Interaction Between Intestinal Microorganisms and Hosts to Regulate Animal Growth

Interaction Between Intestinal Microorganisms and Hosts to Regulate Animal Growth

Guest Editors

Shifeng Pan
Yanhong Zhang
Penggang Liu

MDPI

Basel • Beijing • Wuhan • Barcelona • Belgrade • Novi Sad • Cluj • Manchester

Guest Editors
Shifeng Pan
College of Veterinary Medicine
Yangzhou University
Yangzhou
China

Yanhong Zhang
College of Animal Science and Veterinary Medicine
Henan Institute of Science and Technology
Xinxiang
China

Penggang Liu
College of Veterinary Medicine
Yangzhou University
Yangzhou
China

Editorial Office
MDPI AG
Grosspeteranlage 5
4052 Basel, Switzerland

This is a reprint of the Special Issue, published open access by the journal *Veterinary Sciences* (ISSN 2306-7381), freely accessible at: https://www.mdpi.com/journal/vetsci/special_issues/65387FT46K.

For citation purposes, cite each article independently as indicated on the article page online and as indicated below:

Lastname, A.A.; Lastname, B.B. Article Title. *Journal Name* **Year**, *Volume Number*, Page Range.

ISBN 978-3-7258-5241-3 (Hbk)
ISBN 978-3-7258-5242-0 (PDF)
https://doi.org/10.3390/books978-3-7258-5242-0

Contents

Ying Wang, Jie Tang, Yao Chen, Shuyi Chen, Sumin Chen, Xin Yu, et al.
The Ecological–Evolutionary Game of the Insect Gut Microbiome: Environmental Drivers, Host Regulation, and Prospects for Cross-Cutting Applications
Reprinted from: *Vet. Sci.* **2025**, *12*, 866, https://doi.org/10.3390/vetsci12090866 . . . . **1**

Yanfeng Lu, Lei Zhang, Rui Zhu, Xiujun Duan, Guobo Sun and Yuying Jiang
Chinese Herbal Medicine Compound Microecological Agent (C-MEA) Improves Egg Production Performance in Caged Laying Ducks via Microbiota–Gut–Ovary Axis
Reprinted from: *Vet. Sci.* **2025**, *12*, 808, https://doi.org/10.3390/vetsci12090808 . . . . **21**

Yichen Yu, Fulin Jin, Lijun Wang, Ji Cheng and Shifeng Pan
Role of Gut Microbiota and Metabolite Remodeling on the Development and Management of Rheumatoid Arthritis: A Narrative Review
Reprinted from: *Vet. Sci.* **2025**, *12*, 642, https://doi.org/10.3390/vetsci12070642 . . . . **40**

Xiangtan Su, Ke Wang, Yeqing Liu, Xinyu Lu, Meiru Chen, Jianlong Dang, et al.
Single or Combined Supplementation of *Rhodotorula* Yeast Culture and *Bacillus Subtilis* Enhances Intestinal Barrier Function in Yellow-Feathered Broilers
Reprinted from: *Vet. Sci.* **2025**, *12*, 558, https://doi.org/10.3390/vetsci12060558 . . . . **59**

Yunqiu Yang, Yongfang Chen, Lijun Wang, Min Du, Rui Zhang, Yao Lu and Shifeng Pan
Ssc-miR-130b Enhances Cell Proliferation and Represses Adipogenesis of Primary Cultured Intramuscular Preadipocytes in Pigs
Reprinted from: *Vet. Sci.* **2025**, *12*, 375, https://doi.org/10.3390/vetsci12040375 . . . . **79**

Kehong Deng, Chang Xu, Qing He, Muhammad Safdar, Mudassar Nazar, Xiaocong Li and Kun Li
Angelicae Pubescentis Radix Remitted Intestine Damage in Mice Induced by *Escherichia coli* via Mediating Antioxidant Defense, Inflammatory Mediators, and Restoring Gut Microbiota
Reprinted from: *Vet. Sci.* **2025**, *12*, 354, https://doi.org/10.3390/vetsci12040354 . . . . **94**

Tossaporn Incharoen, Rangsun Charoensook, Wandee Tartrakoon, Sonthaya Numthuam, Yutthana Sunanta, Guillermo Jimenez and Juan J. Loor
Dietary *Bacillus toyonensis* BCT-7112^{T} Supplementation Influences Performance, Egg Quality, Ammonia Emission, and Cecal Microbiome in Laying Ducks
Reprinted from: *Vet. Sci.* **2025**, *12*, 259, https://doi.org/10.3390/vetsci12030259 . . . . **112**

Tian-Ling Pan, Jin-Long Cha, Hao Wang, Jing-Song Zhang, Jin-Long Xiao, Jue Shen, et al.
The CRISPR/Cas9-Mediated Knockout of *VgrG2* in Wild Pathogenic *E. coli* to Alleviate the Effects on Cell Damage and Autophagy
Reprinted from: *Vet. Sci.* **2025**, *12*, 249, https://doi.org/10.3390/vetsci12030249 . . . . **129**

Yang Yang, Xin Zhang, Jin Yang, Ziyan Wu, Junpeng Li, Ruilong Song, et al.
N-(3-Oxododecanoyl)-Homoserine Lactone Induces Intestinal Barrier Damage in Piglets via the Lipid Raft-Mediated Apoptosis Pathway
Reprinted from: *Vet. Sci.* **2025**, *12*, 233, https://doi.org/10.3390/vetsci12030233 . . . . **142**

Xueping Zhang, Hongyu Zhou, Lina Ye, Jiayu Shi, Huiling Zhang and Tangjie Zhang
The Occurrence and Meta-Analysis of Investigations on Intestinal Parasitic Infections Among Captive Wild Mammals in Mainland China
Reprinted from: *Vet. Sci.* **2025**, *12*, 182, https://doi.org/10.3390/vetsci12020182 . . . . **160**

Jiaqiao Zhu, Yonggang Ma, Waseem Ali, Rui Yu, Hui Zou and Zongping Liu
Investigation of Off-Season Breeding Effects on Egg-Laying Performance, Serum Biochemical Parameters, and Reproductive Hormones in Zhedong White
Reprinted from: *Vet. Sci.* **2025**, *12*, 179, https://doi.org/10.3390/vetsci12020179 . . . . **177**

Tao Wang, Xiao Meng, Miao Qian, Shanhao Jin, Ruoyu Bao, Liqi Zhu and Quan Zhang
Helicobacter hepaticus CdtB Triggers Colonic Mucosal Barrier Disruption in Mice via Epithelial Tight Junction Impairment Mediated by MLCK/pMLC2 Signaling Pathway
Reprinted from: *Vet. Sci.* **2025**, *12*, 174, https://doi.org/10.3390/vetsci12020174 . . . . **194**

Aikebaier Reheman, Zhichao Wang, Ruihuan Gao, Jiang He, Juncheng Huang, Changqing Shi, et al.
Effect of Ascites Syndrome on Diversity of Cecal Microbiota of Broiler Chickens
Reprinted from: *Vet. Sci.* **2025**, *12*, 126, https://doi.org/10.3390/vetsci12020126 . . . . **210**

Dan Li, Min Yang, Zhao Ma, Lianqiang Che, Bin Feng, Zhengfeng Fang, et al.
Glycerol Monolaurate Complex Improved Antioxidant, Anti-Inflammation, and Gut Microbiota Composition of Offspring in a Sow–Piglet Model
Reprinted from: *Vet. Sci.* **2025**, *12*, 24, https://doi.org/10.3390/vetsci12010024 . . . . **221**

Yi Wang, Juan Gao, Bojun Fan, Yuemin Hu, Yuefei Yang, Yajie Wu, et al.
Effects of Long-Term Serum Starvation on Autophagy, Metabolism, and Differentiation of Porcine Skeletal Muscle Satellite Cells
Reprinted from: *Vet. Sci.* **2025**, *12*, 11, https://doi.org/10.3390/vetsci12010011 . . . . **238**

veterinary sciences

MDPI

Review

The Ecological–Evolutionary Game of the Insect Gut Microbiome: Environmental Drivers, Host Regulation, and Prospects for Cross-Cutting Applications

Ying Wang [1,†], Jie Tang [1,†], Yao Chen [2], Shuyi Chen [1], Sumin Chen [1], Xin Yu [1], Caijing Wan [1], Guoqi Xiang [3], Yaping Chen [2,*] and Qiang Li [1,*]

1 Key Laboratory of Coarse Cereal Processing, Ministry of Agriculture and Rural Affairs, Sichuan Engineering & Technology Research Center of Coarse Cereal Industrialization, School of Food and Biological Engineering, Chengdu University, Chengdu 610106, China
2 Yunnan Plateau Characteristic Agricultural Industry Research Institute, Yunnan Agricultural University, Kunming 650500, China
3 Key Laboratory of Basic Research and Application Promotion of Solar Energy Technology in Higher Education Institutions of Sichuan Province, Panzhihua University, Panzhihua 617000, China
* Correspondence: cyp83@ynau.edu.cn (Y.C.); liqiang02@cdu.edu.cn (Q.L.); Tel.: +86-15196619794 (Q.L.)
† These authors contributed equally to this work.

Simple Summary: The diversity, function, and complex interactions of insect gut microbiota with their hosts and environments. The composition of insect gut microbiota is influenced by a combination of host genetics, diet, and environmental factors, and is finely regulated by the host immune system. These microorganisms play a critical role in nutrient metabolism, detoxification, immune regulation, and host development and behavior. Recent studies have shown that insect gut microbiota not only hold significant importance in basic theoretical research but also demonstrate broad application potential in fields such as agriculture, environmental protection, industrial biotechnology, and healthcare. In-depth exploration of this field will provide a crucial foundation for understanding host-microbiota interaction mechanisms and developing novel biotechnologies.

Abstract: The insect gut contains a complex and diverse microbial community, and the composition of the insect gut microbial community is influenced by multiple factors such as the host's genetics, dietary habits, and the external environment. The host's immune system maintains the stability and balance of the microbial community through a number of mechanisms. The microorganisms in this community play key roles in the nutrient metabolism, detoxification, immune regulation, development, and behaveior of insects. In recent years, the relevant literature has reported advances in the study of insect gut microbes, indicating the potential applications of insect gut microbes in several fields. The aim of this review is to provide a comprehensive overview of the current information on the structure of insect gut microbial communities and complex host–microbe–environment interactions. The diversity of insects' gut microbial communities and the functions of their gut microbes are revealed. By studying insect gut microbial communities, we can gain insights into the functions of these microbes in the host and explore the causal relationships between them and the host's physiology and behavior. This will not only help us to understand the mechanism of action of the microbiome, but also provide a basis for the development of innovative biotechnology based on insect gut microbes. This research has significant theoretical value in academia and also has a wide range of applications in agriculture, environmental protection, industrial production, and healthcare.

Keywords: insect gut microbes; microecosystems; host–microbe interactions; immune regulation; biotechnology applications

1. Introduction

Recently, there has been increasing attention to the role of the insect gut microbiome in host health, immunity, and metabolism. The gut microbiota is a microbial population inhabiting the gastrointestinal tract and has profound effects on the physiological processes of the host, including digestion and immunity. The study of insects, as a diverse and ecologically significant group of species, provides unique perspectives on host–microbiome interactions and their evolutionary impacts. Studies of the insect gut microbiota have provided valuable insights into microbial stability, immune regulation, and mechanisms of adaptation to environmental changes.

The fruit fly Drosophila melanogaster serves as an important model organism for gut microbiome research, in which intestinal microbes critically regulate both host immunity and metabolism. Buchon et al. pointed out the complex interactions between the gut microbiota and the host immune system, and studies suggest that the microbiota is capable of modulating the intestinal morphology and gene expression and plays an important role under environmental stress [1]. Their study also revealed the role of immune signaling pathways in changes in host physiology, highlighting the relationship between the insect host and its gut microbiota. K.A. Lee and W.J. Lee further explored immune–metabolic interactions in *Drosophila* during systemic and intestinal infections, revealing how the immune system's response to microbial pathogens affects metabolic processes such as energy production and storage [2]. Research shows that the insect gut regulates the composition of the microbial community through selective filtration mechanisms, maintaining beneficial bacteria and eliminating pathogens. This process plays a key role in maintaining the gut microecological balance and host health [3]. It has also been shown that this filtering mechanism acts as a structural barrier to organize the microbiota into different functional regions, thereby improving the host's ability to cope with microbial perturbations. Bai et al. further explored the role of regulatory proteins in maintaining microbial homeostasis and investigated their effects on insect health and development [4]. Research shows that DUOX prevents chronic inflammation by maintaining the intestinal integrity and promoting immune tolerance [5]. The gut microbiota not only participates in nutrient metabolism, but also plays an important role in barrier maintenance, resistance to environmental stress, and pathogen defense. Research shows that environmental factors (such as diet, heavy metals, and microplastics) significantly affect the composition of insect gut microbiota. Microbial signals such as quorum sensing molecules participate in immune regulation [6]. Exposure to pollutants may disrupt the balance of microbial communities, thereby having negative effects on host health and behavior, highlighting the importance of environment–microbe–host interactions.

In summary, the insect gut microbiota is a complex and dynamic system with a crucial role in host health, immunity, and ecological interactions. As research progresses, there is growing evidence that the microbiota is not merely an appendage of the host, but actively participates in shaping the host physiology and behavior. The gut microbiota plays a variety of roles in processes such as immunomodulation, detoxification, microbiota regulation, and adaptation to environmental stresses. Further research will offer new insights and strategies for improving insect health, managing pests, and utilizing microbiota in biotechnology applications.

2. The Structural Features and Microbial Diversity of the Insect Gut

2.1. The Anatomical Structure and Microenvironment of the Insect Gut

The insect gut is a complex and functionally diverse organ system, whose unique anatomy and microenvironment provide an ideal habitat for abundant microbial communities. Lin et al. showed that, from the perspective of its developmental origin and functional features, the insect gut is divided into three major regions: the foregut, midgut, and hindgut. This structural division reflects its functional divergence during evolution [7]. Broderick et al. conducted a thorough study of Drosophila melanogaster and found that the foregut and hindgut originate from the ectoderm. The inner wall of these structures has a chitin layer, which is mainly responsible for food storage and water reabsorption. On the other hand, the midgut originates from the endoderm and is the main site of digestion and absorption. These histological characteristics are highly related to their functions [8]. As shown in Figure 1.

As the core region of the insect digestive system, the midgut has a unique organizational structure. A review by Engel and Moran systematically elaborated on the structural characteristics of the midgut, indicating that its epithelium consists of a single layer of columnar cells. The surface is covered with a special structure called the peritrophic membrane, which has important physiological significance in the insect digestive system [9]. Through morphological and molecular characterization studies, Buchon et al. found that the peritrophic membrane is a translucent network composed of chitin fibers, glycoproteins, and proteins that play multiple roles in food digestion and protection of the intestinal epithelium [10]. Their study specifically highlighted the critical role of the peritrophic membrane in maintaining intestinal homeostasis, not only preventing mechanical damage but also selectively regulating substance exchange.

The characteristics of the intestinal microenvironment are regulated by multiple factors, and a study by Neyen et al. revealed complex interactions of key factors, including the pH, oxygen concentration, osmolality, and nutrient distribution [11]. In a systematic study of Lepidoptera, Paredes et al. found that the pH of different intestinal segments varies significantly, with the anterior portion of the midgut usually being strongly alkaline (pH 10–11) and the posterior portion tending to be neutral. This pH gradient not only affects the activity of digestive enzymes, but also provides a diverse habitat for microorganisms with different pH adaptations, forming a unique microcosm [12]. In terms of the dynamic regulation of the microenvironment, Ryu et al. showed a gradual gradient of the oxygen concentration in the gut from the intestinal wall to the lumen, forming a continuous environment that varied from aerobic to slightly aerobic to anaerobic. This gradient distribution is important for maintaining the diversity of microbial communities and can provide a suitable living space for microorganisms with different oxygen requirements [13]. In particular, Ha et al. found in their Drosophila research that the gut has a unique immune microenvironment. Intestinal epithelial cells can actively secrete various antimicrobial peptides and reactive oxygen species (ROS), as well as other defensive substances. This immune regulatory mechanism plays a central role in maintaining the balance of intestinal microbes [14].

Recent studies have also revealed the distribution characteristics of nutrients in the gut. Cole et al. found that simple sugars were mainly absorbed in the foregut and anterior midgut, while amino acids and fatty acids were absorbed in the posterior gut and posterior midgut. This heterogeneous spatial distribution provided a specific niche for microbial communities with different functions [15]. The role of this material distribution pattern in shaping the microbial community structure was highlighted as an important basis for ensuring the stability of insect gut microecosystems [16].

This highly organized structure provides the basis for maintaining a stable commensal microbial community, while also providing effective defense mechanisms against pathogen invasion. Recent studies have shown that the structural features and microenvironmental regulatory mechanisms of the insect gut are the result of long-term evolution, reflecting the coevolutionary relationship between the host and the microbe. Future studies should further focus on the dynamic changes in microenvironmental factors and their mechanisms regulating the microbial community's structure and function, especially in the context of global environmental change. An in-depth understanding of these mechanisms is important for predicting and regulating insect–microbe interaction relationships. Figure 1. Structure and microenvironment of the insect gut: evolutionary adaptations for microbial colonization and functional maintenance.

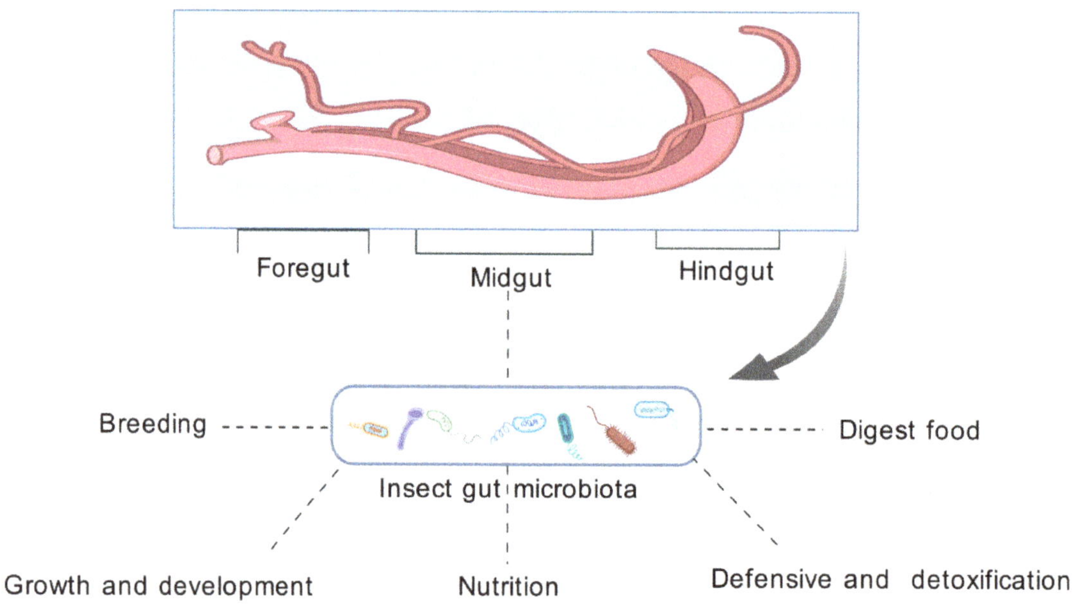

Figure 1. Dynamic regulation and functional evolution of the insect gut microbiota during host development (illustrates the dynamic roles of insect gut microbes at different stages of development). The insect gut is divided into the foregut, midgut, and hindgut, and microbes in each region perform specific functions at different stages. This graphic was generated using the General-Purpose High-Quality Biomedical Graphics Database (GDP), a comprehensive repository of standardized biomedical visualization tools [17].

2.2. The Composition of the Insect Gut Microbial Community

Insect gut microbial communities exhibit unique diversity and complexity. In the study of black soldier fly, *Hermetia illucens*, larvae, Daniele et al. found that the composition of the intestinal microbial community was significantly affected by the diet and the intestinal region. Different sections of the intestine showed specific microbial distribution characteristics [18]. This review study further revealed that these microorganisms mainly included bacteria, fungi, archaea, and viruses, with bacteria being the most dominant and widely studied group [19]. As shown in Table 1. The evolution of insects has led to highly specialized host structures, including specialized bacterial cls, symbiotic organs, and microhabitats, which provide obligate symbionts with a stable living environment and the necessary nutritional support. The main obligate symbiotic microbial communities include the phyla Proteobacteria, Bacteroidetes, Firmicutes, Actinobacteria, Spirochaetes, and Verrucomicrobia [20].

Table 1. Common phyla in gut microbiota isolated from different insects, associated microorganisms found in different insect orders, and their functions.

Insect Classification	**Representative Species**	**Microbial Genus**	**Relative Abundance (%)**
Isoptera	Black-winged termite	Treponema Bacteroides	35–60% 20–45%
Order Lepidoptera	Fall armyworm	Enterococcus	25–50%
Order Hymenoptera	Bees	Gilliamella	15–30%
Order Coleoptera	Red-bellied woodpecker	Lactobacillus	20–40%
Order Diptera	Black-bellied fruit fly	Acetobacter Lactobacillus	10–20% 5–15%
Order Orthoptera	Desert locust	Weissella	30–45%
Order Hemiptera	Aphid	Buchnera	90%
Order Odonata	Green dragonfly larva	Comamonas	20–35%
Order Blattodea	Blattodea	Blattabacterium	70–80%
Order Phthiraptera	Body lice	Riesia	95%
Primary Functions	Representative species		Citation
Wood cellulose degradation, nitrogen fixation Hemicellulose decomposition	Biomass energy conversion, soil improvement		[21]
Biodegradation of microplastics	Environmental management		[22]
Pollen polysaccharide metabolism	Improving agricultural pollination efficiency		[23]
Pesticide (pyrethroid) degradation	Green control of agricultural pests		[24]
Ethanol metabolism, lifespan regulation	Human intestinal disease model		[25]
Cellulose digestion, group pheromone synthesis	Development of locust control strate gies		[26]
Essential amino acid synthesis	Water pollution biological monitor ing		[27]
Aquatic heavy metal (Cd/Pb) chelation	Biological monitoring of water pollution		[28]
Degradation of organochlorine pesticides	Urban pest control		[29]
Vitamin B synthesis	Control of animal and human parasitic disease		[30]

Through the study of herbivorous insects, Shan et al. found that Proteobacteria (Pseudomonas) and thick-walled bacteria (*Bacillus*) are usually the most dominant groups in bacterial communities. This dominance may be closely related to the evolutionary history and ecological adaptation of the host [31]. In particular, Yang et al., in studying plastic-degrading insects, found that certain bacterial taxa with special functions, such as *Pseudomonas* and *Bacillus*, play important roles in degrading complex organic substances [32].

In terms of the dynamic changes in microbial communities, it has been shown that environmental factors such as the temperature, humidity, and pollutants can significantly affect the composition and structure of microbial communities [33]. Insect gut microbiota represent a potential resource for biofuel production, with their unique enzyme systems capable of efficiently degrading lignocellulose (e.g., termite and beetle gut microbiota). Research is focusing on the identification and modification of key microorganisms and enzymes to enhance the biomass conversion efficiency. These findings not only reveal the ecological functions of microorganisms but also provide new insights for sustainable energy development [34]. Future research can further analyse the functional interaction mechanisms of microbial communities through multi-omics technologies and promote their application in synthetic biology [35].

These research results not only deepen our understanding of the composition of insect gut microbial communities, but also provide an important basis for understanding the ecological function and evolutionary mechanisms of microbial communities. Future studies are required to further explore the regulatory mechanisms of the microbial community composition and the interaction relationships among different microbial taxa.

2.3. Factors That Influence the Microbial Community Structure

The assembly of the insect gut microbiota is a complex process driven by multiple factors, including host genetics, developmental stage, and environmental exposure [36]. This finding provides an important basis for understanding the role of nutritional factors in shaping microbial communities. In the study of polystyrene biodegradation, the genetic background of the host is a fundamental factor that influences the microbial community structure. Different insect species, and even different genotypes of the same species, may lead to significant differences in the microbial community composition [37]. Further research suggests that host genetics may exert regulatory effects on the microbial community through pathways such as modulating the intestinal environment, immune response, and metabolic profile [38]. The importance of environmental factors was well demonstrated by Zhang et al. They found that the nanoplastic photoaging process affected the gut microbial community structure of aquatic insect chironomids, *Chironomus kiiensis*, mainly by disrupting their gut integrity and changing the composition of the microbial community [39]. Transitions between developmental stages are another key influencing factor. When studying the pink cotton bollworm, Sun et al. found that the insect's development from the larval stage to an adult was accompanied by significant remodeling of the gut microbial community [40]. Research indicates that environmental factors, such as the temperature and humidity, could indirectly affect microbial communities by altering the physiological state of the host [41]. This remodeling may be related to developmental changes in the intestinal structure, physiological function, and the immune system. Yanchun et al. further showed that these developmentally relevant changes in the microbial community have important implications for the host's physiological functions [42].

Microbial interactions also play an important role in the formation of the community structure. Studying the interactions between plants and microorganisms, Yuxin et al. found that certain microorganisms may affect the structure of other communities by producing metabolites or competing for nutrients [43]. This finding confirms that interactions between microorganisms are a key mechanism for maintaining the stability of community ecological function and structure. It is noteworthy that these influencing factors do not act independently. Existing research indicates that there are complex interactions between the host and the microbial community, and these interactions collectively drive the dynamic succession of the microecosystem [44]. Elucidating the relative contributions of these factors and their interaction mechanisms is of great value for predicting and targeting the regulation of the insect gut microbiota structure, providing a theoretical basis for the development of green pest control strategies based on microbiome management.

3. Functions of Insect Gut Microbes

3.1. Function of Nutrient Metabolism

The gut microbiota of an insect forms interacts closely with its host. By encoding a rich metabolic enzyme system, it directly participates in the host's nutritional metabolism process, significantly enhancing the insect's adaptability to different food sources [45]. This finding provides new insights allowing us to understand the mechanisms of nutritional adaptation in insects. In their study on wild pollinating insects, Jilian et al. revealed that the gut microbial community plays a decisive role in the degradation of complex carbohydrates. They found that certain microorganisms are able to secrete a variety of glycoside hydrolases, such as cellulase and xylanase. The synergistic action of these enzymes can convert indigestible plant polysaccharides into monosaccharides that the host can absorb [46]. Further analysis using metagenomics and metabolomics revealed that this degradation process relies on an interactive network composed of multiple microbial groups, which work together through functional complementarity to break down and convert complex polysaccharides [47]. In terms of nitrogen metabolism, Yaxin et al.'s study of black gadfly, *Hermetia illucens*, larvae found that certain intestinal bacteria have a nitrogen-fixing capacity and are able to convert atmospheric nitrogen into ammonia, providing an additional source of nitrogen for the host [48]. When the nitrogen sources in food are limited, the key role of nitrogen-fixing microorganisms becomes apparent. Research by McMillan's team has confirmed that plants can actively regulate the composition of insect gut microbiota, thereby altering the host's nitrogen metabolism pathways, demonstrating a cross-border nutritional regulation system [49]. Insects and their gut microbiota interact closely to effectively promote the biological conversion of agricultural food waste. The relevant microorganisms not only support the growth and development of insects, but also directly participate in the decomposition and conversion of organic waste [50]. Yuan Qi and colleagues found in a review of recent research that pathogenic fungi and gut microbiota interact to induce host death, and that gut microbiota with significant effects on the host can participate in pest control. Advances in genome editing technology are also new to the field of pest management [51]. The digestibility of cellulose and hemicellulose was determined for four species of grasshoppers, and the relationship between digestibility and gut microbial diversity was analysed. This study provides foundational data for the development of digestible bioreactors for cellulose and hemicellulose, and may offer new insights into straw degradation [52]. This highly efficient interaction mechanism between insects and their gut microbiota also provides both theoretical underpinnings and practical potential for developing agricultural waste recycling processes based on microbiome technology [53].

These findings not only deepen our understanding of microbial trophic interactions in insects, but also provide a theoretical basis for the development of microbial-based biotechnological applications. Understanding these complex metabolic networks is important for improving the nutritional status of beneficial insects or for developing novel bioenergy technologies.

3.2. Detoxification Effect

The detoxification of insect gut microbes is a remarkable area of research. In grasshoppers, *Eucriotettix oculatus*, Xiao-Dong et al. found that gut microbial communities can respond to heavy metal pollution and enhance the host tolerance by changing the community composition and diversity [54]. This adaptive alteration reflects the plasticity of microbial communities under environmental stress. Meiki et al. developed sterile insect culture methods while studying leaf beetles, and their study revealed the critical role of gut microbes in the degradation of plant secondary metabolites [55]. A review by Sandipan et al. further states that certain insect gut microbes are capable of producing specific enzymes that convert toxic compounds into less toxic or non-toxic forms [56]. This detoxification mechanism allows insects to adapt to plant foods containing natural toxins. When studying *Bacillus thuringiensis*, Siyi et al. found that certain intestinal bacteria not only participate in the degradation of environmental pollutants, but also enhance the host's detoxification ability by producing specific metabolites [57]. Lau et al. showed that the detoxification of gut microbes is often closely related to the immune regulation function, and this multiple-action mechanism improves the host's ability to cope with environmental stresses [58]. Through termite transcriptome analysis, YaLing et al. found that pathogenic fungal infection can significantly alter the structure of the intestinal microbiota, and this community restructuring may be an adaptive strategy for the host to enhance their disease resistance [59]. These findings not only deepen our understanding of insect–microbial interactions, but also provide new ideas for the development of microbe-based bioremediation techniques.

3.3. Immunomodulatory Effects

The regulatory role of insect gut microbes in the host immune system is complex and delicate. This immunomodulatory function reflects the result of long-term coevolution between microbes and their host. In their study of plastic-degrading superbugs, Lu et al. found that gut microbial communities are remarkably ecologically robust and able to maintain host immune homeostasis in the face of environmental stresses such as antibiotics and microplastics [60]. This adaptive mechanism is crucial for the survival of insects in complex environments. Further research has revealed that certain gut microbes can regulate the host's metabolism by producing specific microbial proteases [61]. By examining trophic-level delivery of metal oxide nanoparticles, Wang et al. found that changes in the gut microbial community impact oxidative stress and immune responses in the host [62]. Morimura et al. observed in the grapevine pest *Sapolygus spinolae* that the seasonal dynamics of the gut microbial community are closely related to the immune defenses of the host [63]. The role of probiotics in enhancing the immune system of honeybees was particularly emphasized in a study by Robino et al., who found that the addition of specific probiotic strains significantly increased hemolymphocyte activity and phenoloxidase activity [64]. These findings deepen our understanding of insect–microbe immune interactions and provide new ideas for further development of microbe-based biocontrol strategies. This complex immunomodulatory network demonstrates the important role of microbial communities in maintaining host health.

3.4. Developmental and Behavioral Effects

Research indicates that the gut microbiota of insects regulates the development and behavioral patterns of its host through complex mechanisms. Taking honeybees (*Apis mellifera*) as an example, changes in the composition of the gut microbiota are closely related to differentiation in labor behavior, suggesting that microbial communities may regulate the behavioral characteristics of social insects through neuroendocrine or immune pathways [65]. This discovery provides a new research direction for explaining the evolutionary mechanisms of insect social behavior. In terms of host development, environmental stressors such as pesticide exposure can have long-term effects on the gut microbiota. Experimental evidence shows that exposure to sublethal doses of thiamethoxam during the larval stage in bees leads to persistent changes in the gut microbiota structure during the adult stage, and this microbiota disruption is significantly associated with abnormal host growth and development [66]. This phenomenon reveals the cascading interactions between environmental stress, the microbiota, and host development. It is worth noting that the microbiota exhibits significant specificity in regulating the host behavior. Studies have found that although the gut microbiota can significantly influence the frequency of feeding behavior, its impact on related characteristic metabolic markers is relatively limited [67]. Plant-growth-promoting rhizobacteria (PGPR) inhibit pest development by regulating the intestinal microbiota of insects, providing a new approach to green pest control. Further research is needed to elucidate the interaction mechanisms and optimize application technologies [68]. The regulation of development and behavior often requires interactions at multiple levels. These discoveries have advanced our understanding of insect–microbe interactions and set the theoretical basis for developing strategies for behavioral regulation based on that performed by microbes. Future research is essential to uncover the molecular mechanisms of microbe-mediated developmental and behavioral regulation, as well as to investigate the conservation and specificity of these mechanisms across different insect taxa.

4. Interaction of the Insect Gut Immune System and the Microbial Community

4.1. The Evolutionary Immune Function of Insect Pattern Recognition Receptors

Pattern recognition receptors (PRRs) play a central role in the insect immune system. Phylogenetic studies of Lepidoptera indicate that understanding the molecular mechanisms of their evolution holds significant importance for both fundamental biology and the silk industry. This co-evolutionary relationship reflects the process by which insects adapt to diverse pathogens [69]. Through the in-depth study of peptidoglycan recognition proteins (PGRPs), research has shown that these receptors, as regulatory switches of the innate immune system, can accurately identify different types of pathogenic microorganisms and activate the corresponding immune response using a specific binding mode [70]. In terms of model organism research, Pereira et al. systematically elaborated on the mechanism of the role of PRRs in pathogen recognition and immune response initiation by using the large wax moth, *Galleria mellonella*, as an infection model. In particular, they highlighted the differential response patterns that these receptors exhibit in identifying both Gram-positive and Gram-negative bacteria [71]. Further exploring this direction, Wang and his colleagues used insect models to study a new generation of antimicrobial agents, confirming the key role of pattern recognition receptors (PRRs) in pathogen recognition and immune responses, which is of great significance for the development of future antimicrobial treatments [72]. Hoffmann et al. elaborated on the evolutionary features of PRRs and found that these receptors are highly conserved among invertebrates. This conservation implies their important role in innate immunity [73]. Subsequently, Silverman and Maniatis found, through a comparative study, that there are significant similarities between the PRR signaling pathways in insects and mammals, particularly in the mechanism of activation of the NF-κB pathway. This finding provides important clues for understanding the evolution of the immune system [74]. In a functional study, Wand and his team analyzed in detail the complex interactions between *Klebsiella pneumoniae* and the host immune system with the help of a wax moth infection model. It was shown that pattern recognition receptors (PRRs) are critical for pathogen recognition and initiation of immune defense. These receptors not only specifically detect the presence of the pathogen, but also trigger a series of complex, hierarchical immune responses in response to infection [75]. In their study on Aedes aegypti, Wang et al. discovered a close connection between the activation of prephenoloxidase-3 and the recognition of PRRs. This discovery offers new perspectives on how an insect's antifungal immunity functions [76].

These findings deepen our understanding of the functions of insect PRRs and provide a theoretical basis for the development of novel biological control strategies. Future studies are needed to further explore the molecular recognition mechanisms of PRRs, especially how to coordinate the activation of multiple immune response pathways and the interactive regulatory network between these pathways.

4.2. Insect Immunology: Signal Networks in Biological Control

Insect immune signaling pathways are complex and sophisticated regulatory networks. By studying pea aphids, Xu et al. found that the phenol oxidase system plays a key role in resistance to bacterial and fungal infections and that this immune response involves the synergistic activation of multiple signaling pathways [77]. It has been shown that phenoloxidases are not only involved in the recognition and clearance of pathogens, but also play an important role in the amplification of immune responses. The biological significance of blood cells in immune signaling was further explored by Stanley et al. They found that blood cells are not only the executors of immune effects but are also able to regulate the activity of the entire immune network through the secretion of various signaling molecules [78]. A study by Zhang et al. provides insight into the mechanism of action of antimicrobial peptides, revealing how these effector molecules are activated through different signaling pathways and perform multiple functions in immune defense [79]. During the activation of the Toll and IMD signaling pathways, the expression of antimicrobial peptides is intricately regulated to ensure the precision and effectiveness of the immune response. Hetru and Hoffmann delved into the role of NF-κB in the immune response of Drosophila through a detailed study. They found that this transcription factor plays a central role in integrating different immune signaling pathways [80].

A groundbreaking study by Leulier et al. demonstrated that the Drosophila immune system detects bacteria through specific peptidoglycan recognition. This finding provides crucial insights for comprehending how pattern recognition receptors activate downstream signaling pathways [81]. Their study places particular emphasis on the specificity of signal transduction, i.e., different pathogen recognition patterns activate different signaling pathways. Wang et al., in their study of the Taiwan termite, identified a key role for prephenol oxidase in the blackening process and bacterial defense, a finding that demonstrates interactive regulation between signaling pathways [82]. Studies on the Asian corn borer by Prabu et al. revealed the relationship between the phenol oxidase activation mechanism and resistance to *Bacillus thuringiensis* and elucidated how immune signaling pathways regulate insects' resistance to biopesticides [83]. This study deepens our understanding of the function of immune signaling pathways and also provides new ideas for improved pest control strategies. In their review, Wang et al. highlighted the importance of antimicrobial peptides in the post-antibiotic era, pointing out the potential for developing novel antimicrobial strategies based on studies of immune signaling pathways [84].

These results demonstrate the complexity and precision of insect immune signaling pathways and also provide a theoretical basis for the development of novel biological control methods. Future studies are needed to further explore the synergistic mechanisms between different signaling pathways and the influence of environmental factors on immune signaling networks. This has important practical implications for improving agricultural production and disease prevention and control.

4.3. Production and Function of Antimicrobial Peptides (AMPs)

Antimicrobial peptides (AMPs) are important immune effector molecules in insects. Research conducted by Cerenius et al. revealed how the prophenoloxidase system functions in invertebrate immunity and its close connection to the generation of AMPs, forming a robust defense mechanism [85]. This discovery helps us better understand how different aspects of the insect immune system collaborate to defend against threats. By conducting a deep investigation into the mechanism of action of pattern recognition receptors, Takeuchi and Akira revealed the upstream regulatory network of AMP production, demonstrating the close relationship between the inflammatory response and the expression of antimicrobial peptides [86]. When studying fungal immune recognition, Patin et al. found that specific pattern recognition receptors can trigger the production of antifungal antimicrobial peptides. This specific response enhances the host defense against various pathogens [87].

A systematic study by Zhang et al. further elucidates the regulatory molecular and cellular pathways involved in the insect antimicrobial immune response, with particular emphasis on the centrality of antimicrobial peptides in innate immunity [88]. Research has found that there is a synergistic effect between capsular polysaccharide-specific antibody responses and antimicrobial peptides (AMPs), revealing the regulatory function of the antimicrobial peptide system in the complement cascade reaction [89]. Swaminathan et al. found that peptidoglycan recognition proteins regulate the production of antimicrobial peptides through dual regulation. This precise regulation ensures the specificity and effectiveness of the immune response [89]. Jiang et al.'s research on the tobacco moth demonstrated that βGRP-2, an acute-phase protein, recognizes both β-1,3-glucan and lipophosphatidic acid, subsequently triggering prophenoloxidase activation and antimicrobial peptide synthesis [90]. This study elucidated the molecular mechanism of antimicrobial peptide production, while a study by Geijtenbeek et al. revealed how pathogens inhibit the host's antimicrobial peptide response by targeting specific receptors. This finding provides new insights into the immune escape mechanism of pathogens [91].

These findings deepen our understanding of the antimicrobial peptide system in insects, demonstrating its multiple functions in immune defense. Future studies are needed to further explore the mechanism of action of antimicrobial peptides, especially regarding how to optimize their defensive efficacy in the face of novel pathogens. This is important for the development of novel biopesticides and disease control strategies.

4.4. Reactive Oxygen Species (ROS) and Double Oxidase (Duox) Systems

Insects rely on ROS/Duox for immunity, where TLR5 detects flagellin and induces ROS as a frontline defense [92].The Oliveira study further clarified the relationship between nucleic acid perception and ROS production, revealing how pathogenic nucleic acids trigger the oxidative stress response [93]. Jentho and Weis's studies on damage-related molecular patterns (DAMPs) and innate immune training have shown that ROS production not only directly participates in the elimination of pathogens, but also regulates the formation of immune memory [94]. The role of ROS in aseptic inflammation has been investigated, and in one study it was found that reactive oxygen species have important functions in sensing and responding to injury [95]. This study specifically underscores the Duox pathway's vital function in regulating gut homeostasis. Through systematic analysis, Gong et al. demonstrated that in sterile inflammatory conditions, DAMP receptor stimulation is intimately associated with increased ROS production [96]. Matzinger revealed the role of ROS signaling in the recognition of potential threats by the immune system by studying immune tolerance and danger signaling [97]. This finding provides new insights into how insects maintain tissue homeostasis using the ROS-Duox system. Recent studies by Hrithik et al. have shown that damage signaling triggers an immune response caused by the release of DAMP molecules during *Bacillus thuringiensis* infection, and that ROS production is a key link in this process [98].

The progress made in this area has deepened our knowledge of how the ROS-Duox system functions in the insect immune response and has also provided the basis for developing pest control methods that focus on regulating oxidative stress. It is important for future studies to delve deeper into the specific regulatory mechanisms of the ROS signaling network and explore how this knowledge can be used to enhance insects' resistance to disease. This research direction will aid in the creation of more effective and environmentally friendly pest control methods. Figure 2 shows the dynamic regulation and functional evolution of the insect gut microbiota during host development [99].

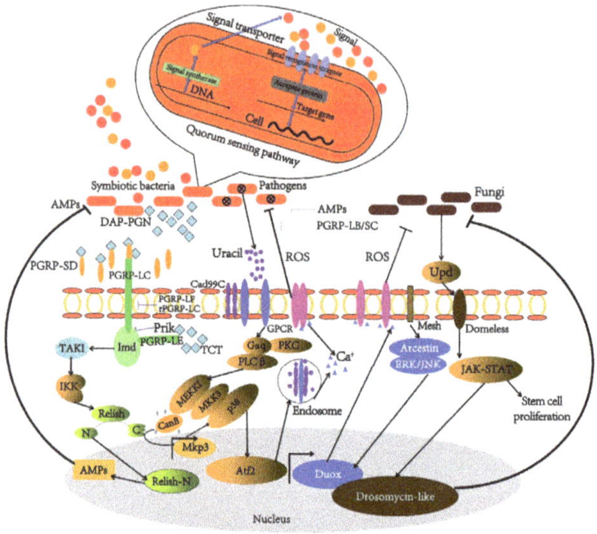

Figure 2. Illustrates multiple immune signalling pathways within the insect gut, including the IMD pathway, Toll pathway, and JAK-STAT pathway. These pathways are activated through various pattern recognition 'receptors (such as PGRP-LC), ultimately inducing the production of antimicrobial peptides (AMPs) to defend against pathogen invasion. Furthermore, the figure depicts the generation of reactive oxygen species (ROS) and their pivotal role in host defence. The upper portion of the diagram elucidates the mechanisms by which symbionts, pathogens, and their interactions influence the host immune response, including microbial community interactions regulated by quorum-sensing signals.

5. Potential Applications of Insect Gut Microbiota: Agriculture

5.1. Agricultural Applications

The insect gut microbiota represents a promising frontier in agricultural biotechnology, offering innovative solutions for sustainable agriculture and crop protection. Recent research has demonstrated the significant potential of insect gut microorganisms for use in various agricultural applications, particularly in pest management, crop protection, and sustainable feed production [100]. The symbiotic relationships between insects and their gut microbiota have prompted the evolution of complex mechanisms that can be harnessed for agricultural enhancement. This mechanism provides scientific justification for the application of insect resources in sustainable animal feed production, helping to reduce traditional agriculture's reliance on chemical additives and promoting the establishment of ecological agricultural systems [101]. In sustainable pest management, the insect gut microbiota has emerged as a powerful tool for developing eco-friendly control strategies. The microbiome-mediated interactions between insects and plants provide novel insights into pest control mechanisms, offering alternatives to conventional chemical pesticides. Recent studies have revealed that manipulating the gut microbiota of pest insects can significantly affect their fitness and host plant adaptation capabilities, presenting opportunities for innovative pest management approaches [102]. This review article systematically outlines how insects—particularly black soldier flies and mealworms—can be utilised to process agricultural and food waste, converting it into high-value products and ultimately achieving sustainable agriculture. The paper emphasises the pivotal role played by insect gut microbiota in this transformation process [103]. This review emphasises that insect bioconversion technology, as a green technology for transforming agricultural waste into valuable resources, holds considerable promise. This technology has successfully established an economic model consistent with the principles of circular agriculture [104].

Moreover, the role of insect gut microbiota in sustainable crop protection extends beyond direct pest control. The vast symbiotic bacterial communities residing within insects represent a largely untapped reservoir. These microorganisms produce an extraordinarily diverse array of secondary metabolites (natural products) possessing unique biological activities. Such compounds hold immense potential for application across agriculture, medicine, and other fields [105].

5.2. Environmental Governance

The application of insect gut microbiota in environmental management represents a significant advancement in sustainable biotechnology solutions. Recent studies have revealed that these microbial communities possess unique enzymatic capabilities that can effectively degrade complex organic pollutants and transform environmental contaminants [106]. The insect gut represents an exceptionally rich and under-explored repository of resources, harbouring microorganisms and enzymes with immense biotechnological potential [107].

The symbiotic microorganisms of insects—including gut microbiota and intracellular symbionts—constitute a vast and largely untapped resource pool, possessing revolutionary potential for the development of next-generation biotechnology products [108]. This study represents an in-depth exploration and validation of this resource repository: a unique alkalophilic bacterium was isolated from the gut of the domesticated silkworm, confirming that this strain can serve as an efficient microbial cell factory for converting biomass into lactic acid. This highlights the immense potential of insect symbionts for application in industrial biotechnology [109].

Moreover, the gut microbiota of insects plays a pivotal role in developing novel biotechnological processes for waste stabilisation. This study comprehensively elucidates the intricate and diverse symbiotic relationships between insects and their internal microorganisms—encompassing both gut microbiota and intracellular symbionts—with particular emphasis on the ecological drivers of these interactions, their maintenance mechanisms, and their profound biological significance for the survival and evolution of insect hosts [110]. The integration of these microbial systems in environmental biotechnology represents a promising approach for addressing contemporary environmental challenges while promoting sustainable industrial practices.

5.3. Industrial Biotechnology

The industrial applications of insect gut microbiota have emerged as a significant frontier in biotechnology, offering innovative solutions for sustainable manufacturing processes. These microorganisms harbor diverse enzymatic systems capable of catalyzing various industrial transformations, particularly in the production of novel biomaterials and specialty chemicals. Recent research has demonstrated their exceptional potential in producing industrial enzymes, including cellulases, lipases, and proteases, which have applications across multiple sectors.

The biotechnological development of insect gut microbiota has revealed its immense application potential in industrial enzymatic processes. Research indicates that these microorganisms possess unique metabolic capabilities to convert complex substrates into high-value products, thereby opening entirely new pathways for industrial bioconversion [111]. The study further indicates that these symbiotic systems and metabolic pathways, optimised through natural evolution, hold immense application potential in the fields of synthetic biology and biotechnology [112].

Furthermore, advances in our understanding of the molecular mechanisms of insect gut microbiota have led to the development of innovative bioreactor systems. The integration of these microbial resources into industrial biotechnology not only enhances the production efficiency but also contributes to the development of sustainable manufacturing practices, marking a significant advancement in bio-based industrial processes.

5.4. Human Health

The exploration of insect gut microbiota offers promising potential for their use in human health applications, particularly in nutrition and therapeutic development. Recent studies have demonstrated that insect-derived products, enriched with beneficial gut microorganisms, can serve as valuable sources of dietary fiber and bioactive compounds that contribute to human nutrition [113]. Yellow mealworms (Tenebrio molitor) are regarded as a highly promising sustainable animal feed source to replace fishmeal and soybean meal, owing to their exceptional nutritional value (high protein content and richness in unsaturated fatty acids) and their capacity to promote a healthy gut microbiome, It has been demonstrated to support multiple aspects of human health [114].

The gut microbiome of edible insects has garnered significant attention for its potential to enhance the nutritional value and safety of insect-based foods. As this review emphasises, the value of edible insects extends far beyond providing basic proteins and nutrients (i.e., "beyond human nutrition"); they also harbour immense potential for promoting health (health benefits) [115]. Furthermore, investigations into the chitin content of insect exoskeletons and their associated gut microbiota have unveiled potential applications in developing functional food ingredients and dietary supplements. These components show promising prebiotic properties and can positively influence the composition and function of the human gut microbiota. The incorporation of insect-derived products and their associated microorganisms into human nutrition is an innovative approach to achieving nutritional security and promoting the achievement of health goals [116].

Figure 3: Interactions between an insect's intestinal immune system and its microbial community (graphical summary).

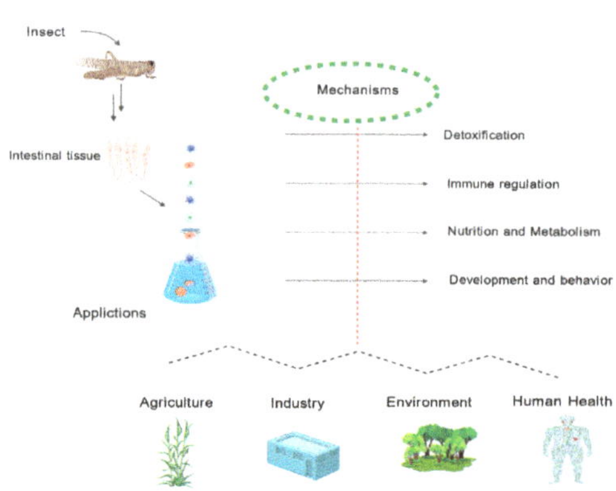

Figure 3. Innate immune defense mechanisms in the insect gut regulate the composition and function of the colonizing microbiota. This graphic was generated using the General-Purpose High-Quality Biomedical Graphics Database (GDP), a comprehensive repository of standardized biomedical visualization tools [17].

6. Conclusions

The gut microbiota of insects demonstrates potential applications across multiple fields: by regulating the microbiota, biological pest control in agriculture and health management of pollinating insects can be achieved; its unique detoxification and degradation capabilities offer new insights into the bioremediation of organic pollutants, heavy metals, and plastic waste; as a valuable source of industrial enzymes and antimicrobial agents, it can drive the development of biomanufacturing technologies; and simultaneously, research into the mechanisms of its interaction with the host provides important references for human gut microbiome studies. Research into insect gut microbiota has revealed a complex and sophisticated microecosystem, demonstrating profound interactions between microorganisms and their hosts Research on insect gut microbiota still faces numerous challenges. For instance, efficiently decodironment interactions remain pressing issues. Additionally, technological limitations pose difficulties in functional validation and application development, necessitating the introduction of more systems biology and synthetic biology approaches for precise control and optimization of microbial function.

Future research directions should focus on (1) employing high-throughput omics technologies and systems biology approaches to comprehensively analyze microbiome functions and explore specific regulatory mechanisms in hosts' physiological processes; (2) utilizing synthetic biology techniques to develop and optimize microbial communities for addressing specific agricultural and environmental challenges; (3) strengthening multidisciplinary collaboration, integrating the latest advances in ecology, microbiology, and biotechnology to jointly promote innovation and breakthroughs in insect gut microbiota research. Other areas for future research include the application of metagenomics/Metatranscriptomics to decode microbial metabolic networks and, when combined with host transcriptomics, to reveal cross-kingdom signaling pathways; synthetic biology applications, utilizing CRISPR-based tools to engineer microbial communities for targeted pollutant degradation, enhanced nutrient acquisition in agricultural systems, development of standardized host strains (e.g., *Bacillus* spp.) for functional validation, and translational medicine based on the comparative responses of insect and human microbiomes to exogenous substances; adapting insect-derived antimicrobial peptides for clinical applications; and establishing a shared database for cross-species microbiome analysis.

In conclusion, research on insect gut microbiota not only deepens our understanding of insect biology, but also provides novel approaches to addressing global challenges in agriculture, the environment, and health. As this field continues to evolve, innovative biotechnology applications based on insect gut microbiota will undoubtedly play an increasingly significant role in the future.

Author Contributions: Conceptualization: Conceptual Design: Y.W. and J.T.; Methodology: C.W.; Software Development: S.C. (Shuyi Chen) and G.X.; Validation: S.C. (Shuyi Chen); Draft Writing: Y.W.; Review & Editing: J.T. and G.X.; Visualization: Q.L.; Supervision: S.C. (Sumin Chen); Project Management: X.Y.; Formal Analysis: Y.C. (Yao Chen); Investigation: Y.C. (Yaping Chen). All authors have read and agreed to the published version of the manuscript.

Funding: Supported by the Open Project Fund of the Key Laboratory of Solar Energy Technology Integration and Application Promotion, Sichuan Province (TYNSYS-2023-Z-03) and The Sichuan Innovation Team Research Program of China Agricultural Research System (SCCXTD-2024-20).

Institutional Review Board Statement: Not applicable.

Informed Consent Statement: Not applicable.

Data Availability Statement: The content of this article is entirely based on a comprehensive synthesis of published literature, with all cited studies duly referenced.

Conflicts of Interest: The authors declare that they have no conflict of interest.

References

1. Buchon, N.; Broderick, N.A.; Lemaitre, B. Gut homeostasis in a microbial world: Insights from Drosophila melanogaster. *Nat. Rev. Microbiol.* **2013**, *11*, 615–626. [CrossRef] [PubMed]
2. Lee, K.A.; Lee, W.J. Immune-metabolic interactions during systemic and enteric infection in Drosophila. *Curr. Opin. Insect Sci.* **2018**, *29*, 21–26. [CrossRef]
3. Lanan, M.C.; Rodrigues, P.A.; Agellon, A.; Jansma, P.; Wheeler, D.E. A bacterial filter protects and structures the gut microbiome of an insect. *ISME J.* **2016**, *10*, 1866–1876. [CrossRef]
4. Buchon, N.; Broderick, N.A.; Poidevin, M.; Pradervand, S.; Lemaitre, B. Drosophila intestinal response to bacterial infection: Activation of host defense and stem cell proliferation. *Cell Host Microbe* **2009**, *5*, 200–211. [CrossRef]
5. Kim, S.H.; Lee, W.J. Role of DUOX in gut inflammation: Lessons from Drosophila model of gut-microbiota interactions. *Front. Cell Infect. Microbiol.* **2014**, *3*, 116. [CrossRef] [PubMed]
6. Coquant, G.; Grill, J.-P.; Seksik, P. Impact of N-Acyl-Homoserine Lactones, Quorum Sensing Molecules, on Gut Immunity. *Front. Immunol.* **2020**, *11*, 1827. [CrossRef]
7. Dar, M.A.; Shaikh, A.F.; Pawar, K.D.; Xie, R.; Sun, J.; Kandasamy, S.; Pandit, R.S. Evaluation of cellulose degrading bacteria isolated from the gut-system of cotton bollworm, Helicoverpa armigera and their potential values in biomass conversion. *PeerJ* **2021**, *9*, e11254. [CrossRef] [PubMed]
8. Broderick, N.A.; Buchon, N.; Lemaitre, B. Microbiota-induced changes in drosophila melanogaster host gene expression and gut morphology. *mBio* **2014**, *5*, e01117-14. [CrossRef]
9. Engel, P.; Moran, N.A. The gut microbiota of insects—Diversity in structure and function. *FEMS Microbiol. Rev.* **2013**, *37*, 699–735. [CrossRef] [PubMed]
10. Buchon, N.; Osman, D.; David, F.P.; Fang, H.Y.; Boquete, J.-P.; Deplancke, B.; Lemaitre, B. Morphological and Molecular Characterization of Adult Midgut Compartmentalization in Drosophila. *Cell Rep.* **2013**, *3*, 1725–1738. [CrossRef]
11. Neyen, C.; Runchel, C.; Schüpfer, F.; Meier, P.; Lemaitre, B. The regulatory isoform rPGRP-LC induces immune resolution via endosomal degradation of receptors. *Nat. Immunol.* **2016**, *17*, 1150–1158. [CrossRef]
12. Paredes, J.C.; Welchman, D.P.; Poidevin, M.; Lemaitre, B. Negative regulation by amidase PGRPs shapes the Drosophila antibacterial response and protects the fly from innocuous infection. *Immunity* **2011**, *35*, 770–779. [CrossRef]
13. Ryu, J.H.; Kim, S.H.; Lee, H.Y.; Bai, J.Y.; Nam, Y.D.; Bae, J.W.; Lee, D.G.; Shin, S.C.; Ha, E.M.; Lee, W.J. Innate immune homeostasis by the homeobox gene caudal and commensal-gut mutualism in Drosophila. *Science* **2008**, *319*, 777–782. [CrossRef]
14. Ha, E.M.; Oh, C.T.; Bae, Y.S.; Lee, W.J. A direct role for dual oxidase in Drosophila gut immunity. *Science* **2005**, *310*, 847–850. [CrossRef]
15. Cole, M.; Lindeque, P.; Halsband, C.; Galloway, T.S. Microplastics as contaminants in the marine environment: A review. *Mar. Pollut. Bull.* **2011**, *62*, 2588–2597. [CrossRef]
16. Hou, Z.; Zhou, Q.; Mo, F.; Kang, W.; Ouyang, S. Enhanced carbon emission driven by the interaction between functional microbial community and hydrocarbons: An enlightenment for carbon cycle. *Sci. Total Environ.* **2023**, *867*, 161402. [CrossRef]
17. Jiang, S.; Li, H.; Zhang, L.; Mu, W.; Zhang, Y.; Chen, T.; Wu, J.; Tang, H.; Zheng, S.; Liu, Y.; et al. Generic Diagramming Platform (GDP): A comprehensive database of high-quality biomedical graphics. *Nucleic Acids Res.* **2025**, *53*, D1670–D1676. [CrossRef]
18. Bruno, D.; Bonelli, M.; De Filippis, F.; Di Lelio, I.; Tettamanti, G.; Casartelli, M.; Ercolini, D.; Caccia, S. The Intestinal Microbiota of *Hermetia illucens* Larvae Is Affected by Diet and Shows a Diverse Composition in the Different Midgut Regions. *Appl. Environ. Microbiol.* **2019**, *85*, e01864-18. [CrossRef]
19. Zhang, X.; Zhang, F.; Lu, X. Diversity and Functional Roles of the Gut Microbiota in Lepidopteran Insects. *Microorganisms* **2022**, *10*, 1234. [CrossRef]
20. Butera, G.; Ferraro, C.; Colazza, S.; Alonzo, G.; Quatrini, P. The culturable bacterial community of frass produced by larvae of *Rhynchophorus ferrugineus* Olivier (Coleoptera: Curculionidae) in the Canary island date palm. *Lett. Appl. Microbiol.* **2012**, *54*, 530–536. [CrossRef]
21. Brune, A. Symbiotic digestion of lignocellulose in termite guts. *Nat. Rev. Microbiol.* **2014**, *12*, 168–180. [CrossRef]
22. Fan, Z.; Khan, M.M.; Wang, K.; Li, Y.; Jin, F.; Peng, J.; Chen, X.; Kong, W.; Lv, X.; Chen, X.; et al. Disruption of midgut homeostasis by microplastics in *Spodoptera frugiperda*: Insights into inflammatory and oxidative mechanisms. *J. Hazard. Mater.* **2025**, *487*, 137262. [CrossRef]

23. Kwong, W.K.; Medina, L.A.; Koch, H.; Sing, K.-W.; Soh, E.J.Y.; Ascher, J.S.; Jaffé, R.; Moran, N.A. Dynamic microbiome evolution in social bees. *Sci. Adv.* **2017**, *3*, e1600513. [CrossRef]
24. Shukla, S.P.; Beran, F. Gut microbiota degrades toxic isothiocyanates in a flea beetle pest. *Mol. Ecol.* **2020**, *29*, 4692–4705. [CrossRef] [PubMed]
25. Broderick, N.A.; Lemaitre, B. Gut-associated microbes of *Drosophila melanogaster*. *Gut Microbes* **2012**, *3*, 307–321. [CrossRef]
26. Dillon, R.; Charnley, K. Mutualism between the desert locust *Schistocerca gregaria* and its gut microbiota. *Res. Microbiol.* **2002**, *153*, 503–509. [CrossRef] [PubMed]
27. Douglas, A.E. Nutritional interactions in insect-microbial symbioses: Aphids and their symbiotic bacteria Buchnera. *Annu. Rev. Entomol.* **1998**, *43*, 17–37. [CrossRef]
28. Sinclair, C.A.; Garcia, T.S.; Vasta, R.; Eagles-Smith, C.A. Mercury trophic transfer to a freshwater biosentinel: Quantifying controlled bioaccumulation in larval dragonflies. *Environ. Toxicol. Chem.* **2025**, *44*, 1824–1834. [CrossRef]
29. Pan, X.; Wang, X.; Zhang, F. New Insights into Cockroach Control: Using Functional Diversity of *Blattella germanica* Symbionts. *Insects* **2020**, *11*, 696. [CrossRef] [PubMed]
30. Kirkness, E.F.; Haas, B.J.; Sun, W.; Braig, H.R.; Perotti, M.A.; Clark, J.M.; Lee, S.H.; Robertson, H.M.; Kennedy, R.C.; Elhaik, E.; et al. Genome sequences of the human body louse and its primary endosymbiont provide insights into the permanent parasitic lifestyle. *Proc. Natl. Acad. Sci. USA* **2010**, *107*, 12168–12173. [CrossRef]
31. Shan, H.W.; Xia, X.J.; Feng, Y.L.; Wu, W.; Li, H.J.; Sun, Z.T.; Li, J.M.; Chen, J.P. The plant-sucking insect selects assembly of the gut microbiota from environment to enhance host reproduction. *NPJ Biofilms Microbiomes* **2024**, *10*, 64. [CrossRef]
32. Yang, X.G.; Wen, P.P.; Yang, Y.F.; Jia, P.P.; Li, W.G.; Pei, D.S. Corrigendum: Plastic biodegradation by in vitro environmental microorganisms and in vivo gut microorganisms of insects. *Front. Microbiol.* **2024**, *15*, 1444678. [CrossRef]
33. Yao, L.; Jiao, J.; Wu, C.; Jiang, B.; Fan, L. Effects of thinning on the structure of soil microbial communities in a subtropical secondary evergreen broad-leaved forest. *Front. Plant Sci.* **2024**, *15*, 1465237. [CrossRef]
34. Zhou, J.; Duan, J.; Gao, M.; Wang, Y.; Wang, X.; Zhao, K. Diversity, Roles, and Biotechnological Applications of Symbiotic Microorganisms in the Gut of Termite. *Curr. Microbiol.* **2019**, *76*, 755–761. [CrossRef] [PubMed]
35. Warnecke, F.; Luginbühl, P.; Ivanova, N.; Ghassemian, M.; Richardson, T.H.; Stege, J.T.; Cayouette, M.; McHardy, A.C.; Djordjevic, G.; Aboushadi, N.; et al. Metagenomic and functional analysis of hindgut microbiota of a wood-feeding higher termite. *Nature* **2007**, *450*, 560–565. [CrossRef] [PubMed]
36. Wang, X.; Sun, S.; Yang, X.; Cheng, J.; Wei, H.; Li, Z.; Michaud, J.P.; Liu, X. Variability of Gut Microbiota Across the Life Cycle of Grapholita molesta (Lepidoptera: Tortricidae). *Front. Microbiol.* **2020**, *11*, 1366. [CrossRef] [PubMed]
37. Xu, L.; Li, Z.; Wang, L.; Xu, Z.; Zhang, S.; Zhang, Q. Progress in polystyrene biodegradation by insect gut microbiota. *World J. Microbiol. Biotechnol.* **2024**, *40*, 143. [CrossRef]
38. Talà, A.; Guerra, F.; Resta, S.C.; Calcagnile, M.; Barca, A.; Tredici, S.M.; De Donno, M.D.; Vacca, M.; Liso, M.; Chieppa, M.; et al. Phenotyping of Fecal Microbiota of Winnie, a Rodent Model of Spontaneous Chronic Colitis, Reveals Specific Metabolic, Genotoxic, and Pro-inflammatory Properties. *Inflammation* **2022**, *45*, 2477–2497. [CrossRef]
39. Zhang, J.; Xia, X.; Huang, W.; Li, Y.; Lin, X.; Li, Y.; Yang, Z. Photoaging of biodegradable nanoplastics regulates their toxicity to aquatic insects (*Chironomus kiinensis*) by impairing gut and disrupting intestinal microbiota. *Environ. Int.* **2024**, *185*, 108483. [CrossRef]
40. Sun, Z.B.; Hu, Y.F.; Song, H.J.; Cong, S.B.; Wang, L. Cry1Ac Mixed with Gentamicin Influences the Intestinal Microbial Diversity and Community Composition of Pink Bollworms. *Life* **2023**, *14*, 58. [CrossRef]
41. Huerta-García, A.; Álvarez-Cervantes, J. The gut microbiota of insects: A potential source of bacteria and metabolites. *Int. J. Trop. Insect Sci.* **2024**, *44*, 13–30. [CrossRef]
42. Deng, Y.; Yang, X.; Chen, J.; Yang, S.; Chi, H.; Chen, C.; Yang, X.; Hou, C. Correction to "Jute (*Corchorus olitorius* L.) Nanocrystalline Cellulose Inhibits Insect Virus via Gut Microbiota and Metabolism". *ACS Nano* **2023**, *17*, 24417. [CrossRef]
43. Zhang, Y.; Zhang, S.; Xu, L. The pivotal roles of gut microbiota in insect plant interactions for sustainable pest management. *NPJ Biofilms Microbiomes* **2023**, *9*, 66. [CrossRef]
44. Chang, C.J.; Chang, C.W.; Lu, H.P.; Hsieh, C.H.; Wu, J.H. Bioenergetically constrained dynamical microbial interactions govern the performance and stability of methane-producing bioreactors. *NPJ Biofilms Microbiomes* **2025**, *11*, 31. [CrossRef]
45. Biasato, I.; Gasco, L.; Schiavone, A.; Capucchio, M.T.; Ferrocino, I. Gut microbiota changes in insect-fed monogastric species: State-of-the-art and future perspectives. *Anim. Front.* **2023**, *13*, 72–80. [CrossRef]
46. Li, J.; Sauers, L.; Zhuang, D.; Ren, H.; Guo, J.; Wang, L.; Zhuang, M.; Guo, Y.; Zhang, Z.; Wu, J.; et al. Divergence and convergence of gut microbiomes of wild insect pollinators. *mBio* **2023**, *14*, e0127023. [CrossRef] [PubMed]

47. Rong, X.; Zhu, L.; Shu, Q. Synergistic gut microbiome-mediated degradation of Astragalus membranaceus polysaccharides and *Codonopsis pilosula* polysaccharides into butyric acid: A metatranscriptomic analysis. *Microbiol. Spectr.* **2025**, *13*, e0303924. [CrossRef] [PubMed]
48. Pei, Y.; Sun, M.; Zhang, J.; Lei, A.; Chen, H.; Kang, X.; Ni, H.; Yang, S. Comparative Metagenomic and Metatranscriptomic Analyses Reveal the Response of Black Soldier Fly (*Hermetia illucens*) Larvae Intestinal Microbes and Reduction Mechanisms to High Concentrations of Tetracycline. *Toxics* **2023**, *11*, 611. [CrossRef]
49. McMillan, H.M. Plants target gut microbes to reduce insect herbivore damage. *Proc. Natl. Acad. Sci. USA* **2023**, *120*, e2308568120. [CrossRef]
50. Siddiqui, S.A.; Harahap, I.A.; Osei-Owusu, J.; Saikia, T.; Wu, Y.S.; Fernando, I.; Perestrelo, R.; Câmara, J.S. Bioconversion of organic waste by insects—A comprehensive review. *Process Saf. Environ. Prot.* **2024**, *187*, 1–25. [CrossRef]
51. Zhao, Y.; Song, Q.; Song, Y. The role of insect intestinal microbes in controlling of Empoasca onukii Matsuda (Hemiptera: Cicadellidae) pest infestations in the production of tea garden: A review. *Arch. Microbiol.* **2023**, *205*, 266. [CrossRef] [PubMed]
52. Bai, J.; Ling, Y.; Li, W.J.; Wang, L.; Xue, X.B.; Gao, Y.Y.; Li, F.F.; Li, X.J. Analysis of Intestinal Microbial Diversity of Four Species of Grasshoppers and Determination of Cellulose Digestibility. *Insects* **2022**, *13*, 432. [CrossRef] [PubMed]
53. Li, J.; Ni, B.; Wu, Y.; Yang, Y.; Mu, D.; Wu, K.; Zhang, A.; Du, Y.; Li, Q. The cultivable gut bacteria *Enterococcus mundtii* promotes early-instar larval growth of *Conogethes punctiferalis* via enhancing digestive enzyme activity. *Pest Manag. Sci.* **2024**, *80*, 6179–6188. [CrossRef]
54. Li, X.D.; Xin, L.; Rong, W.T.; Liu, X.Y.; Deng, W.A.; Qin, Y.C.; Li, X.L. Effect of heavy metals pollution on the composition and diversity of the intestinal microbial community of a pygmy grasshopper (*Eucriotettix oculatus*). *Ecotoxicol. Environ. Saf.* **2021**, *223*, 112582. [CrossRef]
55. Ma, M.; Liu, P.; Yu, J.; Han, R.; Xu, L. Preparing and Rearing Axenic Insects with Tissue Cultured Seedlings for Host-Gut Microbiota Interaction Studies of the Leaf Beetle. *J. Vis. Exp.* **2021**, *176*, e63195.
56. Banerjee, S.; Maiti, T.K.; Roy, R.N. Enzyme producing insect gut microbes: An unexplored biotechnological aspect. *Crit. Rev. Biotechnol.* **2022**, *42*, 384–402. [CrossRef]
57. Wu, S.; Zhong, J.; Lei, Q.; Song, H.; Chen, S.F.; Wahla, A.Q.; Bhatt, K.; Chen, S. New roles for Bacillus thuringiensis in the removal of environmental pollutants. *Environ. Res.* **2023**, *236*, 116699. [CrossRef]
58. Lau, E.; Maccaro, J.; McFrederick, Q.S.; Nieh, J.C. Exploring the interactions between Nosema ceranae infection and the honey bee gut microbiome. *Sci. Rep.* **2024**, *14*, 20037. [CrossRef]
59. Tang, Y.L.; Kong, Y.H.; Qin, S.; Merchant, A.; Shi, J.Z.; Zhou, X.G.; Li, M.W.; Wang, Q. Transcriptomic dissection of termite gut microbiota following entomopathogenic fungal infection. *Front. Physiol.* **2023**, *14*, 1194370. [CrossRef] [PubMed]
60. Lu, B.; Lou, Y.; Wang, J.; Liu, Q.; Yang, S.S.; Ren, N.; Wu, W.M.; Xing, D. Understanding the Ecological Robustness and Adaptability of the Gut Microbiome in Plastic-Degrading Superworms (*Zophobas atratus*) in Response to Microplastics and Antibiotics. *Environ. Sci. Technol.* **2024**, *58*, 12028–12041. [CrossRef]
61. Kannan, M.; Mubarakali, D.; Thiyonila, B.; Krishnan, M.; Padmanaban, B.; Shantkriti, S. Insect gut as a bioresource for potential enzymes-an unexploited area for industrial biotechnology. *Biocatal. Agric. Biotechnol.* **2019**, *18*, 101010.
62. Wang, Z.; Fan, N.; Li, X.; Yue, L.; Wang, X.; Liao, H.; Xiao, Z. Trophic Transfer of Metal Oxide Nanoparticles in the Tomato-Helicoverpa armigera Food Chain: Effects on Phyllosphere Microbiota, Insect Oxidative Stress, and Gut Microbiome. *ACS Nano* **2024**, *18*, 26631–26642. [CrossRef]
63. Morimura, H.; Ishigami, K.; Sato, T.; Sone, T.; Kikuchi, Y. Geographical, Seasonal, and Growth-Related Dynamics of Gut Microbiota in a Grapevine Pest, *Apolygus spinolae* (Heteroptera: Miridae). *Microb. Ecol.* **2024**, *87*, 112. [CrossRef]
64. Robino, P.; Galosi, L.; Bellato, A.; Vincenzetti, S.; Gonella, E.; Ferrocino, I.; Serri, E.; Biagini, L.; Roncarati, A.; Nebbia, P.; et al. Effects of a supplemented diet containing 7 probiotic strains (Honeybeeotic) on honeybee physiology and immune response: Analysis of hemolymph cytology, phenoloxidase activity, and gut microbiome. *Biol. Res.* **2024**, *57*, 50. [CrossRef]
65. Wang, K.; Zheng, M.; Cai, M.; Zhang, Y.; Fan, Y.; Lin, Z.; Wang, Z.; Niu, Q.; Ji, T. Possible interactions between gut microbiome and division of labor in honey bees. *Ecol. Evol.* **2024**, *14*, e11707. [CrossRef]
66. Li, B.; Chen, X.; Ke, L.; Dai, P.; Ge, Y.; Liu, Y.J. Early-Life Sublethal Exposure to Thiacloprid Alters Adult Honeybee Gut Microbiota. *Genes* **2024**, *15*, 1001. [CrossRef] [PubMed]
67. Liberti, J.; Frank, E.T.; Kay, T.; Kesner, L.; Monié-Ibanes, M.; Quinn, A.; Schmitt, T.; Keller, L.; Engel, P. Gut microbiota influences onset of foraging-related behavior but not physiological hallmarks of division of labor in honeybees. *mBio* **2024**, *15*, e0103424. [CrossRef] [PubMed]
68. Adesemoye, A.O.; Antony-Babu, S.; Nagy, E.M.; Kafle, B.D.; Gregory, T.A.; Xiong, C.; Fadamiro, H.Y. Bacteria-based artificial diets modulate larval development, survival and gut microbiota of two insect pests. *Biol. Control* **2025**, *205*, 105769.

69. Ding, X.; Liu, J.; Zheng, L.; Song, J.; Li, N.; Hu, H.; Tong, X.; Dai, F. Genome-Wide Identification and Expression Profiling of Wnt Family Genes in the Silkworm, *Bombyx mori*. *Int. J. Mol. Sci.* **2019**, *20*, 1221. [CrossRef]
70. Steiner, H. Peptidoglycan recognition proteins: On and off switches for innate immunity. *Immunol. Rev.* **2004**, *198*, 83–96. [CrossRef] [PubMed]
71. Pereira, M.F.; Rossi, C.C.; da Silva, G.C.; Rosa, J.N.; Bazzolli, D.M.S. Galleria mellonella as an infection model: An in-depth look at why it works and practical considerations for successful application. *Pathog. Dis.* **2020**, *78*, ftaa056. [CrossRef]
72. Wang, Y.; Li, D.D.; Jiang, Y.Y.; Mylonakis, E. Utility of insects for studying human pathogens and evaluating new antimicrobial agents. *Adv. Biochem. Eng. Biotechnol.* **2013**, *135*, 1–25.
73. Hoffmann, J.A.; Kafatos, F.C.; Janeway, C.A.; Ezekowitz, R.A. Phylogenetic perspectives in innate immunity. *Science* **1999**, *284*, 1313–1318. [CrossRef]
74. Silverman, N.; Maniatis, T. NF-κB signaling pathways in mammalian and insect innate immunity. *Genes Dev.* **2001**, *15*, 2321–2342. [CrossRef]
75. Wand, M.E.; McCowen, J.W.I.; Nugent, P.G.; Sutton, J.M. Complex interactions of Klebsiella pneumoniae with the host immune system in a Galleria mellonella infection model. *J. Med. Microbiol.* **2013**, *62*, 1790–1798. [CrossRef]
76. Wang, Y.; Jiang, H.; Cheng, Y.; An, C.; Chu, Y.; Raikhel, A.S.; Zou, Z. Activation of Aedes aegypti prophenoloxidase-3 and its role in the immune response against entomopathogenic fungi. *Insect Mol. Biol.* **2017**, *26*, 552–563. [CrossRef] [PubMed]
77. Xu, L.; Ma, L.; Wang, W.; Li, L.; Lu, Z. Phenoloxidases are required for the pea aphid's defence against bacterial and fungal infection. *Insect Mol. Biol.* **2019**, *28*, 176–186. [CrossRef]
78. Stanley, D.; Haas, E.; Kim, Y. Beyond Cellular Immunity: On the Biological Significance of Insect Hemocytes. *Cells* **2023**, *12*, 599. [CrossRef]
79. Zhang, Q.Y.; Yan, Z.B.; Meng, Y.M.; Hong, X.Y.; Shao, G.; Ma, J.J.; Cheng, X.R.; Liu, J.; Kang, J.; Fu, C.Y. Antimicrobial peptides: Mechanism of action, activity and clinical potential. *Mil. Med. Res.* **2021**, *8*, 48. [CrossRef] [PubMed]
80. Hetru, C.; Hoffmann, J.A. NF-kappaB in the immune response of Drosophila. *Cold Spring Harb. Perspect. Biol.* **2009**, *1*, a000232. [CrossRef]
81. Leulier, F.; Parquet, C.; Pili-Floury, S.; Ryu, J.H.; Caroff, M.; Lee, W.J.; Mengin-Lecreulx, D.; Lemaitre, B. The Drosophila immune system detects bacteria through specific peptidoglycan recognition. *Nat. Immunol.* **2003**, *4*, 478–484. [CrossRef] [PubMed]
82. Wang, Z.; Luo, J.; Feng, K.; Zhou, Y.; Tang, F. Prophenoloxidase of *Odontotermes formosanus* (Shiraki) (Blattodea: Termitidae) Is a Key Gene in Melanization and Has a Defensive Role during Bacterial Infection. *Int. J. Mol. Sci.* **2022**, *24*, 406. [CrossRef]
83. Prabu, S.; Jing, D.; Shabbir, M.Z.; Yuan, W.; Wang, Z.; He, K. Contribution of phenoloxidase activation mechanism to Bt insecticidal protein resistance in Asian corn borer. *Int. J. Biol. Macromol.* **2020**, *153*, 88–99. [CrossRef]
84. Wang, J.; Dou, X.; Song, J.; Lyu, Y.; Zhu, X.; Xu, L.; Li, W.; Shan, A. Antimicrobial peptides: Promising alternatives in the post feeding antibiotic era. *Med. Res. Rev.* **2019**, *39*, 831–859. [CrossRef]
85. Cerenius, L.; Lee, B.L.; Söderhäll, K. The proPO-system: Pros and cons for its role in invertebrate immunity. *Trends Immunol.* **2008**, *29*, 263–271. [CrossRef] [PubMed]
86. Takeuchi, O.; Akira, S. Pattern recognition receptors and inflammation. *Cell* **2010**, *140*, 805–820. [CrossRef]
87. Patin, E.C.; Thompson, A.; Orr, S.J. Pattern recognition receptors in fungal immunity. *Semin. Cell Dev. Biol.* **2019**, *89*, 24–33. [CrossRef]
88. Zhang, W.; Tettamanti, G.; Bassal, T.; Heryanto, C.; Eleftherianos, I.; Mohamed, A. Regulators and signalling in insect antimicrobial innate immunity: Functional molecules and cellular pathways. *Cell Signal* **2021**, *83*, 110003. [CrossRef]
89. Calzas, C.; Lemire, P.; Auray, G.; Gerdts, V.; Gottschalk, M.; Segura, M. Antibody response specific to the capsular polysaccharide is impaired in Streptococcus suis serotype 2-infected animals. *Infect. Immun.* **2015**, *83*, 441–453. [CrossRef] [PubMed]
90. Swaminathan, C.P.; Brown, P.H.; Roychowdhury, A.; Wang, Q.; Guan, R.; Silverman, N.; Goldman, W.E.; Boons, G.J.; Mariuzza, R.A. Dual strategies for peptidoglycan discrimination by peptidoglycan recognition proteins (PGRPs). *Proc. Natl. Acad. Sci. USA* **2006**, *103*, 684–689. [CrossRef] [PubMed]
91. Jiang, H.; Ma, C.; Lu, Z.Q.; Kanost, M.R. Beta-1,3-glucan recognition protein-2 (betaGRP-2)from Manduca sexta; an acute-phase protein that binds beta-1,3-glucan and lipoteichoic acid to aggregate fungi and bacteria and stimulate prophenoloxidase activation. *Insect Biochem. Mol. Biol.* **2004**, *34*, 89–100. [CrossRef]
92. Geijtenbeek, T.B.; Van Vliet, S.J.; Koppel, E.A.; Sanchez-Hernandez, M.; Vandenbroucke-Grauls, C.M.; Appelmelk, B.; Van Kooyk, Y. Mycobacteria target DC-SIGN to suppress dendritic cell function. *J. Exp. Med.* **2003**, *197*, 7–17. [CrossRef]
93. Hayashi, F.; Smith, K.D.; Ozinsky, A.; Hawn, T.R.; Yi, E.C.; Goodlett, D.R.; Eng, J.K.; Akira, S.; Underhill, D.M.; Aderem, A. The innate immune response to bacterial flagellin is mediated by Toll-like receptor 5. *Nature* **2001**, *410*, 1099–1103. [CrossRef]
94. Wu, J.; Chen, Z.J. Innate immune sensing and signaling of cytosolic nucleic acids. *Annu. Rev. Immunol.* **2014**, *32*, 461–488. [CrossRef]

95. Jentho, E.; Weis, S. DAMPs and Innate Immune Training. *Front. Immunol.* **2021**, *12*, 699563. [CrossRef]
96. Chen, G.Y.; Nuñez, G. Sterile inflammation: Sensing and reacting to damage. *Nat. Rev. Immunol.* **2010**, *10*, 826–837. [CrossRef] [PubMed]
97. Gong, T.; Liu, L.; Jiang, W.; Zhou, R. DAMP-sensing receptors in sterile inflammation and inflammatory diseases. *Nat. Rev. Immunol.* **2020**, *20*, 95–112. [CrossRef]
98. Matzinger, P. Tolerance, danger, and the extended family. *Annu. Rev. Immunol.* **1994**, *12*, 991–1045. [CrossRef]
99. Hrithik, M.T.H.; Ahmed, S.; Kim, Y. Damage signal induced by Bacillus thuringiensis infection triggers immune responses via a DAMP molecule in lepidopteran insect, Spodoptera exigua. *Dev. Comp. Immunol.* **2023**, *139*, 104559. [CrossRef] [PubMed]
100. Chierici, E.; Marchetti, E.; Poccia, A.; Russo, A.; Giannuzzi, V.A.; Governatori, L.; Zucchi, L.; Rondoni, G.; Conti, E. Laboratory and field efficacy of natural products against the invasive pest Halyomorpha halys and side effects on the biocontrol agent Trissolcus japonicus. *Sci. Rep.* **2025**, *15*, 4622. [CrossRef]
101. Hancz, C.; Sultana, S.; Nagy, Z.; Biró, J. The Role of Insects in Sustainable Animal Feed Production for Environmentally Friendly Agriculture: A Review. *Animals* **2024**, *14*, 1009. [CrossRef]
102. Haider, K.; Abbas, D.; Galian, J.; Ghafar, M.A.; Kabir, K.; Ijaz, M.; Hussain, M.; Khan, K.A.; Ghramh, H.A.; Raza, A. The multifaceted roles of gut microbiota in insect physiology, metabolism, and environmental adaptation: Implications for pest management strategies. *World J. Microbiol. Biotechnol.* **2025**, *41*, 75. [CrossRef] [PubMed]
103. Mannaa, M.; Mansour, A.; Park, I.; Lee, D.-W.; Seo, Y.-S. Insect-based agri-food waste valorization: Agricultural applications and roles of insect gut microbiota. *Environ. Sci. Ecotechnol.* **2024**, *17*, 100287. [CrossRef]
104. Crotti, E.; Balloi, A.; Hamdi, C.; Sansonno, L.; Marzorati, M.; Gonella, E.; Favia, G.; Cherif, A.; Bandi, C.; Alma, A. Microbial symbionts: A resource for the management of insect-related problems. *Microb. Biotechnol.* **2012**, *5*, 307–317. [CrossRef]
105. Bode, H.B. Insect-Associated Microorganisms as a Source for Novel Secondary Metabolites with Therapeutic Potential. In *Insect Biotechnology*; Vilcinskas, A., Ed.; Springer: Dordrecht, The Netherlands, 2011; pp. 77–93.
106. Krishnan, M.; Bharathiraja, C.; Pandiarajan, J.; Prasanna, V.A.; Rajendhran, J.; Gunasekaran, P. Insect gut microbiome—An unexploited reserve for biotechnological application. *Asian Pac. J. Trop. Biomed.* **2014**, *4*, S16–S21. [CrossRef]
107. Singh, B.; Mal, G.; Gautam, S.K.; Mukesh, M.; Singh, B.; Mal, G.; Gautam, S.K.; Mukesh, M. Insect Gut—A treasure of microbes and microbial enzymes. In *Advances in Animal Biotechnology*; Springer: Cham, Switzerland, 2019; pp. 51–58.
108. Xie, S.; Lan, Y.; Sun, C.; Shao, Y. Insect microbial symbionts as a novel source for biotechnology. *World J. Microbiol. Biotechnol.* **2019**, *35*, 1–7. [CrossRef]
109. Liang, X.; Sun, C.; Chen, B.; Du, K.; Yu, T.; Luang-In, V.; Lu, X.; Shao, Y. Insect symbionts as valuable grist for the biotechnological mill: An alkaliphilic silkworm gut bacterium for efficient lactic acid production. *Appl. Microbiol. Biotechnol.* **2018**, *102*, 4951–4962. [CrossRef]
110. Mondal, S.; Somani, J.; Roy, S.; Babu, A.; Pandey, A.K. Insect microbial symbionts: Ecology, interactions, and biological significance. *Microorganisms* **2023**, *11*, 2665. [CrossRef] [PubMed]
111. Mika, N.; Zorn, H.; Rühl, M. Insect-derived enzymes: A treasure for industrial biotechnology and food biotechnology. In *Yellow Biotechnology II: Insect Biotechnology in Plant Protection and Industry*; Springer: Cham, Switzerland, 2013; pp. 1–17.
112. Shi, W.; Xie, S.; Chen, X.; Sun, S.; Zhou, X.; Liu, L.; Gao, P.; Kyrpides, N.C.; No, E.-G.; Yuan, J.S. Comparative genomic analysis of the endosymbionts of herbivorous insects reveals eco-environmental adaptations: Biotechnology applications. *PLoS Genet.* **2013**, *9*, e1003131. [CrossRef]
113. Kipkoech, C. Beyond proteins—Edible insects as a source of dietary fiber. *Polysaccharides* **2023**, *4*, 116–128. [CrossRef]
114. Khanal, P.; Pandey, D.; Næss, G.; Cabrita, A.R.; Fonseca, A.J.; Maia, M.R.; Timilsina, B.; Veldkamp, T.; Sapkota, R.; Overrein, H. Yellow mealworms (*Tenebrio molitor*) as an alternative animal feed source: A comprehensive characterization of nutritional values and the larval gut microbiome. *J. Clean. Prod.* **2023**, *389*, 136104. [CrossRef]
115. Aguilar-Toalá, J.E.; Cruz-Monterrosa, R.G.; Liceaga, A.M. Beyond Human Nutrition of Edible Insects: Health Benefits and Safety Aspects. *Insects* **2022**, *13*, 1007. [CrossRef] [PubMed]
116. Baiano, A. Edible insects: An overview on nutritional characteristics, safety, farming, production technologies, regulatory framework, and socio-economic and ethical implications. *Trends Food Sci. Technol.* **2020**, *100*, 35–50. [CrossRef]

veterinary sciences

MDPI

Article

Chinese Herbal Medicine Compound Microecological Agent (C-MEA) Improves Egg Production Performance in Caged Laying Ducks via Microbiota–Gut–Ovary Axis

Yanfeng Lu [1], Lei Zhang [1], Rui Zhu [1,2], Xiujun Duan [1,2], Guobo Sun [1,*] and Yuying Jiang [1]

1 School of Animal Science and Technology, Jiangsu Agri-Animal Husbandry Vocational College, Taizhou 225300, China; luyanfeng@jsnmkjzyxy6.wecom.work (Y.L.); leizhang@jsahvc.edu.cn (L.Z.); rzhu2021@jsahvc.edu.cn (R.Z.); sgb1981@126.com (X.D.); m19517822335@163.com (Y.J.)

2 Jiangsu Key Laboratory for High-Tech Research and Development of Veterinary Biopharmaceuticals, Engineering Technology Research Center for Modern Animal Science and Novel Veterinary Pharmaceutic Development, Taizhou 225300, China

* Correspondence: 2007020169@jsahvc.edu.cn; Tel.: +86-15850851019

Simple Summary: Farming laying ducks and processing their eggs is a traditional industry in China that is gradually transitioning towards modern cage-rearing systems to achieve large-scale and high-efficiency production. However, cage rearing alters ducks' natural behaviors, making them prone to individual stress, which may lead to pathological conditions, ultimately impacting laying performance. To explore antibiotic-free additives, this study investigated the effects of a Chinese herbal medicine compound microecological agent (C-MEA) on the egg production performance, ovarian follicle development, ovary transcriptome, and cecal microbiota of caged laying ducks. The results show increased egg production and follicle development due to up- and downregulated candidate key genes, as well as increased cecal microbiota diversity. Our findings suggest that the C-MEA enhances egg production, ovary function, and microbial diversity, offering new insights into improving the gut health and reproductive performance of caged laying ducks.

Abstract: This study was conducted to investigate the effects of a Chinese herbal medicine compound microecological agent (C-MEA) on the egg production performance, ovarian follicle development, ovary transcriptome, and cecal microbiota of caged laying ducks. A total of 108 black Muscovy ducks (150 days old) were randomly divided into three groups for 30 days in a formal feeding trial. Compared with the control basic diet (Group C) and 16 g/kg C-MEA dosage (Group B), the 8 g/kg C-MEA dosage (Group A) increased egg production (average laying rate 69.35%) and follicle development (5~7 Fs, 6~7 LYFs, 11~13 SYFs) mass ($p < 0.05$). According to RNA-Seq, the ovaries' transcriptome among different dietary groups enriched six key pathways, including neuroactive ligand–receptor interaction, the PPAR signaling pathway, ECM–receptor interaction, focal adhesion, the adherens junction, and the FoxO signaling pathway, as well as 46 candidate key genes. According to 16S-Seq, the microbial diversity was significantly increased in Group A, and the genus abundances of *Sphaerochaeta* and *UCG-004* were significantly changed among different dietary groups ($p < 0.05$). Supplementation with C-MEA may optimize the cecal microflora and the interactions between the intestinal microflora and the host. The results from combining RNA-Seq and 16S-Seq demonstrated that the relationship between *Sphaerochaeta* and the hub gene cluster (*F2, KNG1, C5, PLG, F2RL1, FABP1*, and *GCG*) is the most prominent. In conclusion, the egg performance of caged laying ducks can be modulated through the microbiota–gut–ovary axis. Our findings provide new insights for improving gut health and reproductive performance of caged laying ducks.

Keywords: *Intestinal microbiota*; caged laying ducks; egg production; C-MEA; follicle development

1. Introduction

The farming of laying ducks and duck egg processing are a traditional industry in China, accounting for over 90% of global laying duck production [1]. In 2024, the laying duck population reached approximately 163 million in stock, yielding 2.914 million tons of commercial duck eggs, with a total industry value of CNY 41.0 billion [2]. With the continuous expansion of breeding scale and improvements in standardized raising levels, traditional water- or ground-based rearing systems can no longer meet the needs of industrial development. There is an urgent need to transform duck farming methods to alleviate environmental pressure and further enhance productivity [3]. Compared to conventional free-range or captive breeding systems, cage-rearing systems for laying ducks enable intensive, large-scale production through mechanization with high breeding efficiency. This approach significantly improves farming efficiency while facilitating the centralized management of manure and waste, thereby substantially reducing environmental pollution [4]. However, cage rearing alters ducks' natural behaviors, making them prone to individual stress, which may lead to diarrhea, compromised immunity, increased mortality, and altered microbial diversity, thereby ultimately impacting laying performance [5–7]. Consequently, implementing effective measures to mitigate stress in caged laying ducks holds critical significance.

With the current comprehensive ban on antibiotics in the raising industry and the rising concern over antibiotic resistance, the search for non-toxic, residue-free antibiotic alternatives has become an urgent need. Microecological agents, a novel class of biological additives designed to rebuild and optimize the diversity of animal gut microbiota, may promote nutrient absorption, enhance intestinal barrier function, and modulate the host's immune response [8,9]. Traditional Chinese herbal medicine, derived from natural animals and plants, exhibits antibacterial, anti-inflammatory, stress-reducing, and endocrine-regulating effects, which can significantly improve the growth performance, immune function, and gut microbiota composition of animals. To combine both advantages of microecological agents and traditional Chinese herbal medicine, a specific novel microecologic agent (C-MEA) has gradually become a research hotspot in animal husbandry [10]. Wang et al. [11] compared Chinese herbs, probiotics, and prebiotics with antibiotics on the performance of Pekin ducks, and they demonstrated that the additives of Chinese herbs and probiotics had no side effects on the ducks. Gao et al. [12] reported that combining traditional Chinese medicine with probiotics in broiler chicken feed created a synergistic effect, improving the growth performance, immune function, and intestinal health of broiler chickens. Compared with antibiotics, C-MEA offers significant advantages, including safety, environmental compatibility, residue-free properties, and the absence of drug resistance, and it shows significant value for potential application as a feed additive within animal husbandry. However, the application of C-MEA in animal husbandry production remains in the exploratory stage, with particularly limited research reports on its application in the duck rearing industry.

In this study, three dietary inclusion levels of C-MEA were designed, and the regulatory effects on laying performance, ovarian follicle development, ovary transcriptome, and cecal microbiota were evaluated in caged laying ducks. The aim of this study was to understand in-depth the changes in production properties, ovary function, and intestinal health of caged laying ducks in response to the inclusion of dietary supplements of different levels of C-MEA in order to furnish a reference basis for implementing C-MEA in the duck industry and to facilitate the transition towards environmentally sustainable animal husbandry.

2. Materials and Methods

2.1. Preparation of C-MEA

The C-MEA (Q/01JXDD ST05-2021) used in this experiment was provided by JINXIUDADI Animal Pharmaceutical Co., Ltd. (Shijiazhuang, Hebei, China). The main ratio was 20 g of Huangqi, 15 g of licorice, 15 g of Ciwujia, 10 g of Houttuynia cordata, 10 g of dandelion, 10 g of chrysanthemum, 10 g of lily, 10 g of motherwort, as well as 0.15 g of Enterococcus faecalis, 0.1 g of Bacillus subtilis, and 0.1 g of Clostridium butyricum.

2.2. Experimental Design and Sample Collection

A total of 108 female 150-day-old black Muscovy ducks individually housed in distinct single cages were raised at the National Gene Bank of Waterfowl Resources in Jiangsu, China. The ducks were randomly divided into 3 groups by a single-factor experimental design (36 ducks per group): (1) Group A, basic diet + 8 g/kg C-MEA; (2) Group B, basic diet + 16 g/Kg C-MEA; and (3) Group C, basic diet without C-MEA. The basic diet was the same commercial formula diet (Twin Biochemical Co., Ltd., Hangzhou, Zhejiang, China), which mainly contained corn, wheat, and soya bean meal. The addition of C-MEA was based on our previous study. The total experimental period was 37 days, including 7 days of the pre-trial period (P1–7) and 30 days of the formal period (F1–30). At the end of the trial period, 6 ducks per group were randomly selected for sampling. Ovaries and digesta of the cecum were carefully collected and promptly snap-frozen in liquid nitrogen for subsequent experiments.

2.3. Laying Performance

Eggs were picked up manually at 16:00 every day under the same management conditions. Duck egg production was recorded daily to calculate the daily laying rate. At the end of the experiment, a total of 90 valid individual records were collected.

2.4. Analysis of Ovarian Follicles

A total of 18 ovarian tissues were separated and compared as representative individuals among three groups. According to the follicle diameter, the follicles of different grades were divided into preovulatory follicle (F), large yellow follicle (LYF), and small yellow follicle (SYF) categories, as described by Gilbert [13] and Kun Zou et al. [14]. The Fs, LYFs, and SYFs were counted in three sections from each individual.

2.5. Ovary Transcriptome Data Analysis

Three ovarian stroma samples were selected randomly from each group, and then the qualified total RNA was extracted and transported to Novogene Co., Ltd. (Beijing, China) for transcriptome sequencing (RNA-Seq). The libraries were sequenced on an Illumina HiSeq 2500 system sequencing platform. The clean reads were mapped to the Muscovy duck's reference genome (GCA_048319975.1) using hisat2 (2.0.5) for the downstream analysis as previously described [15]. The differential expressed genes (DEGs) between samples were identified using the DESeq2 R package (1.20.0) [16], and $p < 0.05$ and $\log_2 |\text{foldchange}| \geq 1$ were used as the criteria of significance of DEGs. The GO term and KEGG pathway analyses were performed by clusterProfiler software (3.8.1) [17], and

$p < 0.05$ was considered significantly enriched. The STRING database [18] was used to explore the interaction between DEGs. The sequencing raw data were deposited into the Genome Sequence Archive in the BIG Data Center (https://ngdc.cncb.ac.cn/gsa, accessed on 3 August 2025) with accession number PRJCA044169.

2.6. Quantitative Reverse Transcription Polymerase Chain Reaction (qRT-PCR) Analysis

To validate the RNA-Seq results, the expression of 7 randomly selected DEGs was assessed using qRT-PCR. Total RNA was extracted from ovary samples using the TRIzol Kit (Invitrogen). The primers (Supplementary Table S1) were designed using Primer 5.0 software. qPCR was performed on a CFX96 Real-Time System (Bio-RAD). The relative gene expression levels of selected DEGs were quantified based on *β-actin* gene expression by the $2^{-\Delta\Delta Ct}$ method.

2.7. Cecal Microbiota Analysis

Genomic DNA was extracted from cecum contents using the QIAamp 96 PowerFecal QIAcube HT kit (QIAGEN, Germany) according to the manufacturer's protocols and then evaluated by 1% agarose gels. Then, two DNA samples from each group were pooled as a 16S sequencing sample ($n = 3$ per group, pooled from 6 individuals). The 16S rDNA V3-V4 region of the ribosomal RNA gene was amplified from genomic DNA using forward primer 341F: CCTACGGGNGGCWGCAG, 806R: GGACTACHVGGGTATCTAAT, of which the PCR requirements described in [19] were followed. The PCR product was analyzed on 2% agarose gel electrophoresis, purified with the QIAquick gel extraction kit (QIAGEN, Frankfurt, Germany). Then, the amplicons were subsequently sequenced (16S-Seq) using the Illumina NovaSeq 6000 platform, and bio-informatics analysis was performed by Novogene Co., Ltd. (Beijing, China). The sequencing raw data were deposited into the Genome Sequence Archive in the BIG Data Center (https://ngdc.cncb.ac.cn/gsa) with accession number PRJCA044144.

The raw fastq files were merged by FLASH (1.2.11) to obtain the raw tags, which were then filtered by fastp (0.23.1) to obtain the clean tags. Using the Silva database (https://www.arb-silva.de) and Unite database (https://unite.ut.ee), these tags were compared to obtain the effective tags, which were clustered and classified into operation taxonomic units (OTUs) of ≥95% similarity by utilizing QIIME2 software (202202). The Tax4Fun2 software package (201807) was applied to predict the functional capabilities of microbial communities. The Spearman correlation co-efficient between environmental factors and genera was calculated using the R project psych package (version 1.8.4) [20].

2.8. Statistical Analysis

For 16S sequencing data, the alpha and beta diversity analysis was evaluated by QIME2 (202202). The alpha and beta indices were compared across groups with Kruskal–Wallis and ANOSIM tests using R software (V3.4.3), respectively. All statistical analyses in this study were performed with statistical testing using one-way analysis of variance (ANOVA). The results are expressed as mean ± standard deviation. * signifies a significant difference between means ($p < 0.05$), and ** signifies an extremely significant difference between means ($p < 0.01$).

2.9. Ethics Statement

The protocol was performed after the approval of the Committee on Experimental Animal Management of Jiangsu Agri-Animal Husbandry Vocational College, Taizhou,

China (No. Jsahvc-2023-94), and every effort was made to minimize animal suffering during the experiments.

3. Results

3.1. Effects of C-MEA on Laying Performance of Caged Laying Ducks

To accurately assess the effect of the C-MEA on the egg-laying performance of caged ducks, we collected the 30-day egg production rates of all three groups after a 7-day pre-trial period. Based on the records, an egg production rate $\geq$ 75% was defined as the peak laying period, as these belongs to the top 25% of total egg production, while egg production rate $\leq$ 55% (bottom 25%) was defined as the low-laying period. As shown in Figure 1a, Group A (8 g/kg) maintained the peak laying period from E5 to E30 and subsequently declined slowly to approximately 66.67%, of which the average egg production rate was 69.35%; Group B (16 g/kg) maintained peak production from E4 to E30 and subsequently declined slowly to 58.33%, of which the average egg production rate was 65.00%; and Group C (basic diet control) exhibited the lowest egg production efficiency (59.17%), maintained peak production only from E5 to E24, and then dropped to 47.22% at the end of the trial period. From the results, Group A demonstrated a significant improvement in egg production performance (Figure 1d).

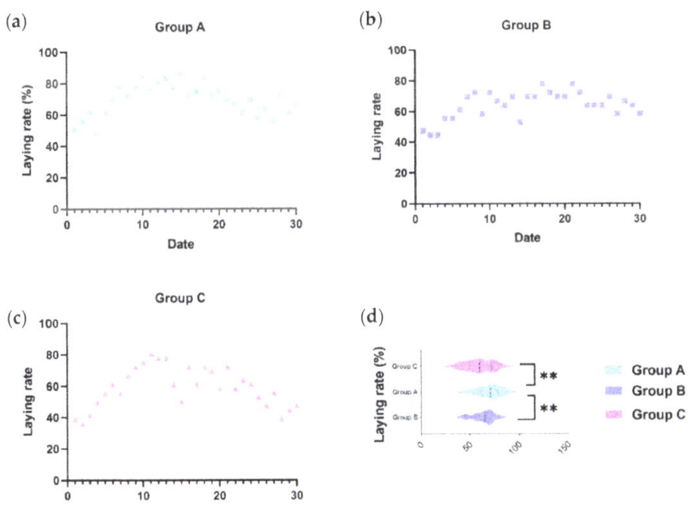

Figure 1. The effects of C-MEA on the laying performance of caged laying ducks, specifically Groups (**a**) A, (**b**) B, and (**c**) C, and laying rate comparasions among three groups (**d**). ** represents an extremely significant difference between means ($p < 0.01$).

3.2. Effects of C-MEA on Follicular Development of Caged Laying Ducks

At E30, F, LYF, and SYF numbers were counted in three groups from 18 individuals (Figure 2a–c). As shown in Figure 2d, there were significantly more Fs, LYFs, and SYFs from Group A than those from Groups B and C. Based on the egg-laying records, the egg-laying efficiency of Group C was lower than 50%, and there were only 1~2 preovulatory follicles on its ovary (Figure 2a). Meanwhile, there were 5~7 and 4~5 preovulatory follicles in Groups A and B, respectively. There were 6~7, 3~4, and 2~3 LYFs in Groups A, B, and C, respectively. There were only 4~5 SYFs in Group C, whereas Group A had 11~13 and

Group B had 6~8. The Fs, LYFs, and SYFs in the C-MEA groups (Groups A and B) were all significantly increased compared to those of the control group (Group C) ($p < 0.05$).

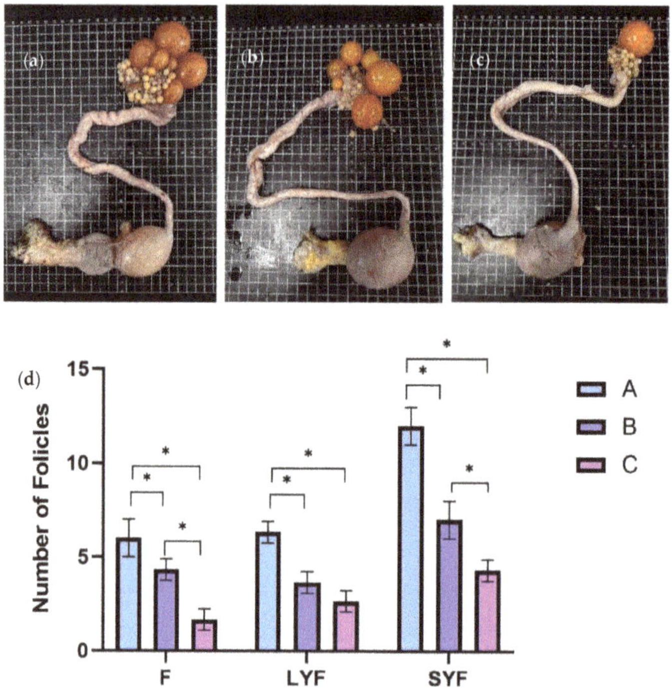

Figure 2. The effects of C-MEA on the follicular development of caged laying ducks, specifically Groups (**a**) A, (**b**) B, and (**c**) C, and the comparasions of follicles' numbers among three groups (**d**). * represents an extremely significant difference between means ($p < 0.05$).

3.3. Analyzing RNA-Seq Data as Compositions

To analyze the mechanism by which the Chinese herbal medicine compound microecological preparation improved the egg production performance of caged laying ducks, transcriptome sequencing was performed on the ovaries of ducks in Groups A, B, and C at E30. Through library construction and platform sequencing, the raw data were filtered by quality control, and nine samples were obtained with 43, 482, 166~49, 194, and 710 clean data (see Supplementary Table S2). The proportion of bases with an accuracy rate > 99.9% (Q30) in the total sequence after filtering exceeded 92.8%. This indicates that the transcriptome data collected in this experiment were of high quality and that the sequencing data obtained were reliable, making them suitable for subsequent analysis. Based on the positional information of gene alignments on the reference genome, the FPKM (fragments per kilobase of exon model per million mapped fragments) of all genes in each sample were quantified. Subsequently, correlation coefficients within and between groups were calculated by FPKM values of all genes across samples. These correlations were visualized in a heatmap, as shown in Figure 3a, and the biological replicates within each group clustered together, indicating a distinct sample grouping situation. The DEGs of the three groups were analyzed by DESeq2 (1.20.0) (Figure 3b). Venn analysis showed that a total of 16,726 DEGs were obtained from the transcriptomes of the three ovarian groups, including 91 DEGs specific to Group A, 174 DEGs specific to Group B, and 287 DEGs specific to Group C (Figure 3c). Pairwise comparison showed 143 up- and 251 downregulated DEGs in A vs. C, 109 up- and 199 downregulated DEGs in B vs. C, and 166 up- and 440 downregulated DEGs in A vs. B (Figure 3d).

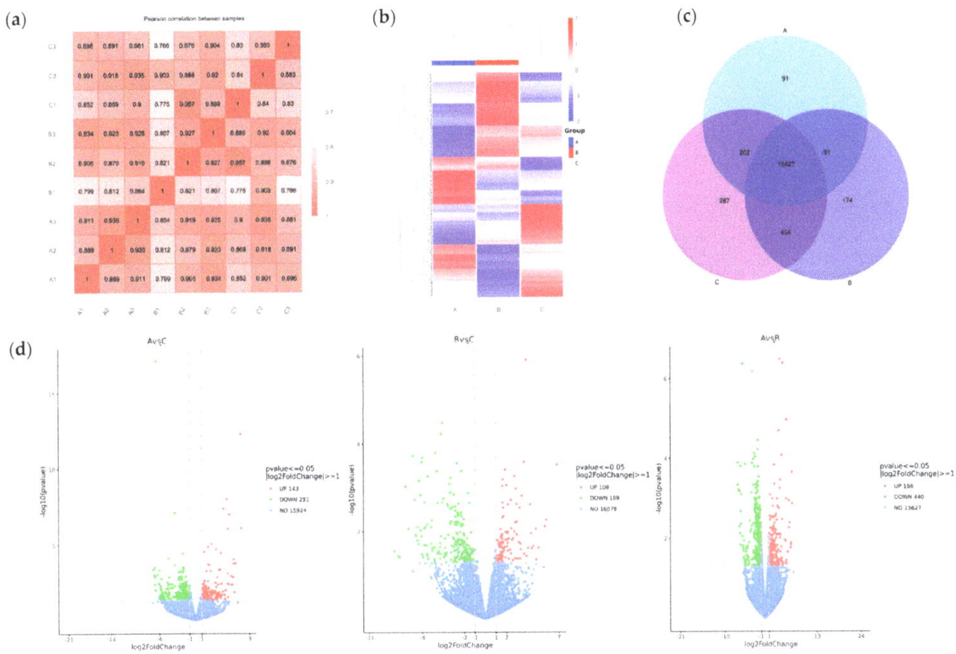

Figure 3. RNA-Seq-based analysis of DEGs in the ovaries of caged laying ducks from different dietary groups. (**a**) The correlation heatmap between samples. (**b**) The cluster analysis result of DEGs in different groups. (**c**) The Venn analysis result of DEGs in different groups. (**d**) The volcano plot of up- and downregulated DEGs of different pairwise comparisons.

GO enrichment analysis was performed to further express the functional roles of DEGs in different feed groups. The top 30 most significant GO terms in the three pairwise comparisons are shown in Figure 4a. The most significant GO terms related to egg production in A vs. C included the G-protein-coupled receptor signaling pathway, calcium ion binding, extracellular matrix, etc. The most significant GO terms related to egg production in B vs. C included lipid transport, the extracellular region, G-protein-coupled receptor activity, etc. The most significant GO terms related to egg production in A vs. B included the oxidoreduction coenzyme metabolic process, lipid transport, extracellular matrix, etc. To better understand the biological functions and interaction of genes, KEGG pathway analysis was performed for the identified DEGs. The top 20 significantly enriched pathways ($p < 0.05$) by DEGs of each pairwise comparison are shown in Figure 4b, including common reproductive-related pathways such as neuroactive ligand–receptor interaction, the PPAR signaling pathway, beta-alanine metabolism, pantothenate and CoA biosynthesis, the hedgehog signaling pathway, and the lysosome in A vs. C; ECM–receptor interaction, neuroactive ligand–receptor interaction, focal adhesion, the adherens junction, the PPAR signaling pathway, and the FoxO signaling pathway in B vs. C; and ECM–receptor interaction, focal adhesion, and neuroactive ligand–receptor interaction in A vs. B.

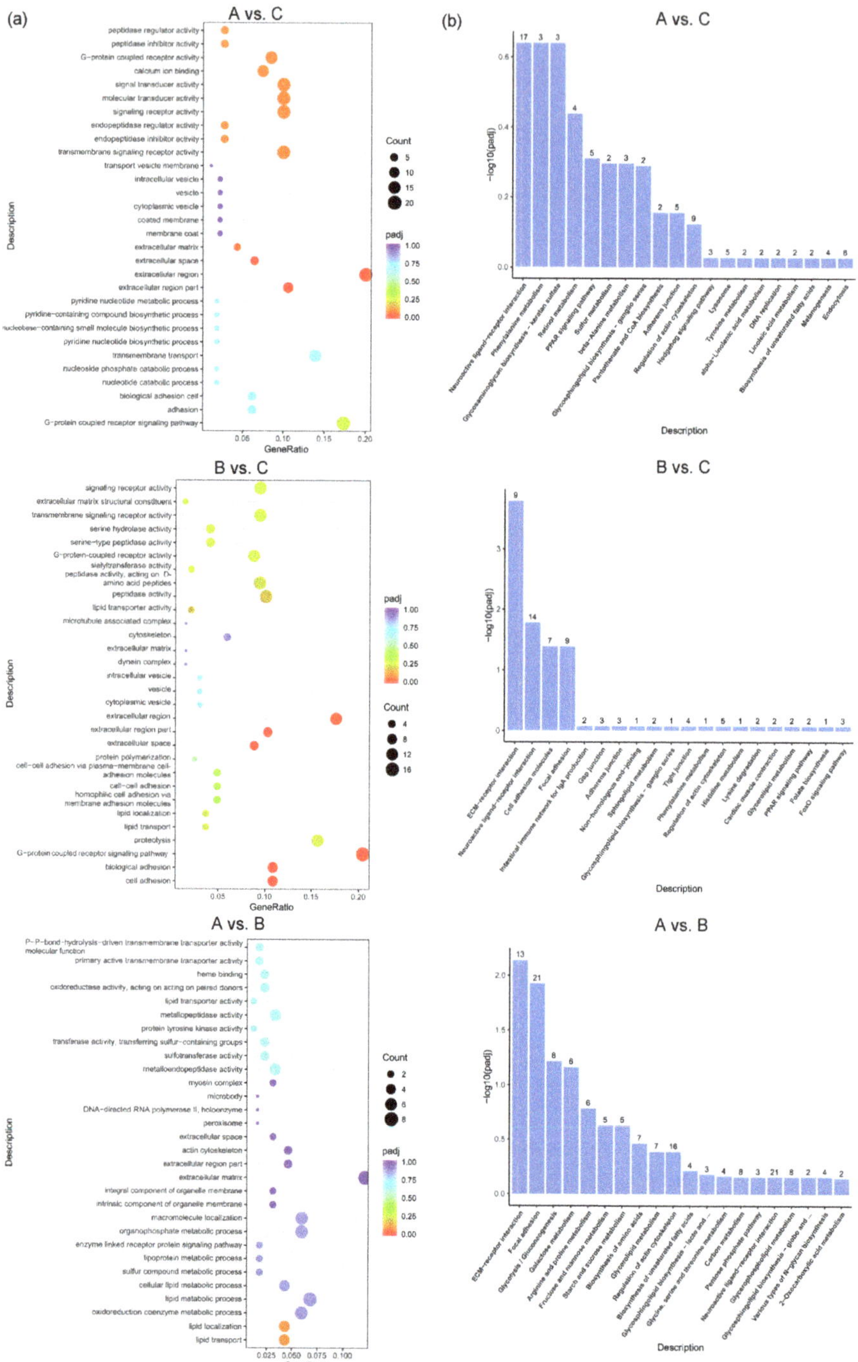

Figure 4. GO and KEGG analysis based on DEGs between different dietary groups of caged laying ducks. (**a**) Top 30 GO classifications and (**b**) top 20 KEGG analysis based on DEGs between different dietary groups of caged laying ducks, the number above the column represents the number of genes enriched in the pathway.

Based on the significance levels of KEGG pathways and the literature reviews of the function of these pathways, six potential pathways were considered related to the laying performance of caged ducks associated with Chinese herbal medicine compound microecological preparations, including neuroactive ligand–receptor interaction, the PPAR signaling pathway, ECM–receptor interaction, focal adhesion, the adherens junction, and the FoxO signaling pathway. Therefore, 73 DEGs were enriched in these six signaling pathways and defined as candidate genes potentially mediating the effect of the microecological preparation of Chinese herbal medicine compound on the differences in egg production performance among the different feed groups of caged laying ducks. To investigate the molecular regulatory mechanisms of these 73 DEGs, a network of mRNA interactions was established by Cytoscape software (3.7.2). Two groups of gene interaction clusters are shown in Figure 5a, and the top cluster had 46 DEGs, which were defined as candidate key genes that may regulate egg production based on the addition of C-MEA. Seven candidate key genes were randomly selected for qRT-PCR validation (Figure 5b). A comparison of RNA-Seq and qRT-PCR data showed that these genes exhibited consistent expression trends, which suggested that transcriptomic sequencing data and candidate key genes from RNA-Seq had high reliability and accuracy.

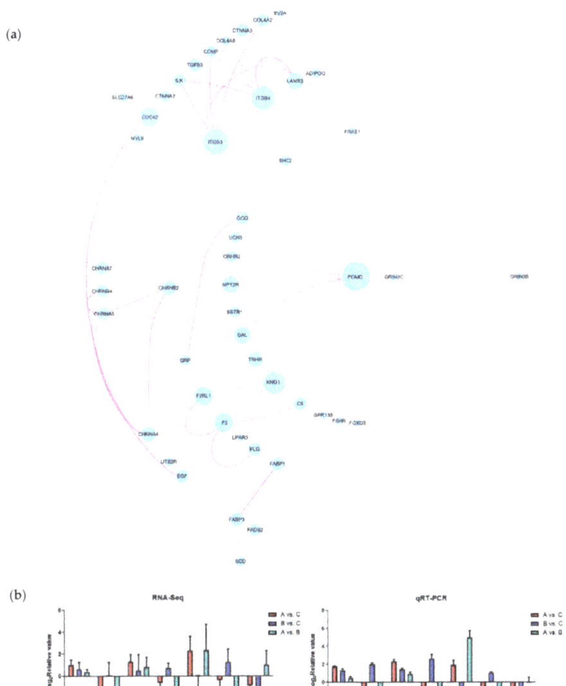

Figure 5. Analysis and verification of candidate key genes in the ovary among different dietary groups of caged laying ducks. (**a**) Interaction network of candidate key genes from RNA-Seq. (**b**) Verification of candidate key genes using qRT-PCR.

3.4. Analyzing 16S rRNA Gene Tag Sequencing Data as Compositions

After the initial filtering and adaptor trimming process, samples from caged Muscovy ducks with different feeds accounted for a total of 933,790 RawPE. After sequence assembly, data filtration, and chimera removal, 839,307 effective tags were obtained, which were classified into 818,289 effective tags (Nochime). The number of reads that passed through each step of assembly and control are presented in Supplementary Table S3. Group A displayed the greatest number of OTUs, and the three groups shared 607 OTUs that were common to the cecum, as shown in Figure 6a. The rarefaction curves for the groups were infinitely close to the saturation plateau, proving that the sequencing results contained enough depth to capture most microbial diversity information and could be used for further analysis (Figure 6b). Based on the 2175 OTUs, the alpha diversity of cecal bacterial communities was compared between different dietary groups. A significant difference in the Chao1, dominance, and Simpson indices was observed between the A and C groups ($p < 0.05$, Figure 7a,b,d), and the dietary A group had a significantly higher richness in the Pielou_e index than the B group ($p < 0.05$, Figure 7c).

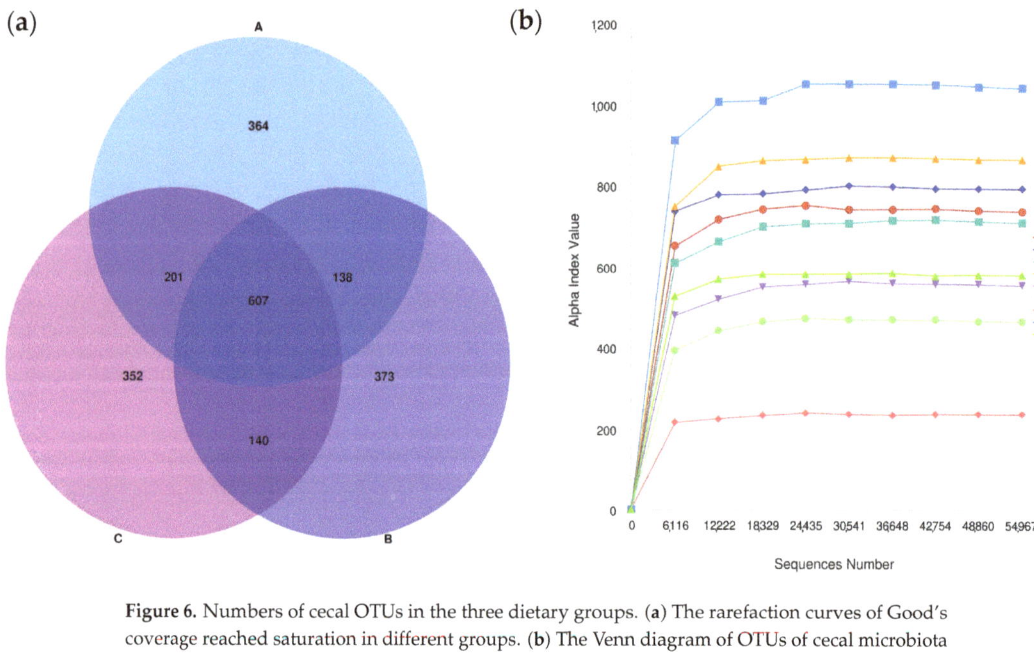

Figure 6. Numbers of cecal OTUs in the three dietary groups. (**a**) The rarefaction curves of Good's coverage reached saturation in different groups. (**b**) The Venn diagram of OTUs of cecal microbiota in different groups of laying ducks.

Based on the analysis of the bacterial composition of different dietary groups at the phylum level, *Bacteroidota* and *Firmicutes* dominated all three groups (Figure 8a). Furthermore, the relative abundance of *Spirochaetota*, *Synergistota*, and *Verrucomicrobiota* was clearly higher than those in the C group ($p < 0.05$, Figure 8d–f), while Chinese medicine combined with treatment with composite microecological preparations decreased the relative abundance of *Bacteroidota* and *Actinobacteriota* ($p < 0.05$, Figure 8b,c).

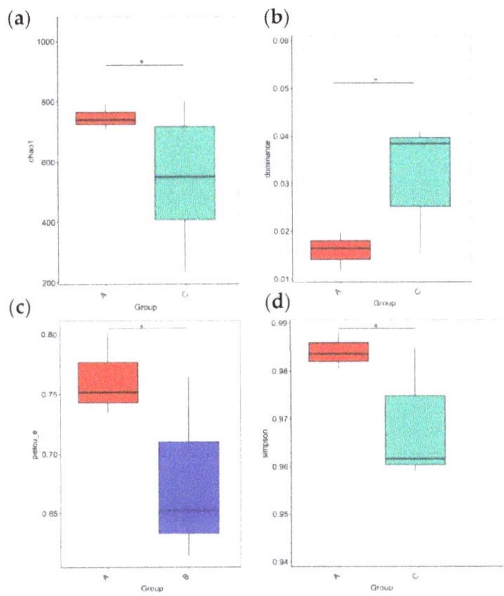

Figure 7. The alpha diversity of cecal microbiota between different dietary groups. (**a**) The Chao1 index of cecal microbiota between Groups A and C. (**b**) The dominance index of cecal microbiota between Groups A and C. (**c**) The Pielou_e index of cecal microbiota between Groups A and B. (**d**) The Simpson index of cecal microbiota between Groups A and C. * indicates statistical significance ($p < 0.05$).

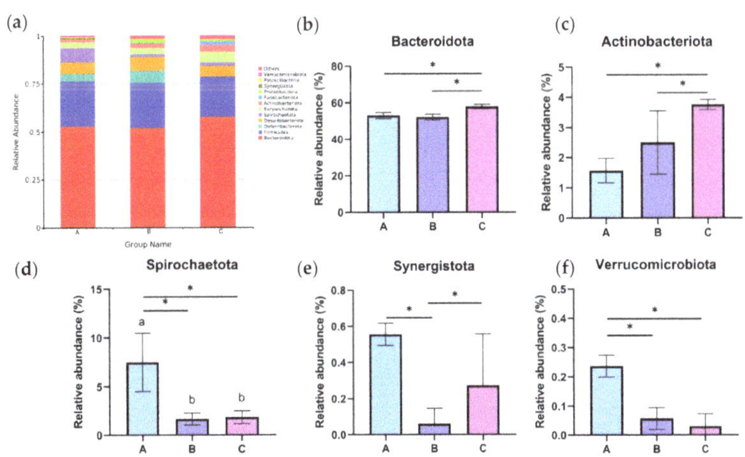

Figure 8. The relative abundance of the cecal microbiota at the phylum level in different dietary groups based on the 16S rDNA gene sequence. (**a**) The stack-column of the cecal microbiota from different groups at the phylum level. (**b**–**f**) The relative abundance of *Bacteroidota*, *Actinobacteriota*, *Spirochaetota*, *Synergistota*, and *Verrucomicrobiota* is expressed as mean ± SEM. * indicates significant differences at $p < 0.05$.

At the genus level, the top 20 bacterial taxa were clustered among the three groups to form heatmaps (Figure 9a). The relative abundances of *Sphaerochaeta* and *UCG-004* were determined (Figure 9b,c). Compared to Groups B and C, the relative abundances of *Sphaerochaeta* and *UCG-004* were significantly changed in Group A ($p < 0.05$).

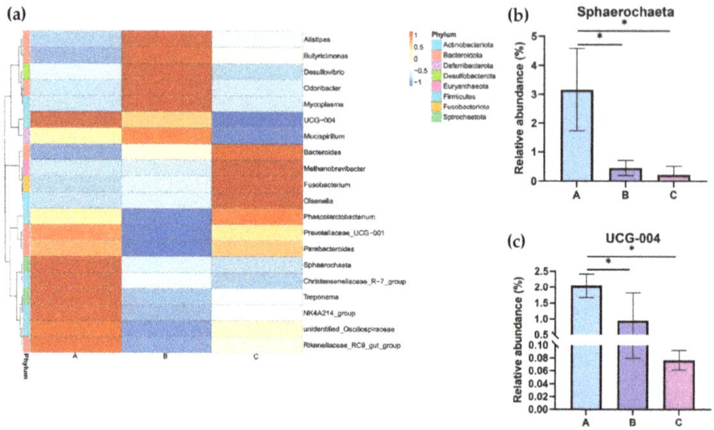

Figure 9. The relative abundance of the cecal microbiota at the genus level in different dietary groups based on the 16S rDNA gene sequence. (**a**) The stack-column of the cecal microbiota from different groups at the genus level. (**b**,**c**) The relative abundance of *Sphaerochaeta* and *UCG-004* is expressed as mean ± SEM. * indicates significant differences at $p < 0.05$.

Microbial contributions to between-group differences among the dietary groups were assessed using the LDA score. The results for the cecum comparisons (A vs. C) are illustrated in Figure 10a; the relative abundance of Coriobacteriales, Coriobacteriia, and Actinobacteriota was highest in the C group, but in the A group, that of *Sphaerochaeta* increased. The main taxa that differed between Groups B and C were the Bacilli, which increased in the B group (Figure 10b). The results from Groups A and B are shown in Figure 10c, where Group B, with the high concentration of C-MEA treatment, promoted the enrichment of *Sphaerochaeta*, *Spirochaetia*, *Spirochaetales*, *Spirochaetaceae*, and *Spirochaetota* but decreased the enrichment of *Clostridium_sp_CAG306* and *Bacilli*.

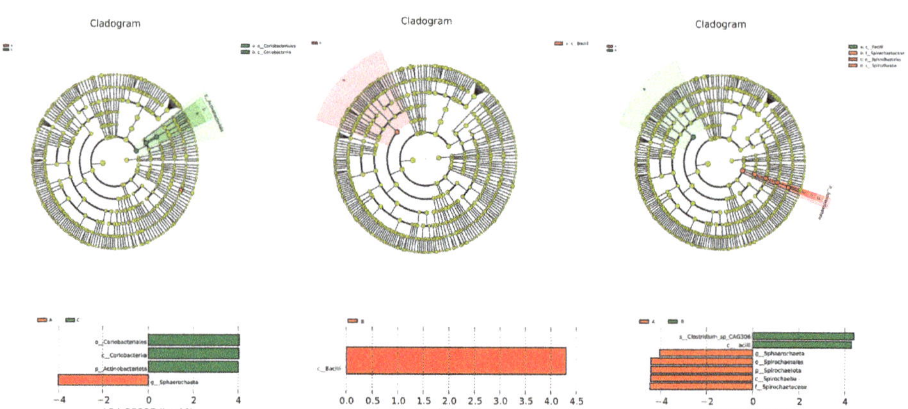

(a) (b) (c)

Figure 10. The main taxa of microbiota that were different in the three dietary groups. A cladogram of the main taxa of microbiota that were different between Groups A and C (**a**), Groups B and C (**b**), and Groups A and B (**c**) on the basis of LEfSe analysis (taxa with LDA score ≥ 4, $p < 0.05$).

3.5. Functional Predictions Based on the Gut Microbiota of Caged Laying Ducks

The physiologic functions of Tax4Fun showed that seven categories of KEGG pathways were classified in all three groups (Figure 11a), including cellular processes, environmental information processing, genetic information processing, human diseases, metabolism, organismal systems, and unclassified. Furthermore, the top 10 KEGGs are listed in Figure 11b. Compared to the control group (Group C), the addition of C-MEA treatment (Groups A and B) inhibited the KEGG pathways of lipid metabolism, biosynthesis of other secondary metabolites, carbohydrate metabolism, and metabolism of cofactors and vitamin, while it promoted the membrane transport pathway.

(**a**)

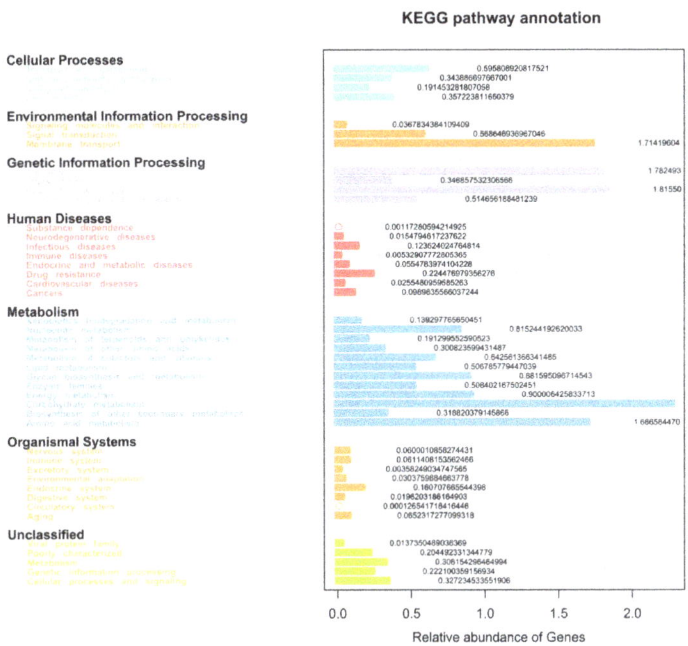

(**b**)

Figure 11. Tax4Fun functional profile of cecal microbiota communities under different dietary groups based on KEGG pathway analysis. (**a**) KEGG pathway prediction based on OTUs among Groups A, B, and C. (**b**) Top 10 KEGG heatmap based on OTUs among Groups A, B, and C.

3.6. Correlation Analysis Between the Microbiota, Laying Rate, and Follicular Status of Caged Laying Ducks

Spearman's correlation analysis (Figure 12) revealed a positive correlation between *Sphaerochaeta* and all four variables, including laying rate and F, LYF, and SYF numbers

($p < 0.05$). In addition, there were also three positive correlations between *UCG-004* and the LYF numbers ($p < 0.05$), *UCG-004* and the SYF numbers ($p < 0.05$), and *Christensenellaceae-R.7-group* and the SYF numbers ($p < 0.05$).

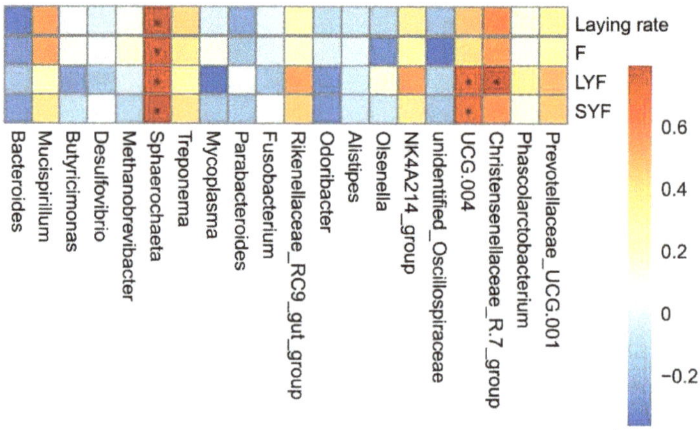

Figure 12. Correlation analysis between the microbiota and variables. Spearman's correlation analysis between the genus level of the gut microbiota (top 20) and the variables. The color (red to blue) indicates the correlation (positive to negative). * represents a significant correlation strength ($p < 0.05$).

3.7. Correlation Analysis Between the Microbiota and DEGs of Caged Laying Ducks

To investigate the potential regulation network between microbiota and mRNA, Spearman's correlation analysis was performed between the top 20 microbiota genera and 46 candidate key DEGs identified earlier. As shown in Figure 13, a bunch of DEGs showed a positive correlation with microbiota genera; for example, key hub gene *POMC* was positively correlated with *Bacteroides*, and *COMP* was positively correlated with *NK4A214group*. Only two DEGs showed negative correlations: *LAMB3-Olsenella* and *CHRNB2-Treponema*. The most attention was focused on *Sphaerochaeta*—a genus previously found to be significantly positively correlated with laying rate among different dietary groups. The analysis revealed positive correlations ($p < 0.01$) between *Sphaerochaeta* and seven genes: *F2*, *KNG1*, *C5*, *PLG*, *F2RL1*, *FABP1*, and *GCG*.

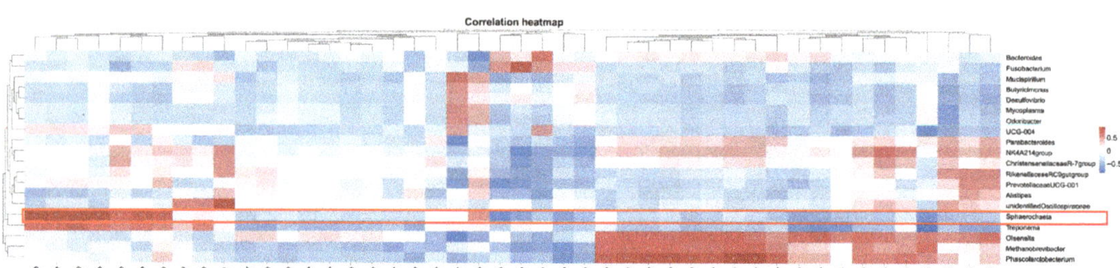

Figure 13. Correlation analysis between the microbiota and key genes. * represents a significant correlation strength ($p < 0.05$), and ** represents an extremely significant correlation strength ($p < 0.01$).

4. Discussion

Numerous studies have demonstrated that dietary supplementation with components of Chinese herbal medicine or a microecological agent has positive effects on ducks. Chen et al. [21] reported that honeycomb (one kind of traditional Chinese medicine) extracts could improve the fatty acid composition and amino acid content of duck eggs, thus improving the eggs' flavor. By use of dietary Artemisia argyi powder, the average laying rate, total PUFA contents of egg yolk, and amino acid levels of eggs had all been improved in laying ducks [22]. Incharoen et al. [23] explored the effects of Bacillus toyonensis BCT-7112^T on laying ducks; the compound was shown to enhance egg production, gut health, and microbial diversity. With the cross-development of different subjects, such as microecology, fermentation engineering, and pharmacy, a kind of novel microecological product, the Chinese herbal medicine compound microecological agent (C-MEA), has sprung up. This formulation can make good use of both the efficacy of Chinese herbal medicine and the microecological agent through their synergistic actions [24,25]. C-MEA may have the potential to develop into a substitute for antibiotics in poultry production, but there are few studies on C-MEA on laying ducks. Our results indicated that feeding ducks with 8 or 16 g/kg of C-MEA was able to clearly increase both egg production and the number of follicles of caged laying ducks during the 30 days of the formal experiment. The above results indicated that cage rearing induces greater stress in laying ducks, potentially causing irreversible reproductive system damage manifested as ovarian dysfunction and significantly reduced egg production. However, the C-MEA promoted grade follicle development in the ovaries of caged ducks and effectively maintained egg production efficiency compared to the control group. One hypothesis for improved production performance is related to the microbiota–gut–ovary axis, in which intestinal microbiota are exploited to enhance the ovary function of laying poultry to improve laying performance. To investigate the potential regulation mechanism of C-MEA on egg performance, we conducted ovary RNA sequencing and cecal 16S sequencing of caged laying ducks among different dietary groups.

The poultry's ovary is a crucial reproductive organ that directly influences the egg production trait. RNA sequencing data analysis comparing three dietary inclusion levels of C-MEA in the ovaries of caged laying ducks revealed a total of 16,726 DEGs. To obtain a preliminary understanding of the function of these DEGs, top 30 GO enrichment analysis was carried out, and their functional roles were mainly in egg production, such as the G-protein-coupled receptor pathway, calcium ion binding, and extracellular matrix. To narrow down the potential candidate key genes that may regulate egg production based on the addition of C-MEA, KEGG enrichment analysis was further carried out. Based on the significance levels of KEGG pathways and the literature reviews of the function of these pathways, this study mainly focused on neuroactive ligand–receptor interaction [26], the PPAR signaling pathway [27], ECM–receptor interaction [28–30], focal adhesion [31–33], the adherens junction [34,35], and the FoxO signaling pathway [36,37], which are involved in egg production performance or follicular development. Therefore, a total of 73 DEGs were enriched from the above six signaling pathways, among which 46 DEGs showed interaction with each other and were defined as candidate key genes in this study, including *POMC*, *FSHR*, *EGF*, *FOXO1*, *FOXO3*, *GCG*, *ITGB3*, *ITGB4*, *F2RL1*, and *FABP1*. The top enriched hub gene was *POMC* (Pro-opiomelanocortin), a polypeptide precursor of several peptide hormones and neuropeptides, which plays an important role in mediating stress, health, survival, and reproduction [38,39]. Liu et al. [40,41] detected the polymorphism of *POMC* in Zhenning yellow chickens and explored the SNP of g.1140C > T that was associated with the E300 egg production strain, while the SNPs of g.958 G > A and g.1817 C > T were significantly associated with E_2 hormone levels and reproduction traits in chickens. Mu et al. [42] compared the ovarian transcriptome of low- and high-yielding Changshun green-shell laying hens, and *POMC* was found to be significantly expressed in high-yielding groups and identified to be the candidate gene for the improvement in egg production. *POMC* mRNA expression was also demonstrated in the hypothalamus of Shaoxing ducks, which was then indicated to initiate the sexual maturation of ducks [43]. Lei et al. [44] immunized laying chickens with LEPR ECD and found that a decreased egg-laying rate and follicle atresia were accompanied by the upregulation of *POMC*. All these studies indicated that *POMC* was involved in follicle development and egg production. Our study indicated that only feeding ducks with 8 g/kg of C-MEA clearly decreased the expression of *POMC*, which was also accompanied by an increase in F, LYF, and SYF numbers. These findings highlight the complex interplay of key pathways and hub genes in modulating reproductive behaviors and genetic regulation mechanisms underlying caged environments in laying ducks.

The cecum plays a crucial role, as it houses a vast and diverse microbial community essential for fermenting and digesting poultry feed [45]. Consequently, investigating how changes in cecal microbiota regulate growth and development in livestock and poultry is of significant importance; however, research on caged laying ducks remains limited. Therefore, this study aims to further explore the effects of C-MEA on cecal microbiota in caged laying ducks by comparing 16S sequencing data between different dietary groups. The experimental results showed that the estimated values of Group A (Chao1 and Simpson) increased significantly compared to the control group (Group C), indicating that the species richness of the cecal digestive tract was higher. Some research concluded that higher intestinal microbe abundance makes animals better able to cope with environmental disturbances, and the distinct microecological spaces critically impact individual nutrition, immunity, disease resistance, and physiological functions, ultimately influencing egg performance [19,46]. *Bacteroidota* was the dominant phylum in all three groups, which is responsible for the fermentation of complex indigestible polysaccharides through its enzymatic capacity for carbohydrate utilization [47]. We also discovered that at the phylum level, the *Spirochaetota*, *Synergistota*, and *Verrucomicrobiota* abundances significantly increased in caged laying ducks supplemented with C-MEA, while the abundance of *Bacteroidota* and *Actinobacteriota* significantly decreased; at the genus level, the relative abundances of *Sphaerochaeta* and *UCG-004* were significantly increased in Group A. Furthermore, correlation analysis showed that *Sphaerochaeta* was positively correlated with E30 laying rate and F, LYF, and SYF numbers among different dietary groups. The genus *Sphaerochaeta* is composed of chemoorganoheterotrophic anaerobes with fermentative metabolism and related to carbohydrate (pentose, hexose, disaccharides, and soluble starch) metabolism [48]. It has been reported that the alteration in *Sphaerochaeta* might affect the metabolism of laying hens to improve their egg quality [49]. Moreover, the abundance of *Sphaerochaeta* was found to be positively related to the chicken egg-laying rate of broiler breeders [50]. It was speculated that the addition of C-MEA increased the abundance of *Sphaerochaeta* and improved the health of the intestine, which might produce the precursors for the synthesis of hormones related to the egg production of ducks. The specific mechanisms by which C-MEA increases host health via their effects on the microbiota–gut–ovary axis need to be further explored. Correlation analysis between the 46 candidate key DEGs (RNA sequencing) and the top 20 microbiota genera (16S sequencing) from all three different dietary groups showed that *Sphaerochaeta* was extremely positively correlated with *F2*, *KNG1*, *C5*, *PLG*, *F2RL1*, *FABP1*, and *GCG* ($p < 0.01$). Due to the short experimental duration and the absence of hormonal analysis, future research will need to be carried out to verify the mechanisms of interaction between hub genes and microbiota on C-MEA and lay the foundation for improving gut health and reproductive performance in caged laying ducks.

5. Conclusions

This study examined the synergistic effect of a C-MEA and confirmed that the effect was most prominent when administering 8 g/kg, significantly increasing the egg-laying rate and F, LYF, and SYF numbers. Combining the RNA-Seq and 16S-Seq results, it can be reasonably deduced that egg production performance can be regulated through the microbiota–gut–ovary axis of caged laying ducks. Notably, the relationship between *Sphaerochaeta* and the hub gene cluster (*F2, KNG1, C5, PLG, F2RL1, FABP1,* and *GCG*) is the most prominent, which requires further experimental validation to confirm the regulation of the microbiota–gut–ovary axis of C-MEA.

Supplementary Materials: The following supporting information can be downloaded at https://www.mdpi.com/article/10.3390/vetsci12090808/s1: Table S1 Primer information for qRT-PCR, Table S2 Quality evaluation and analysis of preprocessing RNA-Seq data, and Table S3 Number of reads that passed through each step of paired-end reads assembly and quality control.

Author Contributions: Conceptualization, G.S. and X.D.; methodology, Y.L. and L.Z.; software, Y.L. and L.Z.; validation, Y.L., R.Z. and Y.J.; formal analysis, Y.L.; resources, Y.L.; data curation, Y.L.; writing—original draft preparation, Y.L.; writing—review and editing, Y.L. and L.Z.; visualization, G.S. and L.Z.; supervision, G.S. and X.D.; project administration, G.S. and X.D.; funding acquisition, G.S. All authors have read and agreed to the published version of the manuscript.

Funding: This research was funded by the Jiangsu Provincial Key Research and Development Program (grant number BE2022359) and the Natural Science Foundation of Jiangsu Agri-animal Husbandry Vocational College (grant number NSF2023ZR03).

Institutional Review Board Statement: All procedures involving the care and use of animals conformed to the U.S. National Institute of Health guidelines (NIH Pub. No. 85-23, revised 1996) and were approved by the Institutional Animal Ethics Committee of Jiangsu Agri-animal Husbandry Vocational College, Taizhou, Jiangsu, China (Approval ID No. Jsahvc-2023-94), date 6 Jan 2023.

Informed Consent Statement: Not applicable.

Data Availability Statement: Data produced in this study are available from the corresponding authors upon reasonable request.

Acknowledgments: The authors sincerely appreciate the valuable contributions of graduate and undergraduate students from Jiangsu Agri-animal Husbandry Vocational College and all the reviewers who participated in the review.

Conflicts of Interest: The authors declare no conflicts of interest.

References

1. Yang, N. Egg production in China: Current status and outlook. *Front. Agric. Sci. Eng.* **2021**, *8*, 25–34. [CrossRef]
2. Hou, S.S. LIULZ 2024 Report on the Development of the Waterfowl Industry and Technology. *Chin. J. Anim. Sci.* **2025**, *61*, 383–387. [CrossRef]
3. Xu, G.; Liu, X.; Wang, Q.; Yu, X.; Hang, Y. Integrated rice-duck farming mitigates the global warming potential in rice season. *Sci. Total Environ.* **2017**, *575*, 58–66. [CrossRef]
4. Chen, B.D.; Zhang, K.J.; Li, L.M.; Zhao, G.M.; Zhou, Y.L.; Dong, J.F.; Shou, X.D.; Zhou, Y.J.; Ma, J.J.; Yang, H. Present status and development trend of egg duck cage rearing. *J. Zhejiang Agric. Sci.* **2022**, *63*, 1868–1870+1882. [CrossRef]
5. Zhang, Y.; Gu, T.; Tian, Y.; Chen, L.; Li, G.; Zhou, W.; Liu, G.F.; Wu, X.S.; Zeng, T.; Xu, Q.; et al. Effects of cage and floor rearing system on the factors of antioxidant defense and inflammatory injury in laying ducks. *BMC Genet.* **2019**, *20*, 103. [CrossRef] [PubMed]
6. Tian, Y.; Liu, X.; Chen, L.; Zeng, T.; Gu, T.T.; Xu, W.X.; Ren, J.D.; Lu, L.Z. Dietary resveratrol alleviates liver and intestine injury in ducks under cage rearing system. *Poult. Sci.* **2025**, *104*, 105330. [CrossRef] [PubMed]

7. Shi, X.F.; Huang, M.Y.; Song, J.L.; Zeng, L.G.; Liang, Q.N.; Qu, Y.Q.; Li, J.Y.; Xu, G.Y.; Zheng, J.Z. Effects of different duck rearing systems on egg flavor and quality and microbial diversity. *Poult. Sci.* **2022**, *101*, 102110. [CrossRef]
8. Shi, S.Q.; Ge, M.G.; Xiong, Y.; Zhang, Y.X.; Li, W.H.; Liu, Z.M.; Wang, J.F.; He, E.H.; Wang, L.M.; Zhou, D.Q. The novel probiotic preparation based on Lactobacillus spp. mixture on the intestinal bacterial community structure of Cherry Valley duck. *World J. Microbiol. Biotechnol.* **2024**, *40*, 194. [CrossRef]
9. Liu, B.H.; Ma, R.Y.; Yang, Q.L.; Yang, Q.L.; Yang, Y.; Fang, Y.J.; Sun, Z.H.; Song, D.J. Effects of traditional chinese herbal feed additive on production performance, egg quality, antioxidant capacity, immunity and intestinal health of laying hens. *Animals* **2023**, *13*, 2510. [CrossRef]
10. Wang, X.; Xie, H.J.; Liu, F.; Wang, Y.H. Production performance, immunity, and heat stress resistance in Jersey cattle fed a concentrate fermented with probiotics in the presence of a Chinese herbal combination. *Anim. Feed Sci. Technol.* **2017**, *228*, 59–65. [CrossRef]
11. Wang, J.; Zhou, H. Comparison of the effects of Chinese herbs, probiotics and prebiotics with those of antibiotics in diets on the performance of meat ducks. *J. Anim. Feed Sci.* **2007**, *16*, 96. [CrossRef]
12. Gao, J.; Wang, R.; Liu Jx Wang, W.L.; Chen, Y.; Cai, W.T. Effects of novel microecologics combined with traditional Chinese medicine and probiotics on growth performance and health of broilers. *Poult. Sci.* **2022**, *101*, 101412. [CrossRef] [PubMed]
13. Gilbert, A.B.; Evans, A.J.; Perry, M.M.; Davidson, M.H. A method for separating the 527 granulosa cells, the basal lamina and the theca of the preovulatory ovarian follicle of 528 the domestic fowl (*Gallus domesticus*). *J. Reprod. Fertil.* **1977**, *50*, 179–181. [CrossRef]
14. Zou, K.; Asiamah, C.A.; Lu, L.; Liu, Y.; Pan, Y.; Chen, T.; Zhao, Z.; Su, Y. Ovarian transcriptomic analysis and follicular development of Leizhou black duck. *Poult. Sci.* **2020**, *99*, 6173–6187. [CrossRef]
15. Zhang, L.; Zhu, R.; Sun, G.B.; Wang, J.; Zuo, Q.S.; Zhu, S.Y. Whole-Transcriptome Sequencing of Ovary Reveals the ceRNA Regulation Network in Egg Production of Gaoyou Duck. *Genes* **2023**, *15*, 9. [CrossRef] [PubMed]
16. Love, M.I.; Huber, W.; Anders, S. Moderated estimation of fold change and dispersion for RNA-seq data with DESeq2. *Genome Biol.* **2014**, *15*, 550. [CrossRef] [PubMed]
17. Yu, G.C.; Wang, L.G.; Han, Y.Y.; He, Q.Y. clusterProfiler: An R package for comparing biological themes among gene clusters. *OMICS* **2012**, *16*, 284–287. [CrossRef]
18. Pertea, M.; Pertea, G.M.; Antonescul, C.M.; Chang, T.S.; Mendell, J.T.; Salzberg, S.L. StringTie enables improved reconstruction of a transcriptome from RNA-seq reads. *Nat. Biotechnol.* **2015**, *33*, 290–295. [CrossRef]
19. Qin, M.; Wang, Z.G.; Liang, M.Z.; Sha, Y.F.; Liu, M.X.; Liu, J.W.; Wang, T.; Zhao, C.X.; Wang, Z.X.; Guo, D.T.; et al. Effects of dietary supplementation with tea polyphenols and probiotics on laying performance, biochemical parameters intestinal morphology and microflora of laying hens. *Int. J. Biol. Macromol.* **2024**, *256 Pt 1*, 128368. [CrossRef]
20. Mohd Shaufi, M.A.; Sieo, C.C.; Chong, C.W.; Gan, H.M.; Ho, Y.W. Deciphering chicken gut microbial dynamics based on high-throughput 16S rRNA metagenomics analyses. *Gut Pathog.* **2015**, *26*, 4. [CrossRef]
21. Chen, X.L.; Xiong, P.W.; Song, W.J.; Song, Q.L.; Zou, Z.H.; Huang, J.N.; Chen, J.; Xu, C.H.; Su, W.D.; Ai, G.X.; et al. Dietary supplementation with honeycomb extracts positively improved egg nutritional and flavor quality, serum antioxidant and immune functions of laying ducks. *Front. Vet. Sci.* **2023**, *10*, 1277293. [CrossRef] [PubMed]
22. Liao, F.Y.; Yao, B.N.; Liu, A.; Zhu, B.G.; Yao, B.Q.; Wang, J.; Yang, S.L.; Boonanuntan, S. Effect of dietary Artemisia argyi powder on egg production, egg quality and nutrients, serum biochemical indices in laying ducks. *J. Appl. Anim. Res.* **2024**, *52*, 2331760. [CrossRef]
23. Incharoen, T.; Charoensook, R.; Tartrakoon, W.; Numthuam, S.; Sunanta, Y.; Jimenez, G.; Loor, J. Dietary Bacillus toyonensis BCT-7112^T supplementation influences performance, egg quality, Ammonia emission, and Cecal microbiome in laying ducks. *Vet. Sci.* **2025**, *12*, 259. [CrossRef] [PubMed]
24. Chen, Z.; Wen, D.; Zhang, Y.; Chen, J.Y.; Pan, F.Q.; Zhang, W.; Zhou, S.S.; Wang, F.; Mu, R. Pituitary transcriptome profile from laying period to incubation period of Changshun green-shell laying hens. *BMC Genom.* **2024**, *25*, 309. [CrossRef] [PubMed]
25. Liang, W.F.; Li, H.T.; Zhou, H.Y.; Wang, M.; Zhao, X.; Sun, X.H.; Li, C.T.; Zhang, X.M. Effects of Taraxacum and Astragalus extracts combined with probiotic Bacillus subtilis and Lactobacillus on Escherichia coli-infected broiler chickens. *Poult. Sci.* **2021**, *100*, 101007. [CrossRef]
26. Chang, Y.G.; Guo, R.B.; Zeng, T.; Sun, H.X.; Tian, Y.; Han, X.; Cao, Y.Q.; Xu, L.G.; Duan, M.C.; Lu, L.Z.; et al. Analysis of transcriptomic differences in the ovaries of high-and low-laying ducks. *Genes* **2024**, *15*, 181. [CrossRef]
27. Komar, C.M. Peroxisome proliferator-activated receptors (PPARs) and ovarian function--implications for regulating steroidogenesis, differentiation, and tissue remodeling. *Reprod. Biol. Endocrinol.* **2005**, *3*, 41. [CrossRef]
28. Sun, T.T.; Xiao, C.; Deng, J.X.; Yang, Z.L.; Zou, L.Q.; Du, W.Y.; Li, S.X.; Huo, X.Q.; Zeng, L.H.; Yang, X.R. Transcriptome analysis reveals key genes and pathways associated with egg production in Nandan-Yao domestic chicken. *Comp. Biochem. Physiol. Part D Genom. Proteom.* **2021**, *40*, 100889. [CrossRef]

29. He, Z.Y.; Chen, Q.L.; Ouyang, Q.Y.; Hu, J.W.; Shen, Z.Y.; Hu, B.; Hu, S.Q.; He, H.; Li, L.; Liu, H.H.; et al. Transcriptomic analysis of the thyroid and ovarian stroma reveals key pathways and potential candidate genes associated with egg production in ducks. *Poult. Sci.* **2023**, *102*, 102292. [CrossRef]
30. Bello, S.F.; Xu, H.P.; Guo, L.J.; Li, K.; Zheng, M.; Xu, Y.B.; Zhang, S.Y.; Bekele, E.J.; Bahareldin, A.A.; Zhu, W.J.; et al. Hypothalamic and ovarian transcriptome profiling reveals potential candidate genes in low and high egg production of white Muscovy ducks (*Cairina moschata*). *Poult. Sci.* **2021**, *100*, 101310. [CrossRef]
31. Yamaguchi, N.; Knaut, H. Focal adhesion-mediated cell anchoring and migration: From in vitro to in vivo. *Development* **2022**, *149*, dev200647. [CrossRef]
32. Wu, X.; Jiang, L.; Xu, F.Q.; Cao, S.M.; Chen, Y.; Zhang, Y.Y.; He, W.F.; Yan, M.J.; Lian, S.Y.; Li, A. Long noncoding RNAs profiling in ovary during laying and nesting in Muscovy ducks (*Cairina moschata*). *Anim. Reprod. Sci.* **2021**, *230*, 106762. [CrossRef] [PubMed]
33. Chen, L.R.; Chao, C.H.; Chen, C.F.; Lee, Y.-P.; Chen, Y.-L.; Shiue, Y.-L. Expression of 25 high egg production related transcripts that identified from hypothalamus and pituitary gland in red-feather Taiwan country chickens. *Anim. Reprod. Sci.* **2007**, *100*, 172–185. [CrossRef] [PubMed]
34. Kabir, M.A.; Ruan, H.; Rong, L.; Horaira, M.A.; Wu, X.; Wang, L. Decoding the duration of fertility of laying chicken through phenotypic and proteomic evaluation. *Br. Poult. Sci.* **2024**, *65*, 677–689. [CrossRef] [PubMed]
35. Du, X.D.; Fu, Y.; Tian, Z.H.; Liu, H.H.; Xin, H.X.; Fu, X.L.; Wang, F.F.; Zhang, H.Z.; Zeng, X. Microcystin-LR accelerates follicular atresia in mice via JNK-mediated adherent junction damage of ovarian granulosa cells. *Ecotoxicol. Environ. Saf.* **2023**, *252*, 114592. [CrossRef]
36. Bai, K.; Hao, E.; Huang, C.X.; Yue, Q.X.; Wang, D.E.; Shi, L.; Chen, H.; Huang, R.L. Melatonin alleviates ovarian function damage and oxidative stress induced by dexamethasone in the laying hens through FOXO1 signaling pathway. *Poult. Sci.* **2023**, *102*, 102745. [CrossRef]
37. Chen, Z.; Wen, D. Ovarian transcriptome profile from egg-laying period to incubation period of Changshun green-shell laying hens. *Genes* **2025**, *16*, 394. [CrossRef]
38. Harno, E.; Gali Ramamoorthy, T.; Coll, A.P.; White, A. POMC: The Physiological Power of Hormone Processing. *Physiol. Rev.* **2018**, *98*, 2381–2430. [CrossRef]
39. Sanders, S.L.; Melner, M.H.; Curry, T.E. Cellular Localization of Ovarian Proopiomelanocortin Messenger RNA during Follicular and Luteal Development in the Rat. *Mol. Endocrinol.* **1990**, *4*, 1311–1319. [CrossRef]
40. Liu, K.; Cao, H.Y.; Dong, X.R.; Liu, H.H.; Wen, Y.Y.; Mao, H.G.; Lu, L.; Yin, Z.Z. Polymorphisms of pro-opiomelanocortin gene and the association with reproduction traits in chickens. *Anim. Reprod. Sci.* **2019**, *210*, 106196. [CrossRef]
41. Liu, K.; Wen, Y.Y.; Liu, H.H.; Cao, H.Y.; Dong, X.Y.; Mao, H.G.; Yin, Z.Z. POMC gene expression, polymorphism, and the association with reproduction traits in chickens. *Poult. Sci.* **2020**, *99*, 2895–2901. [CrossRef]
42. Mu, R.; Yu, Y.Y.; Gegen, T.; Wang, F.; Chen, Z.; Xu, W.B. Transcriptome analysis of ovary tissues from low-and high-yielding Changshun green-shell laying hens. *BMC Genom.* **2021**, *22*, 349. [CrossRef] [PubMed]
43. Ni, Y.; Lu, L.; Chen, J.; Zhao, R. Changes of hypothalamic GnRH-I, POMC and NPY mRNA expression and serum IGF-I and leptin concentrations during maturation of shaoxing ducks (*Anas platyrhynchos*). *Asian-Australas. J. Anim. Sci.* **2011**, *24*, 1211–1216. [CrossRef]
44. Lei, M.M.; Wu, S.Q.; Li, X.W.; Chen, Z.; Shi, Z.D. Leptin receptor signaling inhibits ovarian follicle development and egg laying in chicken hens. *Reprod. Biol. Endocrinol.* **2014**, *12*, 25. [CrossRef] [PubMed]
45. Mu, Q.M. The Study on Microbial Characteristics of the Cecum in Ducks and Its Genetic Relationship with the Host. Master's Thesis, Chinese Academy of Agricultural Sciences, Beijing, China, 2024. [CrossRef]
46. Zhu, Z.H.; Huang, B.B.; Sun, N.N.; Yu, X.C.; Du, Z.Y.; Li, A.; Huang, C.Y. Variations in gut microbiota composition and reproductive hormone levels between laying and broody Muscovy ducks. *Poult. Sci.* **2024**, *103*, 104399. [CrossRef]
47. Shang, H.M.; Zhao, J.C.; Dong, X.Q.; Guo y Zhang, H.X.; Cheng, J.L.; Zhou, H.Z. Inulin improves the egg production performance and affects the cecum microbiota of laying hens. *Int. J. Biol. Macromol.* **2020**, *155*, 1599–1609. [CrossRef]
48. Nazina, T.N.; Bidzhieva, S.K.; Grouzdev, D.S.; Löffler, F.E. Sphaerochaeta. In *Bergey's Manual of Systematics of Archaea and Bacteria*; John Wiley & Sons, Inc.: Hoboken, NJ, USA, 2015; pp. 1–13. [CrossRef]
49. Gong, H.J.; Yang, Z.Q.; Celi, P.; Yan, L.; Ding, X.M.; Bai, S.P.; Zeng, Q.F.; Xu, S.Y.; Su, Z.W.; Zhuo, Y.; et al. Effect of benzoic acid on production performance, egg quality, intestinal morphology, and cecal microbial community of laying hens. *Poult. Sci.* **2021**, *100*, 196–205. [CrossRef]
50. Yang, Z.Q.; Zhang, C.H.; Wang, J.P.; Celi, P.; Ding, X.M.; Bai, S.P.; Zeng, Q.F.; Mao, X.B.; Zhou, Y.; Xu, S.Y.; et al. Characterization of the intestinal microbiota of broiler breeders with different egg laying rate. *Front. Vet. Sci.* **2020**, *7*, 599337. [CrossRef]

veterinary sciences

MDPI

Review

Role of Gut Microbiota and Metabolite Remodeling on the Development and Management of Rheumatoid Arthritis: A Narrative Review

Yichen Yu [1,†], Fulin Jin [1,†], Lijun Wang [1], Ji Cheng [1] and Shifeng Pan [1,2,3,4,*]

1 College of Veterinary Medicine, Yangzhou University, Yangzhou 225009, China; yuyichen1608@163.com (Y.Y.); 13851118926@163.com (F.J.); wlj6379@163.com (L.W.); 008743@yzu.edu.cn (J.C.)
2 Department of Animal Sciences, Washington State University, Pullman, WA 99163, USA
3 Guangling College, Yangzhou University, Yangzhou 225009, China
4 Jiangsu Co-Innovation Center for Prevention and Control of Important Animal Infectious Diseases and Zoonoses, Yangzhou University, Yangzhou 225009, China
* Correspondence: sfpan@yzu.edu.cn
† These authors contributed equally to this work.

Simple Summary: Rheumatoid arthritis is a chronic autoimmune disease that causes joint pain, inflammation, and disability in humans and animals, harming health and livestock productivity. This review explores the growing evidence linking rheumatoid arthritis to imbalances in the gut microbiome—the trillions of bacteria and microbes living in the intestines. The goal is to understand how these microbial changes contribute to rheumatoid arthritis and whether modifying the gut microbiome could offer new treatment strategies. Studies show RA patients and animals have fewer beneficial gut bacteria and more harmful bacteria. These imbalances weaken the gut barrier, allowing harmful substances to enter the bloodstream and spark systemic inflammation. The gut microbiome also disrupts immune balance, promoting damaging inflammation while reducing protective responses. Interventions like probiotics and fecal microbiota transplantation improved gut health and reduced rheumatoid arthritis symptoms in animal studies. By highlighting the gut–rheumatoid arthritis connection, this review suggests targeting gut health could provide affordable, low-risk treatments, especially for animals with limited therapies. The findings may inspire new probiotic products or personalized microbial treatments to prevent rheumatoid arthritis progression, benefiting both human patients and global livestock health.

Abstract: Rheumatoid arthritis (RA) is a chronic autoimmune disease that has a serious impact on both human health and animal production. The gut microbiota is a large and complex symbiotic ecosystem in animals, and the imbalance of gut microbiota is closely related to the pathogenesis of numerous diseases, including RA. The interactions among the gut microbiota, intestinal barrier, and immune system play key roles in maintaining intestinal homeostasis and affecting the development of RA. Regulating intestinal flora and metabolites provides new ideas for the prevention and treatment of RA. Probiotics can regulate the balance of intestinal flora and metabolites, improve the immune environment, and provide novel therapeutic strategies against RA. In order to summarize the role of gut microbiota and metabolite remodeling in the development and management of RA, this review will elaborate on the role of intestinal flora imbalance in the pathogenesis of RA and assess prospective therapeutic approaches that target the gut flora. Understanding the interaction among intestinal flora, metabolites, and RA will help to clarify the pathogenesis of RA and develop innovative and personalized therapeutic interventions against chronic autoimmune diseases.

Keywords: rheumatoid arthritis; intestinal microbiota; animal production; intestinal barrier; immune response; probiotic therapy

1. Introduction

Rheumatoid arthritis (RA) is a chronic autoimmune disease characterized by joint stiffness, pain, and limited mobility. It can lead to a great loss of joint function and, in severe cases, disability or involvement of other tissues and organs. The occurrence of RA seriously affects both human health and the development of livestock and poultry.

At present, RA research in animals is relatively scarce, which seriously hinders the development of effective treatment strategies against RA in animals and causes a lack of specific therapeutic drugs and methods. Additionally, due to the differences in the immune systems of animals from those of humans [1], the direct application of human drugs to animals may result in many difficulties in determining drug doses and preventing adverse reactions [2]. Furthermore, the treatment of RA in animals faces challenges in terms of economic cost and breeder awareness [3].

An increasing number of studies have demonstrated that multiple therapies based on altering gut microbiota diversity can play a vital role in treating RA, suggesting that gut microbiota remodeling can be a more cost-effective option in RA prevention and treatment [4,5]. At the preventive level, the balance of the intestinal microbial community can be used as a target for early intervention in RA, with the advantages of being relatively inexpensive and easy to implement. In terms of disease relief, the regulatory effect of gut microbial remodeling on metabolism helps to reduce joint inflammation and improve feed conversion and has relatively few side effects and adverse reactions, such as drug residue [6]. In terms of long-term benefits, the plasticity of gut microbes offers the possibility of sustainable RA management. Through continuous microbial community monitoring and targeted interventions, intestinal homeostasis may be maintained and RA recurrence may be prevented.

This article aims to provide a synthesis of recent research findings on the relationship between RA and the gut microbiota and will explore the role of gut microbiota imbalance in the pathogenesis of RA and the potential applications of gut microbiota regulation in the treatment of RA, to better aid the development of novel therapeutic approaches against RA.

2. The Compositional and Functional Imbalance of the Gut Microbiota in Patients and Animals with RA

The gut microbiota represents an extremely large and complex symbiotic microbial ecosystem in animals. These microorganisms not only participate in the digestive and absorptive processes of food but also exert a profound impact on the health of the host through their metabolites [7,8]. The gut microbiota produces a wide variety of metabolites, including short-chain fatty acids (SCFAs), amino acids, and so forth. With the advancement of microbiome research, these metabolites have been regarded as a crucial link between gut microbes and host health [9].

In recent years, an increasing number of studies have demonstrated that the gut microbiota and derived metabolites play a pivotal role in the pathogenesis and/or progression of arthritis [10,11]. The compositional and functional imbalance of the gut microbiota has been summarized in animals with inflammatory arthritis (Table 1) and RA patients (Table 2).

Table 1. The compositional and functional imbalance of the gut microbiota in arthritis animals.

Animals Included	Types of Arthritis	Classification Levels	Changes in Microbiota in Animals with Inflammatory Arthritis
Rhesus monkeys [12]	Pro-inflammatory arthritis (CHIKV infected)	Genus	*Prevotella* and *Sarcina* ↑ *Anaerostipes* and *Lactobacillus* ↓
		Species	*Prevotella copri* and *Bacteroides fragilis* ↑ *Butyricicoccus intestinisimiae* and *Streptococcus lutetiensis* ↓
Pigs [13]	Diet-induced exudative arthritis	Species	*Clostridium perfringens* ↑
Chicks [14]	Hypervirulent arthritis (*Salmonella Pullorum* infected)	Phylum	*Proteobacteria* ↑ *Firmicutes* ↓
		Genus	*Escherichia-Shigella* and *Klebsiell* ↑ *Lachnoclostridium* and *Blautia* ↓

↑ represents an increase in abundance, and ↓ represents a decrease in abundance.

Table 2. The compositional and functional imbalance of gut microbiota in RA patients.

Patients Included	Classification Levels	Changes in Microbiota in RA Patients
RA patients [4]	Genus	*Klebsiella* and *Escherichia* ↑ *Fusicatenibacter* and *Megamonas* ↓
Female patients with early RA [5]	Phylum	*Bacteroidetes* ↑ *Actinobacteria* ↓
	Genus	*Collinsella* ↓
Female RA patients [15]	Genus	*Bacteroides* and *Megamonas* ↑ *Prevotella* and *Gemmiger* ↓
RA patients [16]	Phylum	*Verrucomicrobia* and *Proteobacteria* ↑ *Bacteroidetes* ↓
	Genus	*Lactobacillus* and *Streptococcus* ↑ *Bacteroides* and *Faecalibacterium* ↓
RA patients [17]	Species	*Bifidobacterium longum* and *Dorea formicigenerans* ↑ *Faecalibacterium prausnitzii* and *Bacteroides* spp. ↓
RA patients [18]	Genus	*Eubacterium* and *Escherichia-Shigella* ↑
RA patients grouped into stages I–IV [19]	Species	*Escherichia coli* ↑ *Bacteroides uniformis* and *Bacteroides plebeius* ↓

↑ represents an increase in abundance, and ↓ represents a decrease in abundance.

3. The Role of Intestinal Flora Imbalance in the Pathogenesis of RA

The mutualism between microbiota and their host is a symbiotic relationship beneficial to both parties. An imbalance between intestinal microbiota and the host can potentially lead to the onset of serious diseases that pose a significant threat to human and animal health [20].

3.1. Relationship Between RA and Gut Microbiota, Gut Microbiota Metabolites, and Gut Barrier

There is a complex and delicate equilibrium between the gut microbiota and its metabolites and the intestinal barrier. On the one hand, the intestinal flora produces beneficial metabolites through the fermentation of indigestible carbohydrates, including dietary fiber [21,22]. These metabolites can provide energy for intestinal epithelial cells and promote the repair and regeneration of the intestinal mucosa, as well as enhance the integrity of the intestinal barrier [23,24]. Concurrently, metabolites can regulate the activity of intestinal immune cells and inhibit the inflammatory response, thereby maintaining intestinal health [25,26]. On the other hand, gut integrity is of paramount importance in preventing the infiltration of harmful substances into the bloodstream. A disruption to the intestinal barrier function, such as an increase in permeability, will give rise to a systemic inflammatory and immune response in the host, the persistence of which will not only exacerbate the damage to the intestine itself but also affect the function of other organ systems, such as triggering autoimmune diseases [27].

Equilibrium disruptions of the gut microbiota are frequently observed in patients and animals with RA, accompanied by an increase in intestinal permeability and a reduction in the integrity of the intestinal barrier. ZO-1, Cluadin-1, and Occludin are key tight junction proteins and have important effects on the integrity and function of the intestinal barrier. A cross-sectional study demonstrated a correlation between fecal tight junction protein levels in RA patients and RA clinical parameters. Among these parameters, Claudin-1 level was found to be significantly decreased, resulting in altered intestinal permeability in patients [28]. The study by Wen et al. found that at the onset of RA, there was a peak in the expression of intestinal hypoxia-inducible factor (HIF)-2α in intestinal epithelial cells, which in turn induced the transcription of Claudin-15, thereby disrupting the intestinal barrier and promoting the development of arthritis [29]. Zonulin is a peptide hormone secreted by intestinal epithelial cells, whose core function is to regulate the dynamic opening and closing of intestinal tight junctions to maintain selective permeability of the intestinal barrier. Elevated levels of Zonulin are often observed in the serum of individuals at risk for RA and in RA patients, and this alteration is accompanied by leakage of the intestinal barrier, microbial community imbalance, and inflammation [30]. Therefore, maintaining the balance of intestinal flora and the integrity of the intestinal barrier function is of great importance in the prevention and treatment of RA.

3.2. Interactions Among RA, Gut Microbiota, and Immune System

The immune system plays a pivotal role in the pathogenesis of RA. Dendritic cells (DCs) present antigens to naive T cells and induce them to differentiate into different cell subtypes (Th1, Th17, and Treg cells, etc.) under the action of antigen-stimulating signal factors and produce related cytokines or immune tolerance. Th1 cells activate macrophages and are responsible for cell-mediated immunity and phagocyte-dependent protective responses. The prevailing theory posits that the manifestation of RA is contingent on the over-expansion and expression of Th1 cells and its hallmark factor, IFN-γ.

The discovery of a new subset of T cells, Th17 cells, has led to the establishment of a link between Th17 cell plasticity and autoimmune diseases [31]. Research has demonstrated that Th17 cells, the sole osteoclast-derived Th cell subset, along with its hallmark factor, IL-17A, assume a pivotal role in bone destruction in RA [32]. The release of inflammatory cytokines (TNF-α, IL-1β, and IL-6), chemokines, and matrix metalloproteinases are promoted by IL-17A, which can also promote the formation and activation of osteoclasts [33]. TNF-α has been demonstrated to up-regulate the expression of secreted proteins in synovial fibroblasts, thereby inhibiting Wnt signaling, which is required for bone formation in osteoblasts [34].

Activation of the TNF-α pathway in CD4+ T cells is critical for the development of RA [35]. These cytokines induce a high responsiveness of osteoclast precursor cells to the receptor activator for nuclear factor-κB ligand (RANKL) and up-regulate RANKL expression in synovial fibroblasts [36]. This process culminates in the amplification of osteoclastogenesis.

In contrast, the function of Treg cells is characterized by their ability to impede RANKL-induced osteoclastogenesis and increase bone volume in vivo [37], thereby exerting a pivotal role in the initiation and management of RA [38,39]. This function is facilitated by the involvement of TGF-β/IL-4, TGF-β/IL-10, or CTLA-4, operating through either a contact-dependent or -independent cell-to-cell interaction. CTLA-4 has been shown to bind to the osteoclast precursor CD80/CD86, inducing osteoclast apoptosis. These findings suggest that the balance between Treg cells and Th17 cells may be a key factor in the inhibition of bone destruction in RA.

It is noteworthy that an imbalance in the gut microbiota and the production of harmful metabolites may be a significant contributing factor to the aberrant immune system response. Fecal microbiota transplantation (FMT) from patients at high risk of developing RA results in an intestinal muco-immune imbalance in mice, characterized by an increase in Th17 cells and elevated mRNA levels of IL-17a, IL-22, and TNF-α in the small intestine [40]. Maeda et al. observed that mice with the microbiota of RA patients exhibited elevated numbers of intestinal Th17 cells and developed severe arthritis following zymosan injections. Furthermore, these mice exhibited augmented IL-17 responses in lymphocytes [41]. The study conducted by Jubair et al. observed marked dysbiosis and mucosal inflammation at an early stage of collagen-induced arthritis (CIA). In addition, depletion of the microbiota was observed to have a mitigating effect on both the severity of inflammation and the progression of disease in both the early and late stages of CIA. This provides further confirmation of the significant role played by the gut microbiota in the pathogenesis of RA [42].

3.3. Relationship Between Specific Bacterial Species and RA

A substantial body of research has demonstrated that particular bacterial species, including *Escherichia coli*, are linked to the onset of RA. It has been demonstrated that the elevation of *E. coli* enhances the arginine succinyl transferase pathway in RA stages II and III, which is associated with the elevation of the rheumatoid factor and can induce bone loss [19]. It is possible that this effect of *E. coli* is closely related to the production of lipopolysaccharide (LPS). As early as 1994, it was demonstrated that LPS produced by *E. coli* induces arthritis in rats at a similar rate to that of *E. coli* itself. Furthermore, local stimulation of IL-1 production may play an important role in this process [43]. In the study conducted by Kitamura et al., the administration of LPS produced by *E. coli* resulted in the destruction of articular cartilage and the infiltration of inflammatory cells in mice, leading to the development of severe arthritis [44]. Consequently, *E. coli* or its LPS is frequently employed in the construction of RA models both in vitro and in vivo [45].

In contrast to LPS from *E. coli*, LPS from *Bacteroides fragilis* demonstrated a diminished capacity to induce arthritis. Furthermore, it exhibited the ability to impede the inflammatory response initiated by LPS from *E. coli*, thereby inhibiting the progression of arthritis [44]. It is hypothesized that this effect may be achieved by promoting the development of intestinal Treg cells and inhibiting the immune responses of intestinal Th17 cells [46]. Furthermore, the cell wall lipoprotein of *Bifidobacterium longum* has been demonstrated to exert a significant influence on the prevention of arthritis via the gut–joint axis [47]. *Lactobacillus rhamnosus* has been demonstrated to alleviate RA symptoms by reducing inflammation mediated by CD8+ T cells [48].

3.4. Molecular Simulation Is a Potential Mechanism for Connecting RA with Intestinal Flora

Molecular mimicry refers to the phenomenon whereby antigens produced by specific pathogens or gut microbes that share sequence homology with autoantigens may lead to cross-activation of autoreactive T or B cells, thereby triggering an autoimmune response [49]. At present, the roles of several molecular mimicry mechanisms related to the intestinal microbiota in the occurrence of RA have been revealed.

Protein citrullination is a post-translational modification process that requires peptidylarginine deiminase (PAD). While this is a normal physiological process and not unique to RA, citrullination of RA autoantigen proteins by PAD is a key event in the pathogenesis of RA [50]. Furthermore, PADI4 is also found to be highly expressed in the synovial membrane of RA joints [51]. A variety of anti-citrullinated protein antibodies can be detected in the serum of patients prior to the onset of RA, and these antibodies form immune complexes with corresponding antigens in the joints and deposit in the synovium, causing synovial inflammation. It has been hypothesized that in individuals with a predisposing genetic background, citrullinated protein antigens disrupt immune tolerance, thereby contributing to the pathogenesis of RA [52]. The structure of cholesterol-dependent cytolysin secreted by some Gram-positive bacteria is similar to that of perforin-membrane attack complex family proteins, which can promote Ca^{2+} influx through the formation of perforin to activate endogenous PAD enzymes and participate in citrullination [50].

A significant number of studies have recently concentrated on the role of autoantigen citrullination mediated by *Porphyromonas gingivalis*. Unlike other PAD, which requires Ca^{2+} to catalyze the conversion of peptidyl arginine to citrulline [53], PAD secreted by *P. gingivalis* can catalyze the citrullination in the absence of Ca^{2+}, resulting in the production of RA-specific autoantigens [54]. These findings suggest that *P. gingivalis* may break the host immune tolerance to citrullinated proteins and induce local immune responses in the joints.

Neutrophils represent the most prevalent cell type in RA synovial fluid and have the capacity to infiltrate synovial tissue. Certain bacterial species have been observed to induce the formation of neutrophil extracellular traps (NETs) [55], a process that involves the release of the PAD enzyme to trigger PAD activation and modify the histone arginine residues in the neutrophil nucleus to citrulline [56,57]. Studies have revealed that RA patients exhibit increased levels of citrullinated histone H3 in their neutrophil populations. Antibodies present in the serum of RA patients have been observed to bind to citrullinated histone H4, a by-product of NETs [58]. Furthermore, it has been determined that the activation of NADPH oxidase (NOX) is a prerequisite for the formation of NETs [59]. It has been established that the Gram-positive bacteria *Lactobacillus* spp. can stimulate the activation of NOX1 [60], while certain bacteria such as *E. coli* have been observed to activate AMPK and inhibit NOX activation [61,62]. Consequently, these bacteria may possess the capacity to affect NET formation.

In RA patients and animals, disruption of the gut microbiota may result in aberrant expression of specific microbial antigens that cross-react with RA-related autoantigens through molecular mimicry mechanisms. Specific phages that are enriched in the gut of RA patients and animals, particularly those derived from phages *Prevotella* and *Oscillibacter*, were found to have a high degree of homology with the epitopes of the RA autoantigen BiP. These phage peptides have the capacity to activate CD4+ T cells and B cells through the mechanism of molecular mimicry, thereby triggering an autoimmune response and promoting the occurrence and development of RA [63]. Additionally, the disruption of the intestinal phagome in RA patients is significantly associated with the presence of disease-specific autoantibodies and disease-susceptibility genes (such as HLA-DR shared epitopes). The phage composition of RA patients with an HLA-DR susceptibility gene differs from that of patients without, indicating a potential correlation between intestinal phage and disease susceptibility genes in RA [63]. These findings offer new insights into the pathogenesis of RA and may provide new avenues for immunotherapy in the future.

4. Application of Intestinal Flora Regulation in the Treatment of RA

In recent years, there have been encouraging indications that oral probiotics and FMT may prove to be effective as adjunctive treatments for RA.

4.1. Therapeutic Effects of Probiotics on RA

Probiotics, as modulators of gut microbiota, have demonstrated efficacy in the treatment of RA (Figure 1).

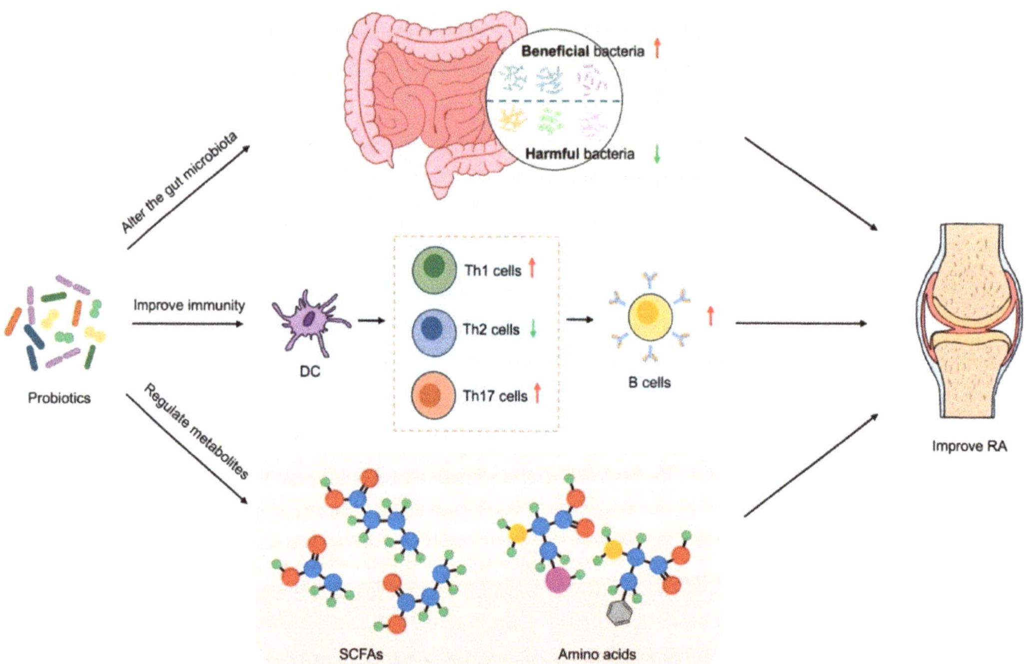

Figure 1. Therapeutic effects of probiotics on RA. Probiotics modulate the gut microbiota by promoting the proliferation of beneficial bacteria and inhibiting harmful bacteria, thereby restoring intestinal flora balance. This modulation regulates immune responses by reducing the differentiation of pro-inflammatory Th1/Th17 cells and enhancing anti-inflammatory Th2 cells and Tregs. Concurrently, probiotics ferment dietary fiber to produce SCFAs and other metabolites, which improve intestinal barrier integrity and suppress systemic inflammation. The combined effects of microbiota–immune–metabolite interactions ultimately alleviate RA symptoms. Red arrows represent promotive effects, and green arrows represent inhibitory effects. PowerPoint was used to draw this figure.

Probiotics have the capacity to regulate the equilibrium of intestinal flora by augmenting the number of beneficial bacteria and impeding the proliferation of harmful bacteria. In the study by Pan et al., treatment with *Lactobacillus casei* increased the abundance of *Lactobacillus acidophilus*, *Lactobacillus hominis*, and *Lactobacillus vaginalis* and then improved arthritis [64]. In addition to competing for nutrients and living space, probiotics are known to produce antibacterial substances, which serve to inhibit the growth and reproduction of harmful bacteria. It has been demonstrated that probiotics entering the gut will automatically aggregate under the mediation of extracellular polysaccharides and hydrophobic surface proteins. They will then adhere to epithelial cells with a stronger adhesion force through competitive repulsion, thereby reducing the adhesion of pathogenic bacteria [65]. For example, *Bacillus coagulans* may eliminate the antagonistic microorganisms that cause inflammatory responses by producing bacteriocins and L(+) lactic and competing for the adhesion sites of the mucosa, thereby exerting a protective effect on RA patients [66].

Probiotics have been demonstrated to stimulate the activation and proliferation of immune cells, including macrophages, DCs, and T lymphocytes, as well as regulate the production of cytokines within the intestine. Reduced production of pro-inflammatory cytokines, such as TNF-α, IL-1β, and IL-6, and increased secretion of anti-inflammatory cytokines, including IL-10 and IL-4, can help to reduce the inflammatory response in the gut and the level of systemic inflammatory response, exerting a beneficial influence on the treatment of RA. Qin et al. found that *Lactiplantibacillus plantarum* was effective in improving autoantibody levels, cytokine imbalance, and oxidative stress in RA mice [67]. *Saccharomyces boulardii* has been demonstrated to reduce the levels of pro-inflammatory cytokines, inhibit the proliferation of Th17 and type 3 innate lymphocytes (ILC3), and regulate immune disorder in arthritis rats [68]. Furthermore, *L. rhamnosus* GR-1 and *Lactobacillus* reuteri RC-14 reduce the serum levels of IL-1α, IL-6, IL-12p70, and TNF-α in RA patients and have a positive anti-inflammatory effect at the cellular level [69]. Similarly, *L. casei* treatment can also regulate the production of cytokines and reduce lymphocyte infiltration at the same time, thereby inhibiting the development of arthritis [70].

Probiotics have also been demonstrated to exert a beneficial impact on the management of RA by modulating the production of metabolites within the gut microbiota (Table 3).

Table 3. Effects of gut microbiota-derived metabolites on RA.

Metabolites of Intestinal Microorganisms	Associated Gut Microbes	Biological Functions
SCFAs [71,72]	*Clostridium*, *Bacteroides*, and *Eubacterium*	Decrease serum Zonulin concentration, increase the expression of tight junction protein, and restore the intestinal permeability
Tryptophan [73,74]	*Bifidobacterium*, *Clostridium*, *Escherichia coli*, and *Enterococcus*	Activate the AHR pathway, inhibit the activation and inflammatory response of intestinal immune cells, regulate the proliferation and differentiation of intestinal mucosal cells, and promote the repair and stability of mucosal barrier
Bile acid [75,76]	*Clostridium*, *Bacteroides*, and *Alistipes*	Promote the expression of tight junction proteins and blocking proteins and reduce the secretion of inflammatory cytokines (TNF-α and IL-6) to protect the intestinal barrier
Indole-3-aldehyde, Indole-3-acetic acid [77,78]	*Lactobacillus*, *Bifidobacterium*, and *Clostridium*	Affect the activity of Tregs cells, thereby affecting the balance of immune responses
Trimethylamine N-Oxide [79,80]	*Anaerocccus hydrongenalis*, *Clostridium asparagiforme*, and *Clostridium hathewayi*	Induce IL-1β, TNF-α, and IL-6, produce chemokines, and hinder bile acid synthesis and metabolism

Probiotics are capable of fermenting dietary fiber to produce SCFAs, including acetic acid, propionic acid, and butyric acid. This process reduces the pH value in the gut and inhibits the growth and reproduction of harmful bacteria. Concurrently, SCFAs can facilitate the restoration of the intestinal mucosal barrier and augment the defensive capabilities of the intestinal mucosal barrier. SCFAs play an immunomodulatory role mainly by promoting the differentiation of Treg cells; in particular, butyric acid regulates the balance of Treg/Th17 cells [81–83], so it also has a protective effect on joint inflammation and damage. A reduction in butyric acid-producing bacteria, such as *Faecalibacterium*, was observed in RA patients. The exogenous intake of butyric acid-producing probiotics has been demonstrated to supplement butyrate, promote the differentiation of Treg cells, inhibit that of Th17 cells and T follicular helper cells, and regulate the Th17/Treg balance, which in turn mitigates the severity of arthritis [84].

It is estimated that approximately 95% of human serotonin (5-hydroxytryptamine, 5-HT) is produced in the gastrointestinal tract. Rodent and human innate immune cells have been shown to express 5-HT receptors, and intra-articular injection of 5-HT in mice has been observed to cause joint inflammation and pain, whereas depletion of 5-HT has been demonstrated to reduce disease severity [85]. Research has demonstrated that the ingestion of probiotics exerts an influence on 5-HT production and function. *B. longum* subsp. *infantis* B6MNI modulates the gut microbiota and fecal metabolites, including 5-hydroxyindole-3-acetic acid (5-HIAA), to affect Pim-1 expression and immune cell differentiation. This then affects joint inflammation, regulates osteoclast differentiation factors, and delays the progression of RA through the JAK/STAT3 pathway [86].

It is worth noting that the safety of using probiotics remains a potential issue. Theoretically, there could be four kinds of side effects: systemic infections, deleterious metabolic activities, excessive immune stimulation in susceptible individuals, and gene transfer [87]. Therefore, more research is needed to confirm these points.

4.2. Intestinal Flora Transplantation and RA Treatment

FMT is a therapeutic modality that involves the transfer of the intestinal flora of healthy individuals into the intestinal tract of patients. FMT has demonstrated considerable efficacy in the treatment of intestinal disorders, such as *Clostridium difficile* infection [88], and is increasingly being employed in the management of other conditions, including RA. In 2020, Zeng and colleagues reported the first successful case of FMT for RA [89].

Although there are still many issues to be explored and solved in practical application, such as the possible occurrence of adverse reactions like abdominal distension, diarrhea and constipation [90,91], the prospect of FMT in treating RA remains optimistic. For example, the efficacy and safety of FMT can be further improved by optimizing the donor screening criteria and improving the accuracy [92]. Moreover, FMT can be used in combination with other treatment modalities, such as immunosuppressants and non-steroidal anti-inflammatory drugs [93], and this comprehensive treatment plan is expected to further improve the treatment effect of RA. In addition to treating RA, FMT treatment in high-risk groups may restore the balance of intestinal flora and normal immune function, thereby reducing the incidence of RA.

4.3. Interaction Between Intestinal Flora and RA Therapeutics

There are numerous interactions between the gut microbiota and various drugs used in the treatment of RA (Figure 2).

On the one hand, pharmaceutical agents exert regulatory effects on the gut microbiota. Studies have demonstrated that certain drugs can exert a therapeutic effect by enhancing the composition of the intestinal flora and reinforcing the intestinal mucosal barrier. Jingfang Granules have been observed to augment the prevalence of intestinal microbiota, consequently elevating the concentration of SCFAs in the intestine and serum of RA rats, which has been shown to activate AMPK, regulate fatty acid metabolism and biosynthesis, and thereby mitigate RA-induced tissue damage [94]. Methotrexate (MTX) is currently the primary synthetic drug used in the treatment of RA. It has been demonstrated that MTX alters the human gut microbiota, resulting in reduced host immune activation and consequently providing protection against RA [95].

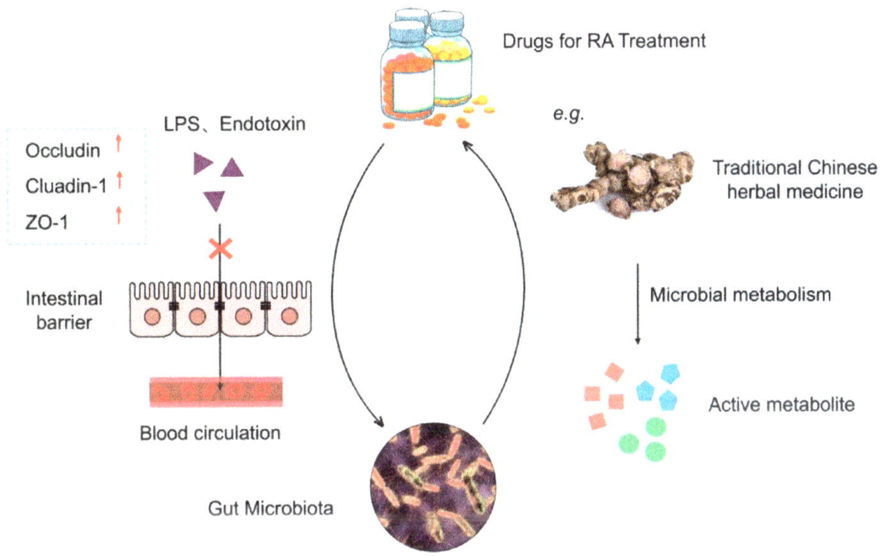

Figure 2. Interactions between gut microbiota and therapeutic drugs for RA. **Left**: Drugs regulate gut microbiota composition and enhance intestinal barrier function by upregulating tight junction proteins (Occludin, ZO-1, Claudin-1). **Right**: Gut microbiota metabolize drugs into active compounds or influence drug bioavailability. The bidirectional arrow indicates dynamic interactions between drugs and microbiota. PowerPoint was used to draw this figure.

On the other hand, the gut microbiota exerts a significant influence on the metabolism of drugs. Some pharmaceutical ingredients, particularly traditional Chinese herbs, are metabolized by intestinal flora into active compounds in the gut, thereby exerting therapeutic effects. For example, O-desmethylangolensin, 3-hydroxydodecanedioic acid, and other metabolites produced by the fermentation of *Lycium barbarum* polysaccharide by gut flora may be the key to alleviating RA [96]. Similarly, *L. casei* 3260-fermented *Gleditsia sinensis* thorn extract can significantly reduce cartilage damage caused by RA [97]. Furthermore, the gut microbiota affects the metabolism and excretion of drugs, which in turn impacts the bioavailability and efficacy of drugs and, ultimately, their therapeutic effect. Zhou et al. observed that the absence of *B. fragilis* resulted in the absence of a significant therapeutic effect of MTX. Conversely, the transplantation of *B. fragilis* restored the efficacy of MTX [98].

It is worth noting that diet, as a non-pharmaceutical intervention method, is an effective approach for the long-term management of diseases. The role of anti-inflammatory diets characterized by low calorie levels and rich in omega-3 fatty acids in the prevention and treatment of RA is receiving increasing attention. Studies conducted among obese and non-obese subjects have shown that calorie restriction can reduce the level of TNF—α in serum [99], as well as the expression of NLRP3 inflammasome and il-1b in adipose tissue [100]. Omega-3 polyunsaturated fatty acids in the diet can directly regulate the types and abundance of intestinal microbiota, change the levels of pro-inflammatory mediators, and regulate the levels of microbiota metabolites. For example, it can reduce the growth of Enterobacteria, increase the growth of *Bifidobacteria*, *Blautia*, etc., and subsequently inhibit the inflammatory response related to endotoxemia and increase the content of SCFA [101,102]. Correspondingly, certain microorganisms, such as *Bifidobacterium*, can also promote the absorption of Omega-3 polyunsaturated fatty acids [103]. This interaction may have a positive effect on reducing inflammation and RA management.

4.4. Summary of Therapeutic Interventions Targeting Gut Microbiota in RA Management

Table 4 summarizes the effects of the intervention measures mentioned herein for RA. These findings highlight the translational potential of microbiota-based therapies for veterinary applications.

Table 4. Summary of therapeutic interventions targeting gut microbiota in RA management.

Intervention Type	Specific Approach	Animal Model/Study Cohort	Observed Effects	Mechanism of Action
Probiotics	*Lactobacillus casei* [64]	Adjuvant-induced arthritis rats	Inhibited joint swelling, lowered arthritis scores, and prevented bone destruction	Increases *Lactobacillus* abundance and reduces pro-inflammatory cytokines
	Lactiplantibacillus plantarum [67]	CIA mice	Reduced autoantibody levels and alleviated joint damage	Reduces intestinal permeability and corrects microbial imbalance
	Bacillus coagulans [66]	Adult RA patients	Relieved pain, reduced total C-reactive protein, and improved patients' self-assessment	Inhibits pathogenic bacteria and promotes SCFA production
FMT	Healthy donor fecal transfer [89]	A female RA patient	Decreased DAS28 score and dropped titer of RA	Restores the microbiota composition and inhibits Th17 cell activation
Dietary Interventions	Anti-inflammatory diet [104]	RA patients	Alleviated swelling and pain of the joint	Reduces pro-inflammatory cytokines
Pharmacological–Microbial Interactions	MTX [95]	RA patients	Altered *Bacteroides* and *Faecalibacterium* abundance and affected the metabolism of the microbial community	Modulates gut microbiota to reduce immune activation
	Traditional Chinese medicine (Jingfang Granules) [94]	RA rats	Restored intestinal tight junction proteins and mitigated tissue damage	Increases SCFA production and activates AMPK signaling

4.5. Safety Considerations and Practical Suggestions for Microbial Therapy in Veterinary RA Treatment

Microbial intervention measures have shown promising therapeutic potential in the treatment of animal RA by regulating the intestinal microbiota and the immune system. However, the potential safety risks associated with them still warrant special attention.

Intestinal barrier damage often exacerbates the inflammatory cascade through complex mechanisms, which are of great significance for the safety of microbial therapies in veterinary RA treatment. Species-specific factors are crucial, as the structure and immune response of the intestinal barrier vary among different species, which affects the risk of systemic inflammation. For example, the higher immune sensitivity of dogs makes them more susceptible to inflammation as caused by barrier damage [105]. Similarly, the fragile intestinal barrier of poultry and the complex rumen ecosystem of ruminants require targeted microbial intervention measures to minimize the risks of bacterial dysbiosis and barrier dysfunction. Therefore, although microbial therapies have the potential to regulate immune responses and reduce inflammation, precise control is necessary to avoid unintended immune activation [106,107].

In the treatment of RA, microbial therapies present a paradoxical phenomenon in treatment: they can both alleviate and exacerbate immune imbalance, potentially leading to symptom deterioration. In particular, probiotic therapy may overstimulate the immune system, especially in individuals with existing immune imbalance, resulting in the production of autoantibodies and the activation of autoreactive T cells, further exacerbating the autoimmune response and worsening RA symptoms. This therapeutic contradiction factor becomes even more complex due to the specific responses of different species to microbial intervention [108,109]. Some microbial strains may enhance regulatory immunity in one species but trigger adverse immune responses in another species. Current methods cannot precisely regulate microbial activity, which further exacerbates this challenge and limits the precision of microbial therapies.

Furthermore, there are serious safety issues with the interaction between microbial therapies and conventional drugs, as there may be cross-toxicity and adverse reactions. Microbial intervention affects the pharmacokinetics and pharmacodynamics of conventional drugs. When microbial therapies (such as probiotics) interact with drugs, cross-toxicity may occur, leading to changes in drug efficacy, increased toxicity, or triggering of unexpected immune reactions. These interactions are complicated by the differences in the composition of microbial communities as affected by diet, environment, and species-specific factors, thereby influencing the results of microbial therapy and drug interaction [105]. Therefore, understanding these interactions is crucial for optimizing treatment strategies and improving the management outcomes of rheumatoid arthritis [110,111].

5. Translational Strategies from Human/Rodent Research to Veterinary Medicine

At present, there are still relatively few studies on RA in livestock and poultry animals. However, the existing findings have a positive promoting effect on veterinary research and practice.

The translational potential of human/rodent RA research to veterinary medicine is rooted in the evolutionary conservation of interactions between the gut microbiota and the immune system. Comparative microbiome analysis reveals shared dysbiosis patterns between human RA and animal joint lesions. As mentioned in Tables 1 and 2, in human RA patients, we observed the enrichment of pro-inflammatory microbiota (such as *Escherichia*) and the reduction of SCFA-producing bacteria (such as *Faecalibacterium*), and similar changes were observed in chicks with inflammatory arthritis. This cross-species microbial signature suggests that therapeutic methods targeting the microbiota that are effective in humans may produce similar benefits in animals. For example, SCFAs produced by commensal bacteria such as *Bacteroides* and *Clostridium* exert anti-inflammatory effects in different species by promoting Treg cell differentiation and inhibiting Th17 cell-mediated inflammation. This mechanism has been verified in the rodent CIA model [112] and is considered promising for application in the intervention of inflammatory diseases in more species, such as cattle [113]. Similarly, the disruption of tight junction proteins in human RA patients is consistent with the observed digestive tract damage in rhesus monkeys with inflammatory arthritis [12], suggesting that there may be a conserved pathway between intestinal flora imbalance and systemic inflammation.

However, translational challenges arise from differences in immune responses and gut microbial ecology among species. There are significant differences in the immune systems and the degree of immune responses among different animals [114], so the dose and mode of intervention need to be adjusted. In addition, ruminants, such as cattle, carry unique gut microbiota, whose effects on fiber fermentation and SCFA production differ from those of humans [115]. This requires the development of ruminant-specific probiotics to optimize anti-inflammatory metabolite production in arthropathy-affected livestock.

To narrow the gap in translational research, future studies can focus on establishing standardized animal RA models to characterize the microbial-immune interactions in RA, and utilize cross-species proteomics and metabolomics to identify conserved biomarkers (such as SCFAs, Zonulin) related to the pathological mechanisms of RA, achieving the transformation of human diagnostic tools to veterinary medicine. At the same time, rodent models can be used for pre-screening the anti-RA effects of therapies targeting the microbiota (such as probiotics and FMT) and for further conducting clinical trials of veterinary-specific intervention measures to develop feasible measures for animal RA management, thereby meeting production needs and improving animal welfare.

6. Conclusions

In conclusion, the gut microbiota and the associated metabolites play an important role in the pathogenesis of RA. Modulating intestinal flora may provide a novel strategy for the prevention and treatment of RA. The information presented in this review will contribute to the elucidation of the pathogenesis of RA, enhance the efficacy of RA treatment in both humans and animals, and improve the life quality of patients and animals with RA to the extent possible.

Author Contributions: Conceptualization, S.P.; writing—original draft preparation, Y.Y., F.J., and L.W.; writing—review and editing, J.C. and S.P. All authors have read and agreed to the published version of the manuscript.

Funding: This work was funded by the National Natural Science Foundation of China, grant number 32072809; the Natural Science Foundation of Jiangsu Province, grant number BK20211119; the National Foreign Experts Project (High-end Foreign Experts Introduction Programs of Ministry of Science and Technology), grant number G2023014067L; the Open Fund of Meat Processing Key Laboratory of Sichuan Province, grant number 22-R-17; the Research and Practice Innovation Plan for Postgraduates in Jiangsu Province, grant number SJCX23_2016; the Innovation and Entrepreneurship Training Program for College Students of Higher Education Institutions in Jiangsu Province, grant number 202411117163Y, the Special Project of Cross-cooperation of Northern Jiangsu People's Hospital, grant number SBJC23007; and the Priority Academic Program Development of Jiangsu Higher Education Institutions (PAPD).

Institutional Review Board Statement: Not applicable.

Informed Consent Statement: Not applicable.

Data Availability Statement: Not applicable.

Acknowledgments: We thank Min Du of Washington State University for the critical reading and editing of the manuscript.

Conflicts of Interest: The funders had no role in the design of the study; in the collection, analyses, or interpretation of data; in the writing of the manuscript; or in the decision to publish the results. The authors declare no conflict of interest.

Abbreviations

The following abbreviations are used in this manuscript:

CIA	Collagen-induced arthritis
DCs	Dendritic cells
FMT	Fecal microbiota transplantation
HIF	Hypoxia-inducible factor
ILC3	Type 3 innate lymphocytes
LPS	Lipopolysaccharide
MTX	Methotrexate
NETs	Neutrophil extracellular traps
NOX	NADPH oxidase
PAD	Peptidylarginine deiminase
RA	Rheumatoid arthritis
RANKL	Nuclear factor-κB ligand
SCFAs	Short-chain fatty acids
5-HIAA	5-hydroxyindole-3-acetic acid
5-HT	5-hydroxytryptamine

References

1. Jiao, A.; Zhang, C.; Wang, X.; Sun, L.; Liu, H.; Su, Y.; Lei, L.; Li, W.; Ding, R.; Ding, C.; et al. Single-cell sequencing reveals the evolution of immune molecules across multiple vertebrate species. *J. Adv. Res.* **2024**, *55*, 73–87. [CrossRef] [PubMed]
2. Aalbers, C.J.; Bevaart, L.; Loiler, S.; de Cortie, K.; Wright, J.F.; Mingozzi, F.; Tak, P.P.; Vervoordeldonk, M.J. Preclinical Potency and Biodistribution Studies of an AAV 5 Vector Expressing Human Interferon-β (ART-I02) for Local Treatment of Patients with Rheumatoid Arthritis. *PLoS ONE* **2015**, *10*, e0130612. [CrossRef] [PubMed]
3. Nixon, E.J.; Rose Vineer, H.; Wall, R. Treatment strategies for sheep scab: An economic model of farmer behaviour. *Prev. Vet. Med.* **2017**, *137*, 43–51. [CrossRef]
4. Yu, D.; Du, J.; Pu, X.; Zheng, L.; Chen, S.; Wang, N.; Li, J.; Chen, S.; Pan, S.; Shen, B. The Gut Microbiome and Metabolites Are Altered and Interrelated in Patients with Rheumatoid Arthritis. *Front. Cell Infect. Microbiol.* **2021**, *11*, 763507. [CrossRef]
5. Jeong, Y.; Kim, J.W.; You, H.J.; Park, S.J.; Lee, J.; Ju, J.H.; Park, M.S.; Jin, H.; Cho, M.L.; Kwon, B.; et al. Gut Microbial Composition and Function Are Altered in Patients with Early Rheumatoid Arthritis. *J. Clin. Med.* **2019**, *8*, 693. [CrossRef]
6. Shi, R.; Yu, Z.; Wu, W.; Ho, H.; Wang, J.; Wang, Y.; Han, R. A Survey of 61 Veterinary Drug Residues in Commercial Liquid Milk Products in China. *J. Food Prot.* **2020**, *83*, 1227–1233. [CrossRef] [PubMed]
7. Martinez-Guryn, K.; Hubert, N.; Frazier, K.; Urlass, S.; Musch, M.W.; Ojeda, P.; Pierre, J.F.; Miyoshi, J.; Sontag, T.J.; Cham, C.M.; et al. Small Intestine Microbiota Regulate Host Digestive and Absorptive Adaptive Responses to Dietary Lipids. *Cell Host Microbe* **2018**, *23*, 458–469.e455. [CrossRef]
8. El Aidy, S.; van den Bogert, B.; Kleerebezem, M. The small intestine microbiota, nutritional modulation and relevance for health. *Curr. Opin. Biotechnol.* **2015**, *32*, 14–20. [CrossRef]
9. Zhang, Z.; Fang, Y.; He, Y.; Farag, M.A.; Zeng, M.; Sun, Y.; Peng, S.; Jiang, S.; Zhang, X.; Chen, K.; et al. Bifidobacterium animalis Probio-M8 improves sarcopenia physical performance by mitigating creatine restrictions imposed by microbial metabolites. *NPJ Biofilms Microbiomes* **2024**, *10*, 144. [CrossRef]
10. Mei, L.; Yang, Z.; Zhang, X.; Liu, Z.; Wang, M.; Wu, X.; Chen, X.; Huang, Q.; Huang, R. Sustained Drug Treatment Alters the Gut Microbiota in Rheumatoid Arthritis. *Front. Immunol.* **2021**, *12*, 704089. [CrossRef]
11. Dong, Y.; Yao, J.; Deng, Q.; Li, X.; He, Y.; Ren, X.; Zheng, Y.; Song, R.; Zhong, X.; Ma, J.; et al. Relationship between gut microbiota and rheumatoid arthritis: A bibliometric analysis. *Front. Immunol.* **2023**, *14*, 1131933. [CrossRef] [PubMed]

12. Chen, H.; Shi, J.; Tang, C.; Xu, J.; Li, B.; Wang, J.; Zhou, Y.; Yang, Y.; Yang, H.; Huang, Q.; et al. CHIKV infection drives shifts in the gastrointestinal microbiome and metabolites in rhesus monkeys. *Microbiome* **2024**, *12*, 161. [CrossRef] [PubMed]
13. Månsson, I.; Norberg, R.; Olhagen, B.; Björklund, N.E. Arthritis in pigs induced by dietary factors. Microbiologic, clinical and histologic studies. *Clin. Exp. Immunol.* **1971**, *9*, 677–693. [PubMed]
14. Huang, J.; Liang, L.; Cui, K.; Li, P.; Hao, G.; Sun, S. Salmonella phage CKT1 significantly relieves the body weight loss of chicks by normalizing the abnormal intestinal microbiome caused by hypervirulent Salmonella Pullorum. *Poult. Sci.* **2022**, *101*, 101668. [CrossRef]
15. Yun, H.; Wang, X.; Wei, C.; Liu, Q.; Li, X.; Li, N.; Zhang, G.; Cui, D.; Liu, R. Alterations of the intestinal microbiome and metabolome in women with rheumatoid arthritis. *Clin. Exp. Med.* **2023**, *23*, 4695–4706. [CrossRef]
16. Chen, Y.; Ma, C.; Liu, L.; He, J.; Zhu, C.; Zheng, F.; Dai, W.; Hong, X.; Liu, D.; Tang, D.; et al. Analysis of gut microbiota and metabolites in patients with rheumatoid arthritis and identification of potential biomarkers. *Aging (Albany NY)* **2021**, *13*, 23689–23701. [CrossRef]
17. Xing, Y.; Liu, Y.; Sha, S.; Zhang, Y.; Dou, Y.; Liu, C.; Xu, M.; Zhao, L.; Wang, J.; Wang, Y.; et al. Multikingdom characterization of gut microbiota in patients with rheumatoid arthritis and rheumatoid arthritis-associated interstitial lung disease. *J. Med. Virol.* **2024**, *96*, e29781. [CrossRef]
18. Zhu, J.; Wang, T.; Lin, Y.; Xiong, M.; Chen, J.; Jian, C.; Zhang, J.; Xie, H.; Zeng, F.; Huang, Q.; et al. The change of plasma metabolic profile and gut microbiome dysbiosis in patients with rheumatoid arthritis. *Front. Microbiol.* **2022**, *13*, 931431. [CrossRef]
19. Cheng, M.; Zhao, Y.; Cui, Y.; Zhong, C.; Zha, Y.; Li, S.; Cao, G.; Li, M.; Zhang, L.; Ning, K.; et al. Stage-specific roles of microbial dysbiosis and metabolic disorders in rheumatoid arthritis. *Ann. Rheum. Dis.* **2022**, *81*, 1669–1677. [CrossRef]
20. Scher, J.U.; Sczesnak, A.; Longman, R.S.; Segata, N.; Ubeda, C.; Bielski, C.; Rostron, T.; Cerundolo, V.; Pamer, E.G.; Abramson, S.B.; et al. Expansion of intestinal Prevotella copri correlates with enhanced susceptibility to arthritis. *eLife* **2013**, *2*, e01202. [CrossRef]
21. De Vadder, F.; Kovatcheva-Datchary, P.; Zitoun, C.; Duchampt, A.; Bäckhed, F.; Mithieux, G. Microbiota-Produced Succinate Improves Glucose Homeostasis via Intestinal Gluconeogenesis. *Cell Metab.* **2016**, *24*, 151–157. [CrossRef]
22. Sinha, A.K.; Laursen, M.F.; Brinck, J.E.; Rybtke, M.L.; Hjørne, A.P.; Procházková, N.; Pedersen, M.; Roager, H.M.; Licht, T.R. Dietary fibre directs microbial tryptophan metabolism via metabolic interactions in the gut microbiota. *Nat. Microbiol.* **2024**, *9*, 1964–1978. [CrossRef] [PubMed]
23. Farhadipour, M.; Arnauts, K.; Clarysse, M.; Thijs, T.; Liszt, K.; Van der Schueren, B.; Ceulemans, L.J.; Deleus, E.; Lannoo, M.; Ferrante, M.; et al. SCFAs switch stem cell fate through HDAC inhibition to improve barrier integrity in 3D intestinal organoids from patients with obesity. *iScience* **2023**, *26*, 108517. [CrossRef]
24. Tang, G.; Du, Y.; Guan, H.; Jia, J.; Zhu, N.; Shi, Y.; Rong, S.; Yuan, W. Butyrate ameliorates skeletal muscle atrophy in diabetic nephropathy by enhancing gut barrier function and FFA2-mediated PI3K/Akt/mTOR signals. *Br. J. Pharmacol.* **2022**, *179*, 159–178. [CrossRef] [PubMed]
25. Ruan, S.; Zhai, L.; Wu, S.; Zhang, C.; Guan, Q. SCFAs promote intestinal double-negative T cells to regulate the inflammatory response mediated by NLRP3 inflammasome. *Aging (Albany NY)* **2021**, *13*, 21470–21482. [CrossRef] [PubMed]
26. Iraporda, C.; Errea, A.; Romanin, D.E.; Cayet, D.; Pereyra, E.; Pignataro, O.; Sirard, J.C.; Garrote, G.L.; Abraham, A.G.; Rumbo, M. Lactate and short chain fatty acids produced by microbial fermentation downregulate proinflammatory responses in intestinal epithelial cells and myeloid cells. *Immunobiology* **2015**, *220*, 1161–1169. [CrossRef]
27. Mouries, J.; Brescia, P.; Silvestri, A.; Spadoni, I.; Sorribas, M.; Wiest, R.; Mileti, E.; Galbiati, M.; Invernizzi, P.; Adorini, L.; et al. Microbiota-driven gut vascular barrier disruption is a prerequisite for non-alcoholic steatohepatitis development. *J. Hepatol.* **2019**, *71*, 1216–1228. [CrossRef]
28. Mucientes, A.; Lisbona-Montañez, J.M.; Mena-Vázquez, N.; Ruiz-Limón, P.; Manrique-Arija, S.; García-Studer, A.; Ortiz-Márquez, F.; Fernández-Nebro, A. Intestinal Dysbiosis, Tight Junction Proteins, and Inflammation in Rheumatoid Arthritis Patients: A Cross-Sectional Study. *Int. J. Mol. Sci.* **2024**, *25*, 8649. [CrossRef]
29. Wen, J.; Lyu, P.; Stolzer, I.; Xu, J.; Gießl, A.; Lin, Z.; Andreev, D.; Kachler, K.; Song, R.; Meng, X.; et al. Epithelial HIF2α expression induces intestinal barrier dysfunction and exacerbation of arthritis. *Ann. Rheum. Dis.* **2022**, *81*, 1119–1130. [CrossRef]
30. Tajik, N.; Frech, M.; Schulz, O.; Schälter, F.; Lucas, S.; Azizov, V.; Dürholz, K.; Steffen, F.; Omata, Y.; Rings, A.; et al. Targeting zonulin and intestinal epithelial barrier function to prevent onset of arthritis. *Nat. Commun.* **2020**, *11*, 1995. [CrossRef]
31. Park, H.; Li, Z.; Yang, X.O.; Chang, S.H.; Nurieva, R.; Wang, Y.H.; Wang, Y.; Hood, L.; Zhu, Z.; Tian, Q.; et al. A distinct lineage of CD4 T cells regulates tissue inflammation by producing interleukin 17. *Nat. Immunol.* **2005**, *6*, 1133–1141. [CrossRef]
32. Gallego, A.; Vargas, J.A.; Castejón, R.; Citores, M.J.; Romero, Y.; Millán, I.; Durántez, A. Production of intracellular IL-2, TNF-alpha, and IFN-gamma by T cells in B-CLL. *Cytom. B Clin. Cytom.* **2003**, *56*, 23–29. [CrossRef]

33. Sato, K.; Suematsu, A.; Okamoto, K.; Yamaguchi, A.; Morishita, Y.; Kadono, Y.; Tanaka, S.; Kodama, T.; Akira, S.; Iwakura, Y.; et al. Th17 functions as an osteoclastogenic helper T cell subset that links T cell activation and bone destruction. *J. Exp. Med.* **2006**, *203*, 2673–2682. [CrossRef] [PubMed]
34. Diarra, D.; Stolina, M.; Polzer, K.; Zwerina, J.; Ominsky, M.S.; Dwyer, D.; Korb, A.; Smolen, J.; Hoffmann, M.; Scheinecker, C.; et al. Dickkopf-1 is a master regulator of joint remodeling. *Nat. Med.* **2007**, *13*, 156–163. [CrossRef] [PubMed]
35. Ishigaki, K.; Kochi, Y.; Suzuki, A.; Tsuchida, Y.; Tsuchiya, H.; Sumitomo, S.; Yamaguchi, K.; Nagafuchi, Y.; Nakachi, S.; Kato, R.; et al. Polygenic burdens on cell-specific pathways underlie the risk of rheumatoid arthritis. *Nat. Genet.* **2017**, *49*, 1120–1125. [CrossRef] [PubMed]
36. Ikebuchi, Y.; Aoki, S.; Honma, M.; Hayashi, M.; Sugamori, Y.; Khan, M.; Kariya, Y.; Kato, G.; Tabata, Y.; Penninger, J.M.; et al. Coupling of bone resorption and formation by RANKL reverse signalling. *Nature* **2018**, *561*, 195–200. [CrossRef]
37. Zaiss, M.M.; Sarter, K.; Hess, A.; Engelke, K.; Böhm, C.; Nimmerjahn, F.; Voll, R.; Schett, G.; David, J.P. Increased bone density and resistance to ovariectomy-induced bone loss in FoxP3-transgenic mice based on impaired osteoclast differentiation. *Arthritis Rheum.* **2010**, *62*, 2328–2338. [CrossRef]
38. Wang, T.; Li, J.; Jin, Z.; Wu, F.; Li, Y.; Wang, X.; Zhou, H.; Zhou, Q. Dynamic Frequency of Blood CD4+CD25+ Regulatory T Cells in Rats with Collagen-induced Arthritis. *Korean J. Physiol. Pharmacol.* **2015**, *19*, 83–88. [CrossRef]
39. Bozec, A.; Zaiss, M.M.; Kagwiria, R.; Voll, R.; Rauh, M.; Chen, Z.; Mueller-Schmucker, S.; Kroczek, R.A.; Heinzerling, L.; Moser, M.; et al. T cell costimulation molecules CD80/86 inhibit osteoclast differentiation by inducing the IDO/tryptophan pathway. *Sci. Transl. Med.* **2014**, *6*, 235ra260. [CrossRef]
40. Luo, Y.; Tong, Y.; Wu, L.; Niu, H.; Li, Y.; Su, L.C.; Wu, Y.; Bozec, A.; Zaiss, M.M.; Qing, P.; et al. Alteration of Gut Microbiota in Individuals at High-Risk for Rheumatoid Arthritis Associated With Disturbed Metabolome and the Initiation of Arthritis Through the Triggering of Mucosal Immunity Imbalance. *Arthritis Rheumatol.* **2023**, *75*, 1736–1748. [CrossRef]
41. Maeda, Y.; Kurakawa, T.; Umemoto, E.; Motooka, D.; Ito, Y.; Gotoh, K.; Hirota, K.; Matsushita, M.; Furuta, Y.; Narazaki, M.; et al. Dysbiosis Contributes to Arthritis Development via Activation of Autoreactive T Cells in the Intestine. *Arthritis Rheumatol.* **2016**, *68*, 2646–2661. [CrossRef] [PubMed]
42. Jubair, W.K.; Hendrickson, J.D.; Severs, E.L.; Schulz, H.M.; Adhikari, S.; Ir, D.; Pagan, J.D.; Anthony, R.M.; Robertson, C.E.; Frank, D.N.; et al. Modulation of Inflammatory Arthritis in Mice by Gut Microbiota Through Mucosal Inflammation and Autoantibody Generation. *Arthritis Rheumatol.* **2018**, *70*, 1220–1233. [CrossRef]
43. Noyori, K.; Okamoto, R.; Takagi, T.; Hyodo, A.; Suzuki, K.; Koshino, T. Experimental induction of arthritis in rats immunized with Escherichia coli 0:14 lipopolysaccharide. *J. Rheumatol.* **1994**, *21*, 484–488.
44. Kitamura, K.; Sasaki, M.; Matsumoto, M.; Shionoya, H.; Iida, K. Protective effect of Bacteroides fragilis LPS on Escherichia coli LPS-induced inflammatory changes in human monocytic cells and in a rheumatoid arthritis mouse model. *Immunol. Lett.* **2021**, *233*, 48–56. [CrossRef]
45. Koppejan, H.; Jansen, D.; Hameetman, M.; Thomas, R.; Toes, R.E.M.; van Gaalen, F.A. Altered composition and phenotype of mucosal-associated invariant T cells in early untreated rheumatoid arthritis. *Arthritis Res. Ther.* **2019**, *21*, 3. [CrossRef] [PubMed]
46. Telesford, K.M.; Yan, W.; Ochoa-Reparaz, J.; Pant, A.; Kircher, C.; Christy, M.A.; Begum-Haque, S.; Kasper, D.L.; Kasper, L.H. A commensal symbiotic factor derived from Bacteroides fragilis promotes human CD39(+)Foxp3(+) T cells and Treg function. *Gut Microbes* **2015**, *6*, 234–242. [CrossRef]
47. Piva, F.; Gervois, P.; Karrout, Y.; Sané, F.; Romond, M.B. Gut-Joint Axis: Impact of Bifidobacterial Cell Wall Lipoproteins on Arthritis Development. *Nutrients* **2023**, *15*, 4861. [CrossRef] [PubMed]
48. Tripathy, A.; Swain, N.; Padhan, P.; Raghav, S.K.; Gupta, B. Lactobacillus rhamnosus reduces CD8(+)T cell mediated inflammation in patients with rheumatoid arthritis. *Immunobiology* **2023**, *228*, 152415. [CrossRef]
49. Lule, S.; Colpak, A.I.; Balci-Peynircioglu, B.; Gursoy-Ozdemir, Y.; Peker, S.; Kalyoncu, U.; Can, A.; Tekin, N.; Demiralp, D.; Dalkara, T. Behçet Disease serum is immunoreactive to neurofilament medium which share common epitopes to bacterial HSP-65, a putative trigger. *J. Autoimmun.* **2017**, *84*, 87–96. [CrossRef]
50. Romero, V.; Fert-Bober, J.; Nigrovic, P.A.; Darrah, E.; Haque, U.J.; Lee, D.M.; van Eyk, J.; Rosen, A.; Andrade, F. Immune-mediated pore-forming pathways induce cellular hypercitrullination and generate citrullinated autoantigens in rheumatoid arthritis. *Sci. Transl. Med.* **2013**, *5*, 209ra150. [CrossRef]
51. Suzuki, A.; Yamada, R.; Chang, X.; Tokuhiro, S.; Sawada, T.; Suzuki, M.; Nagasaki, M.; Nakayama-Hamada, M.; Kawaida, R.; Ono, M.; et al. Functional haplotypes of PADI4, encoding citrullinating enzyme peptidylarginine deiminase 4, are associated with rheumatoid arthritis. *Nat. Genet.* **2003**, *34*, 395–402. [CrossRef]
52. Kinloch, A.; Lundberg, K.; Wait, R.; Wegner, N.; Lim, N.H.; Zendman, A.J.; Saxne, T.; Malmström, V.; Venables, P.J. Synovial fluid is a site of citrullination of autoantigens in inflammatory arthritis. *Arthritis Rheum.* **2008**, *58*, 2287–2295. [CrossRef] [PubMed]

53. Vossenaar, E.R.; Zendman, A.J.; van Venrooij, W.J.; Pruijn, G.J. PAD, a growing family of citrullinating enzymes: Genes, features and involvement in disease. *Bioessays* **2003**, *25*, 1106–1118. [CrossRef]
54. McGraw, W.T.; Potempa, J.; Farley, D.; Travis, J. Purification, characterization, and sequence analysis of a potential virulence factor from Porphyromonas gingivalis, peptidylarginine deiminase. *Infect. Immun.* **1999**, *67*, 3248–3256. [CrossRef] [PubMed]
55. Konig, M.F.; Andrade, F. A Critical Reappraisal of Neutrophil Extracellular Traps and NETosis Mimics Based on Differential Requirements for Protein Citrullination. *Front. Immunol.* **2016**, *7*, 461. [CrossRef]
56. Sohn, D.H.; Rhodes, C.; Onuma, K.; Zhao, X.; Sharpe, O.; Gazitt, T.; Shiao, R.; Fert-Bober, J.; Cheng, D.; Lahey, L.J.; et al. Local Joint inflammation and histone citrullination in a murine model of the transition from preclinical autoimmunity to inflammatory arthritis. *Arthritis Rheumatol.* **2015**, *67*, 2877–2887. [CrossRef] [PubMed]
57. Neeli, I.; Khan, S.N.; Radic, M. Histone deimination as a response to inflammatory stimuli in neutrophils. *J. Immunol.* **2008**, *180*, 1895–1902. [CrossRef]
58. Pratesi, F.; Dioni, I.; Tommasi, C.; Alcaro, M.C.; Paolini, I.; Barbetti, F.; Boscaro, F.; Panza, F.; Puxeddu, I.; Rovero, P.; et al. Antibodies from patients with rheumatoid arthritis target citrullinated histone 4 contained in neutrophils extracellular traps. *Ann. Rheum. Dis.* **2014**, *73*, 1414–1422. [CrossRef]
59. Hakkim, A.; Fuchs, T.A.; Martinez, N.E.; Hess, S.; Prinz, H.; Zychlinsky, A.; Waldmann, H. Activation of the Raf-MEK-ERK pathway is required for neutrophil extracellular trap formation. *Nat. Chem. Biol.* **2011**, *7*, 75–77. [CrossRef]
60. Jones, R.M.; Luo, L.; Ardita, C.S.; Richardson, A.N.; Kwon, Y.M.; Mercante, J.W.; Alam, A.; Gates, C.L.; Wu, H.; Swanson, P.A.; et al. Symbiotic lactobacilli stimulate gut epithelial proliferation via Nox-mediated generation of reactive oxygen species. *Embo J.* **2013**, *32*, 3017–3028. [CrossRef]
61. Liu, N.; Li, Y.; Dong, C.; Xu, X.; Wei, P.; Sun, W.; Peng, Q. Inositol-Requiring Enzyme 1-Dependent Activation of AMPK Promotes Brucella abortus Intracellular Growth. *J. Bacteriol.* **2016**, *198*, 986–993. [CrossRef] [PubMed]
62. Liu, X.; Wang, N.; Zhu, Y.; Yang, Y.; Chen, X.; Chen, Q.; Zhou, H.; Zheng, J. Extracellular calcium influx promotes antibacterial autophagy in Escherichia coli infected murine macrophages via CaMKKβ dependent activation of ERK1/2, AMPK and FoxO1. *Biochem. Biophys. Res. Commun.* **2016**, *469*, 639–645. [CrossRef] [PubMed]
63. Hu, F.; Li, X.; Liu, K.; Li, Y.; Xie, Y.; Wei, C.; Liu, S.; Song, J.; Wang, P.; Shi, L.; et al. Rheumatoid arthritis patients harbour aberrant enteric bacteriophages with autoimmunity-provoking potential: A paired sibling study. *Ann. Rheum. Dis.* **2024**, *83*, 1677–1690. [CrossRef]
64. Pan, H.; Guo, R.; Ju, Y.; Wang, Q.; Zhu, J.; Xie, Y.; Zheng, Y.; Li, T.; Liu, Z.; Lu, L.; et al. A single bacterium restores the microbiome dysbiosis to protect bones from destruction in a rat model of rheumatoid arthritis. *Microbiome* **2019**, *7*, 107. [CrossRef] [PubMed]
65. Albuquerque-Souza, E.; Balzarini, D.; Ando-Suguimoto, E.S.; Ishikawa, K.H.; Simionato, M.R.L.; Holzhausen, M.; Mayer, M.P.A. Probiotics alter the immune response of gingival epithelial cells challenged by Porphyromonas gingivalis. *J. Periodontal Res.* **2019**, *54*, 115–127. [CrossRef]
66. Mandel, D.R.; Eichas, K.; Holmes, J. Bacillus coagulans: A viable adjunct therapy for relieving symptoms of rheumatoid arthritis according to a randomized, controlled trial. *BMC Complement. Altern. Med.* **2010**, *10*, 1. [CrossRef]
67. Qin, Q.; Hu, G.; Zhou, X.; Zhu, R.; Chen, J.; Zeng, K.; Wu, J.; Wei, J.; Chen, T.; Guo, X. Therapeutic potential of the probiotic Lactiplantibacillus plantarum BX 62 and its postbiotics in alleviating rheumatoid arthritis in mice. *Curr. Res. Food Sci.* **2024**, *9*, 100915. [CrossRef]
68. Ahmad; Zhang, C.; Wang, Y.; Ullah, H.; Rahman, A.U.; Wei, J.; Qin, Y.H.; Wang, G.; Wang, B.; Li, X. Saccharomyces boulardii (CNCM I-745) alleviates collagen-induced arthritis by partially maintaining intestinal mucosal integrity through TLR2/MYD88/NF-κB pathway inhibition. *Int. Immunopharmacol.* **2024**, *139*, 112738. [CrossRef]
69. Pineda Mde, L.; Thompson, S.F.; Summers, K.; de Leon, F.; Pope, J.; Reid, G. A randomized, double-blinded, placebo-controlled pilot study of probiotics in active rheumatoid arthritis. *Med. Sci. Monit.* **2011**, *17*, Cr347–Cr354. [CrossRef]
70. Huang, Z.Y.; Stabler, T.; Pei, F.X.; Kraus, V.B. Both systemic and local lipopolysaccharide (LPS) burden are associated with knee OA severity and inflammation. *Osteoarthr. Cartil.* **2016**, *24*, 1769–1775. [CrossRef]
71. Schauber, J.; Svanholm, C.; Termén, S.; Iffland, K.; Menzel, T.; Scheppach, W.; Melcher, R.; Agerberth, B.; Lührs, H.; Gudmundsson, G.H. Expression of the cathelicidin LL-37 is modulated by short chain fatty acids in colonocytes: Relevance of signalling pathways. *Gut* **2003**, *52*, 735–741. [CrossRef]
72. Haase, S.; Mäurer, J.; Duscha, A.; Lee, D.H.; Balogh, A.; Gold, R.; Müller, D.N.; Haghikia, A.; Linker, R.A. Propionic Acid Rescues High-Fat Diet Enhanced Immunopathology in Autoimmunity via Effects on Th17 Responses. *Front. Immunol.* **2021**, *12*, 701626. [CrossRef] [PubMed]
73. Scott, S.A.; Fu, J.; Chang, P.V. Microbial tryptophan metabolites regulate gut barrier function via the aryl hydrocarbon receptor. *Proc. Natl. Acad. Sci. USA* **2020**, *117*, 19376–19387. [CrossRef]

74. Yan, X.; Yan, J.; Xiang, Q.; Wang, F.; Dai, H.; Huang, K.; Fang, L.; Yao, H.; Wang, L.; Zhang, W. Fructooligosaccharides protect against OVA-induced food allergy in mice by regulating the Th17/Treg cell balance using tryptophan metabolites. *Food Funct.* **2021**, *12*, 3191–3205. [CrossRef]
75. Guo, Q.; Tang, Y.; Li, Y.; Xu, Z.; Zhang, D.; Liu, J.; Wang, X.; Xia, W.; Xu, S. Perinatal High-Salt Diet Induces Gut Microbiota Dysbiosis, Bile Acid Homeostasis Disbalance, and NAFLD in Weanling Mice Offspring. *Nutrients* **2021**, *13*, 2135. [CrossRef] [PubMed]
76. Xiao, F.; Dong, F.; Li, X.; Li, Y.; Yu, G.; Liu, Z.; Wang, Y.; Zhang, T. Bifidobacterium longum CECT 7894 Improves the Efficacy of Infliximab for DSS-Induced Colitis via Regulating the Gut Microbiota and Bile Acid Metabolism. *Front. Pharmacol.* **2022**, *13*, 902337. [CrossRef] [PubMed]
77. Langan, D.; Perkins, D.J.; Vogel, S.N.; Moudgil, K.D. Microbiota-Derived Metabolites, Indole-3-aldehyde and Indole-3-acetic Acid, Differentially Modulate Innate Cytokines and Stromal Remodeling Processes Associated with Autoimmune Arthritis. *Int. J. Mol. Sci.* **2021**, *22*, 2017. [CrossRef]
78. Su, X.; Wang, X.; Zhang, X.; Sun, Y.; Jia, Y. β-Indole-3-acetic acid attenuated collagen-induced arthritis through reducing the ubiquitination of Foxp3 via the AhR-TAZ-Tip60 pathway. *Immunol. Res.* **2024**, *72*, 741–753. [CrossRef]
79. Romano, K.A.; Vivas, E.I.; Amador-Noguez, D.; Rey, F.E. Intestinal microbiota composition modulates choline bioavailability from diet and accumulation of the proatherogenic metabolite trimethylamine-N-oxide. *mBio* **2015**, *6*, e02481. [CrossRef]
80. Chen, M.L.; Yi, L.; Zhang, Y.; Zhou, X.; Ran, L.; Yang, J.; Zhu, J.D.; Zhang, Q.Y.; Mi, M.T. Resveratrol Attenuates Trimethylamine-N-Oxide (TMAO)-Induced Atherosclerosis by Regulating TMAO Synthesis and Bile Acid Metabolism via Remodeling of the Gut Microbiota. *mBio* **2016**, *7*, e02210–e02215. [CrossRef]
81. Arpaia, N.; Campbell, C.; Fan, X.; Dikiy, S.; van der Veeken, J.; deRoos, P.; Liu, H.; Cross, J.R.; Pfeffer, K.; Coffer, P.J.; et al. Metabolites produced by commensal bacteria promote peripheral regulatory T-cell generation. *Nature* **2013**, *504*, 451–455. [CrossRef] [PubMed]
82. Furusawa, Y.; Obata, Y.; Fukuda, S.; Endo, T.A.; Nakato, G.; Takahashi, D.; Nakanishi, Y.; Uetake, C.; Kato, K.; Kato, T.; et al. Commensal microbe-derived butyrate induces the differentiation of colonic regulatory T cells. *Nature* **2013**, *504*, 446–450. [CrossRef]
83. Smith, P.M.; Howitt, M.R.; Panikov, N.; Michaud, M.; Gallini, C.A.; Bohlooly, Y.M.; Glickman, J.N.; Garrett, W.S. The microbial metabolites, short-chain fatty acids, regulate colonic Treg cell homeostasis. *Science* **2013**, *341*, 569–573. [CrossRef]
84. He, J.; Chu, Y.; Li, J.; Meng, Q.; Liu, Y.; Jin, J.; Wang, Y.; Wang, J.; Huang, B.; Shi, L.; et al. Intestinal butyrate-metabolizing species contribute to autoantibody production and bone erosion in rheumatoid arthritis. *Sci. Adv.* **2022**, *8*, eabm1511. [CrossRef] [PubMed]
85. Tominaga, K.; Alstergren, P.; Kurita, H.; Kopp, S. Serotonin in an antigen-induced arthritis of the rabbit temporomandibular joint. *Arch. Oral. Biol.* **1999**, *44*, 595–601. [CrossRef]
86. Li, B.; Ding, M.; Chen, C.; Zhao, J.; Shi, G.; Ross, P.; Stanton, C.; Chen, W.; Yang, B. Bifidobacterium longum subsp. infantis B6MNI Alleviates Collagen-Induced Arthritis in Rats via Regulating 5-HIAA and Pim-1/JAK/STAT3 Inflammation Pathways. *J. Agric. Food Chem.* **2023**, *71*, 17819–17832. [CrossRef] [PubMed]
87. Doron, S.; Snydman, D.R. Risk and safety of probiotics. *Clin. Infect. Dis.* **2015**, *60* (Suppl. 2), S129–S134. [CrossRef] [PubMed]
88. Benech, N.; Cassir, N.; Alric, L.; Barbut, F.; Batista, R.; Bleibtreu, A.; Briot, T.; Davido, B.; Galperine, T.; Joly, A.C.; et al. Impact of Clinical and Pharmacological Parameters on Faecal Microbiota Transplantation Outcome in Clostridioides difficile Infections: Results of a 5-Year French National Survey. *Aliment. Pharmacol. Ther.* **2025**, *61*, 159–167. [CrossRef]
89. Zeng, J.; Peng, L.; Zheng, W.; Huang, F.; Zhang, N.; Wu, D.; Yang, Y. Fecal microbiota transplantation for rheumatoid arthritis: A case report. *Clin. Case Rep.* **2021**, *9*, 906–909. [CrossRef]
90. Minkoff, N.Z.; Aslam, S.; Medina, M.; Tanner-Smith, E.E.; Zackular, J.P.; Acra, S.; Nicholson, M.R.; Imdad, A. Fecal microbiota transplantation for the treatment of recurrent Clostridioides difficile (*Clostridium difficile*). *Cochrane Database Syst. Rev.* **2023**, *4*, Cd013871. [CrossRef]
91. Rossen, N.G.; MacDonald, J.K.; de Vries, E.M.; D'Haens, G.R.; de Vos, W.M.; Zoetendal, E.G.; Ponsioen, C.Y. Fecal microbiota transplantation as novel therapy in gastroenterology: A systematic review. *World J. Gastroenterol.* **2015**, *21*, 5359–5371. [CrossRef] [PubMed]
92. Jochumsen, E.A.; Kragsnaes, M.S.; Nilsson, A.C.; Rasmussen, K.F.; Ellingsen, T.; Juul, M.A.; Kjeldsen, J.; Holm, D.K. 'Does this fecal microbiota transplant work?' Quality assurance of capsule based fecal microbiota transplant production. *Scand. J. Gastroenterol.* **2024**, *59*, 1234–1239. [CrossRef]
93. Gefen, R.; Dourado, J.; Emile, S.H.; Wignakumar, A.; Rogers, P.; Aeschbacher, P.; Garoufalia, Z.; Horesh, N.; Wexner, S.D. Fecal microbiota transplantation for patients with ulcerative colitis: A systematic review and meta-analysis of randomized control trials. *Tech Coloproctol.* **2025**, *29*, 103. [CrossRef] [PubMed]
94. Wang, X.; Pan, L.; Niu, D.; Zhou, J.; Shen, M.; Zeng, Z.; Gong, W.; Yang, E.; Tang, Y.; Cheng, G.; et al. Jingfang Granules alleviates the lipid peroxidation induced ferroptosis in rheumatoid arthritis rats by regulating gut microbiota and metabolism of short chain fatty acids. *J. Ethnopharmacol.* **2025**, *339*, 119160. [CrossRef]

95. Nayak, R.R.; Alexander, M.; Deshpande, I.; Stapleton-Gray, K.; Rimal, B.; Patterson, A.D.; Ubeda, C.; Scher, J.U.; Turnbaugh, P.J. Methotrexate impacts conserved pathways in diverse human gut bacteria leading to decreased host immune activation. *Cell Host Microbe* **2021**, *29*, 362–377.e11. [CrossRef]
96. Liu, Y.; Liu, L.; Luo, J.; Peng, X. Metabolites from specific intestinal bacteria in vivo fermenting Lycium barbarum polysaccharide improve collagenous arthritis in rats. *Int. J. Biol. Macromol.* **2023**, *226*, 1455–1467. [CrossRef] [PubMed]
97. Eor, J.Y.; Park, N.; Son, Y.J.; Kim, S.H. Therapeutic Effects of Gleditsia sinensis Thorn Extract Fermented by Lactobacillus casei 3260 in a Type II Collagen-Induced Rheumatoid Arthritis Mouse Model. *Food Sci. Anim. Resour.* **2021**, *41*, 497–508. [CrossRef]
98. Zhou, B.; Dong, C.; Zhao, B.; Lin, K.; Tian, Y.; Zhang, R.; Zhu, L.; Xu, H.; Yang, L. Bacteroides fragilis participates in the therapeutic effect of methotrexate on arthritis through metabolite regulation. *Front. Microbiol.* **2022**, *13*, 1015130. [CrossRef]
99. Ravussin, E.; Redman, L.M.; Rochon, J.; Das, S.K.; Fontana, L.; Kraus, W.E.; Romashkan, S.; Williamson, D.A.; Meydani, S.N.; Villareal, D.T.; et al. A 2-Year Randomized Controlled Trial of Human Caloric Restriction: Feasibility and Effects on Predictors of Health Span and Longevity. *J. Gerontol. A Biol. Sci. Med. Sci.* **2015**, *70*, 1097–1104. [CrossRef]
100. Vandanmagsar, B.; Youm, Y.H.; Ravussin, A.; Galgani, J.E.; Stadler, K.; Mynatt, R.L.; Ravussin, E.; Stephens, J.M.; Dixit, V.D. The NLRP3 inflammasome instigates obesity-induced inflammation and insulin resistance. *Nat. Med.* **2011**, *17*, 179–188. [CrossRef]
101. Cao, W.; Wang, C.; Chin, Y.; Chen, X.; Gao, Y.; Yuan, S.; Xue, C.; Wang, Y.; Tang, Q. DHA-phospholipids (DHA-PL) and EPA-phospholipids (EPA-PL) prevent intestinal dysfunction induced by chronic stress. *Food Funct.* **2019**, *10*, 277–288. [CrossRef] [PubMed]
102. Ochoa-Repáraz, J.; Kasper, L.H. The Second Brain: Is the Gut Microbiota a Link Between Obesity and Central Nervous System Disorders? *Curr. Obes. Rep.* **2016**, *5*, 51–64. [CrossRef] [PubMed]
103. Wall, R.; Ross, R.P.; Shanahan, F.; O'Mahony, L.; Kiely, B.; Quigley, E.; Dinan, T.G.; Fitzgerald, G.; Stanton, C. Impact of administered bifidobacterium on murine host fatty acid composition. *Lipids* **2010**, *45*, 429–436. [CrossRef]
104. Adam, O.; Beringer, C.; Kless, T.; Lemmen, C.; Adam, A.; Wiseman, M.; Adam, P.; Klimmek, R.; Forth, W. Anti-inflammatory effects of a low arachidonic acid diet and fish oil in patients with rheumatoid arthritis. *Rheumatol. Int.* **2003**, *23*, 27–36. [CrossRef] [PubMed]
105. Jia, X. Host-Microbe Interactions in Pets: Implications for Immune System Dynamics. *Int. J. Mol. Vet. Res.* **2024**, *14*, 46–55.
106. Mohammed, A.T.; Khattab, M.; Ahmed, A.M.; Turk, T.; Sakr, N.; Khalil, A.M.; Abdelhalim, M.; Sawaf, B.; Hirayama, K.; Huy, N.T. The therapeutic effect of probiotics on rheumatoid arthritis: A systematic review and meta-analysis of randomized control trials. *Clin. Rheumatol.* **2017**, *36*, 2697–2707. [CrossRef]
107. Piraner, D.I.; Abedi, M.H.; Moser, B.A.; Lee-Gosselin, A.; Shapiro, M.G. Tunable thermal bioswitches for in vivo control of microbial therapeutics. *Nat. Chem. Biol.* **2017**, *13*, 75–80. [CrossRef]
108. Swartwout, B.; Luo, X.M. Implications of Probiotics on the Maternal-Neonatal Interface: Gut Microbiota, Immunomodulation, and Autoimmunity. *Front. Immunol.* **2018**, *9*, 2840. [CrossRef]
109. Lindenberg, F.; Krych, L.; Fielden, J.; Kot, W.; Frøkiær, H.; van Galen, G.; Nielsen, D.S.; Hansen, A.K. Expression of immune regulatory genes correlate with the abundance of specific Clostridiales and Verrucomicrobia species in the equine ileum and cecum. *Sci. Rep.* **2019**, *9*, 12674. [CrossRef]
110. Aarts, J.; Boleij, A.; Pieters, B.C.H.; Feitsma, A.L.; van Neerven, R.J.J.; Ten Klooster, J.P.; M'Rabet, L.; Arntz, O.J.; Koenders, M.I.; van de Loo, F.A.J. Flood Control: How Milk-Derived Extracellular Vesicles Can Help to Improve the Intestinal Barrier Function and Break the Gut-Joint Axis in Rheumatoid Arthritis. *Front. Immunol.* **2021**, *12*, 703277. [CrossRef]
111. Strzelec, M.; Detka, J.; Mieszczak, P.; Sobocińska, M.K.; Majka, M. Immunomodulation-a general review of the current state-of-the-art and new therapeutic strategies for targeting the immune system. *Front. Immunol.* **2023**, *14*, 1127704. [CrossRef] [PubMed]
112. Lv, J.; Hao, P.; Zhou, Y.; Liu, T.; Wang, L.; Song, C.; Wang, Z.; Liu, Z.; Liu, Y. Role of the intestinal flora-immunity axis in the pathogenesis of rheumatoid arthritis-mechanisms regulating short-chain fatty acids and Th17/Treg homeostasis. *Mol. Biol. Rep.* **2025**, *52*, 617. [CrossRef] [PubMed]
113. He, Z.; Dong, H. The roles of short-chain fatty acids derived from colonic bacteria fermentation of non-digestible carbohydrates and exogenous forms in ameliorating intestinal mucosal immunity of young ruminants. *Front. Immunol.* **2023**, *14*, 1291846. [CrossRef] [PubMed]
114. Wlaźlak, S.; Pietrzak, E.; Biesek, J.; Dunislawska, A. Modulation of the immune system of chickens a key factor in maintaining poultry production-a review. *Poult. Sci.* **2023**, *102*, 102785. [CrossRef]
115. Tapio, I.; Snelling, T.J.; Strozzi, F.; Wallace, R.J. The ruminal microbiome associated with methane emissions from ruminant livestock. *J. Anim. Sci. Biotechnol.* **2017**, *8*, 7. [CrossRef] [PubMed]

veterinary sciences

MDPI

Article

Single or Combined Supplementation of *Rhodotorula* Yeast Culture and *Bacillus Subtilis* Enhances Intestinal Barrier Function in Yellow-Feathered Broilers

Xiangtan Su, Ke Wang, Yeqing Liu, Xinyu Lu, Meiru Chen, Jianlong Dang, Gaowei Zhang, Guang Yang, Aiqin Gao * and Yuanqing Xu *

College of Animal Science, Inner Mongolia Agricultural University, Hohhot 010018, China; fs646464@163.com (X.S.); wangke99658@163.com (K.W.); jiangseruyan@163.com (Y.L.); 15134666337@163.com (X.L.); jjjjtmmjt@163.com (M.C.); perter1733@formail.com (J.D.); 17637324277@163.com (G.Z.); yangguang_94@126.com (G.Y.)

* Correspondence: gaoaiqin999@imau.edu.cn (A.G.); xuyuanqing@imau.edu.cn (Y.X.)

Simple Summary: Yellow-feathered broilers, a distinctive breed in China, are more susceptible to immune dysfunction and intestinal diseases due to intensive farming practices. To address these challenges, enhancing the intestinal barrier function is a critical strategy. This study investigated the effects of *Rhodotorula* yeast culture (RYC) and *Bacillus subtilis* (BS), either alone or in combination, on the intestinal health of yellow-feathered broilers. RYC has been shown to alleviate immune stress and strengthen intestinal barrier function, while BS can regulate intestinal microbiota and maintain gut health. The experimental results revealed that the combination of RYC and BS improved jejunal structure, as evidenced by an increased villus height-to-crypt depth (VH/CD) ratio. However, the two supplements exhibited antagonistic interactions in modulating immune responses and the expression of tight junction proteins. Despite these antagonistic effects, the study highlighted the potential of RYC and BS to enhance intestinal barrier function and modulate immune activity in yellow-feathered broilers. These findings provide valuable insights into the scientific application of RYC and BS as dietary supplements, offering strategies to improve gut health and overall well-being in poultry farming.

Abstract: This study aimed to investigate the effects of dietary supplementation with *Rhodotorula* yeast cultures (RYC) and *Bacillus subtilis* (BS), alone or in combination, on growth performance and intestinal barrier function in yellow-feathered broilers. A 2 × 2 factorial interaction study was conducted, in which a total of 192 one-day-old yellow-feathered broilers were randomly assigned into four treatment groups: CON group, fed a basal diet; BS group, supplemented with 5×10^9 CFU/kg of BS; RYC group, supplemented with 5000 mg/kg of RYC; and RYC + BS group, supplemented with both. During the 56-day experimental period, body weight, average daily feed intake, average daily gain, and feed conversion ratio were not significantly affected by RYC, BS, or their interaction ($p > 0.05$). In the jejunum, the villus height-to-crypt depth ratio was significantly increased by the interaction of RYC and BS ($p < 0.05$). mRNA expression of tight junction proteins (*JAM2*, *TJP1*) was significantly upregulated by BS alone ($p < 0.05$), but this effect was diminished when RYC and BS were combined, indicating an antagonistic interaction between the two supplements. Mucin-2 (*MUC2*) expression was significantly increased by RYC or BS alone ($p < 0.05$). In immune function analysis, IgM levels were significantly increased by RYC alone but decreased when RYC and BS were combined ($p < 0.05$), further highlighting their antagonistic interaction. BS supplementation significantly increased IgG and pro-inflammatory gene expression (*TNFA*, *IL1B*, and *NFKB1*)

($p < 0.05$), while RYC supplementation reduced IFN-γ and increased anti-inflammatory gene expression (*IL10* and *MyD88*). Cecal microbial analysis revealed increased abundance of *g_Bacillus* in the BS group, *g_norank_f__norank_o__Clostridia_UCG-014* in the RYC group, and *g_norank_f__norank_o__norank_c_Clostridia* in the BS + RYC group were significantly increased compared to the CON group ($p < 0.05$). These results suggest that RYC and BS, as dietary supplements, may enhance intestinal health and barrier function in yellow-feathered broilers with minimal effects on growth performance. However, the antagonistic interactions between RYC and BS in modulating immune responses and tight junction protein expression highlight the need for careful consideration when combining these supplements in poultry nutrition strategies.

Keywords: yellow-feathered broilers; *Rhodotorula* yeast culture; *Bacillus subtilis*; intestinal immunity; intestinal barrier function

1. Introduction

Yellow-feathered broilers, a distinctive breed in China, are highly valued for their distinctive flavor profiles and superior nutritional composition [1]. This growing consumer preference has driven their expanding market share in China's meat sector [2]. However, modern intensive breeding practices aimed at accelerating growth rates—particularly through high-fat diets—impose substantial intestinal stress, compromising barrier integrity in broilers [3]. Such barrier dysfunction can trigger a cascade of metabolic dysregulation, manifesting as suppressed growth metrics, amplified stress susceptibility, and impaired immunocompetence [4]. These challenges are exacerbated by the breed's extended production cycle, necessitating more rigorous gut health management compared to commercial broiler strains [5].

The gastrointestinal tract serves as the primary interface between external environments and internal homeostasis, with its multilayered barrier system providing a critical defense against luminal pathogens and antigens [6]. The intestinal barrier system consists of a tightly connected epithelial cell layer, tight junction protein (TJP) expression, mucus secretion, intestinal immune cells, and the resident symbiotic microbiota [7]. The system is categorized into four distinct types based on their functions: the physical barrier, the chemical barrier, the immune barrier, and the microbial barrier. TJP, specialized structures situated between intestinal epithelial cells, function as a critical regulatory barrier that governs paracellular permeability, maintaining gut barrier integrity [8].

Emerging evidence shows that synbiotic interventions (combined prebiotics and probiotics) can optimize the intestinal ecosystem by reshaping microbial composition, enhancing α-diversity, and competitively excluding enteropathogens [9,10]. Within this paradigm, yeast cultures—a class of multifunctional probiotic preparations—deliver a bioactive matrix comprising fermentation substrates, microbial proteins, yeast metabolites, and cell wall components like β-glucans and mannooligosaccharides [11]. Notably, yeast-derived components have been shown to increase the ileal villus height-to-crypt depth (VH/CD) ratio in broilers [12]. Yeast cell walls also enhance intestinal barrier integrity in broilers through the upregulation of TJP gene expression, thereby improving gut health [13,14]. Additionally, dietary yeast culture supplementation has been found to modulate the composition and diversity of the cecal microbiota in broilers [11]. Beyond structural benefits, red yeast enhances gastrointestinal health in poultry by adsorbing mycotoxins, reducing toxin-induced intestinal dysfunction [11,15]. Importantly, studies have shown that *Rhodotorula* yeast culture (RYC) is rich in metabolites such as carotenoids, which provide significant benefits in reducing immune stress and enhancing intestinal barrier function [16].

Bacillus subtilis (BS), a prominent member of the animal intestinal microbiota, produces bioactive compounds such as lipopeptides, surfactants, and bacteriocins, which can be used as a viable alternative to traditional feed antibiotics [17,18]. Beyond its antimicrobial effects, BS enhances intestinal health in broilers by improving VH/CD, upregulating TJ expression, and reducing pathogenic colonization in the cecum, thereby synergistically improving gut barrier integrity and microbiota homeostasis [19,20]. Additionally, BS facilitates nutrient release, modulates digestive enzyme activity, and optimizes microbial community structure, further supporting intestinal function [21,22].

Recent studies have demonstrated that combining dietary supplementation with BS and *Saccharomyces cerevisiae* can significantly improve growth performance in *Salmonella Typhimurium*-challenged broilers, primarily by enhancing intestinal morphology [23]. Further research suggests that BS achieves its maximal viable biomass when co-administered with yeast, indicating a potential microbial synergy [24].

Despite these findings, most studies have focused on the isolated application of yeast products or BS, leaving a critical gap in understanding their combined effects—particularly in yellow-feathered broilers. Specifically, systematic studies evaluating the impact of RYC and BS cosupplementation on intestinal barrier function are lacking. This study hypothesizes that the synergistic use of RYC and BS will enhance intestinal barrier function through complementary mechanisms: RYC primarily supports intestinal morphology via nutrient provision, while BS exerts its benefits through cecal microbiota modulation.

2. Materials and Methods

This study was conducted in accordance with the ethical guidelines and protocols of the People's Republic of China. All experimental procedures were assessed and approved by the Animal Care and Use Committee at Inner Mongolia Agricultural University (Hohhot, China; Approval No. NND2024124).

2.1. Feed Additives

The *Rhodotorula* yeast culture (β-glucan $\geq$ 1.45 mg/g, mannan oligosaccharides $\geq$ 0.33 mg/g, carotenoids $\geq$ 1.60 mg/g) was provided by the State Key Laboratory of Animal Nutrition at the Beijing Institute of Animal Husbandry and Veterinary Medicine, Chinese Academy of Agricultural Sciences (Beijing, China). The RYC was produced by fermenting *Rhodotorula mucilaginosa* in a liquid broth with soybean meal as the solid substrate. The final product contained yeast cell walls, cellular metabolites, and residual medium components [25,26]. The *Bacillus subtilis* was provided by Beihai Yiqiang Biotechnology Co., Ltd. (Beihai, China) and had a viable count of 1×10^{11} CFU/g. Both the RYC and BS were used in powder form for this study.

The additives were incorporated into the feed in a three-stage mixing process. In the initial mixing stage, the probiotics (RYC or BS) were thoroughly blended with 100 g of mixed feed according to the experimental design. During the secondary mixing stage, this 100 g mixture was then combined with 900 g of standard feed, producing 1 kg of supplemented feed. In the final dilution stage, the 1 kg supplemented feed was mixed with 9 kg of standard feed, resulting in 10 kg of intermediate feed. For complete feed production, the 10 kg intermediate feed was blended with 30 kg of standard feed, yielding 40 kg of final feed. To ensure uniform distribution, the feed was mixed thoroughly at each stage to prevent ingredient separation [27].

2.2. Animals, Research Protocol, and Nutritional Regimens

A total of 192 one-day-old yellow-feathered broilers with similar initial body weights (40.00 ± 0.64 g) were randomly assigned to four treatment groups. Each group consisted of six replicate pens, with eight chicks per pen (four males and four females) to ensure sex balance. Based on previous research [28,29], the optimal inclusion levels of RYC and BS in the broiler diets were determined. The experiment was designed as a 2 × 2 factorial interaction study to evaluate the effects of RYC and BS, alone or in combination, on yellow-feathered broilers. The four experimental groups were as follows: Control group (basal diet only), BS group (basal diet + 5 × 10^9 CFU/kg BS), RYC group (basal diet + 5000 mg/kg RYC), and BS + RYC group (basal diet + both additives: 5 × 10^9 CFU/kg BS + 5000 mg/kg of RYC). The basal diet was formulated according to the Nutrition Requirement for Yellow-Feather Broilers (NY/T 3645-2020) [30]. The detailed composition and nutritional profile of the basal diet are provided in Table 1.

Table 1. Composition and nutrient levels of the basal diet (%, as fed).

Items	1 to 28 Days of Age	29 to 56 Days of Age
Ingredients		
Maize	58.95	60.90
Soybean oil	3.00	3.00
Soybean meal	23.00	20.00
Cottonseed cake	6.00	6.00
Rape seed cake	5.00	6.00
Limestone	1.07	1.22
$CaHPO_4$	1.90	1.80
NaCl	0.37	0.37
Choline	0.11	0.11
Vitamin premix [1]	0.10	0.10
Mineral premix [2]	0.50	0.50
Total	100.00	100.00
Nutrient levels		
ME (MJ/kg)	12.26	12.64
CP (%)	21.00	20.00
CF (%)	7.00	6.00
Ca (%)	1.05	0.95
P (%)	0.52	0.47
Met + Cys (%)	0.82	0.71

Abbreviations: Each kilogram of the diet contains one premix [1], which includes the following concentrations: vitamin A at 9000 IU, vitamin B1 at 3 mg, vitamin B2 at 8 mg, vitamin B6 at 4.4 mg, vitamin B12 at 0.012 mg, vitamin D3 at 3000 IU, vitamin E at 26 IU, vitamin K3 at 1.20 mg, D-pantothenic acid at 1 mg, nicotinic acid at 45 mg. Mineral premix [2] copper (Cu) at 10 mg, iron (Fe) at 100 mg, manganese (Mn) at 120 mg, zinc (Zn) at 108 mg, iodine (I) at 108 mg, and selenium (Se) at 0.35 mg. Metabolic energy (ME) was calculated based on data from the Chinese Raw Material Database on animal nutrition.

2.3. Assessment of Growth Performance and Acquisition of Samples

Feed intake and residual feed were recorded daily throughout the trial. Body weights were measured at two time points: the initial weight was recorded after a 2-h fasting period posthatching, and the final weight was measured at 56 days following a 12-h fast to ensure an empty gastrointestinal tract. Average daily gain (ADG), average daily feed intake (ADFI), and feed-to-gain ratio (F/G) were calculated per replicate using the following formulas:

$$ADG = (\text{final body weight} - \text{initial body weight})/\text{trial duration (days)}$$

$$ADFI = feed\ intake/trial\ duration\ (days)$$

$$F/G = ADFI/ADG$$

At 56 days, six broilers per group (three males and three females, representing average body weight per sex) were selected for sampling. Following the administration of anesthesia, the selected yellow-feathered broilers were euthanized via carotid artery transection. A longitudinal surgical section was made along the ventral median axis to access the coelomic compartment. The duodenum, jejunum, ileum, and cecum were sequentially removed. During sampling, the contents of the distal end of the right cecum were collected uniformly, and 1.5 mL of the contents were rapidly transferred into a 2 mL freezing tube and flash-frozen in liquid nitrogen. A 2 cm segment of intestinal tissue from the midileum was excised, and the luminal digesta were flushed with 4 °C normal saline. The tissue was then placed into a 2 mL cryogenic storage tube and flash-frozen in liquid nitrogen. Duodenal, jejunal, and ileal samples were taken from approximately 1 cm of tissue 3 cm from the pylorus, 5 cm proximal to the yolk sac diverticulum, and 10 cm proximal to the ileocecal junction, respectively, for morphometric analysis. The samples were fixed in a 4% paraformaldehyde solution. All procedures were conducted in accordance with the methods described by Liu et al. [31].

2.4. The Structural Characteristics of Enteric Tissue

Fixed intestinal tissue samples were processed through progressive ethanol dehydration, paraffin embedding, sectioning, and hematoxylin and eosin (H&E) staining. Morphometric measurements, including villus height (VH), crypt depth (CD), and the VH/CD ratio, were performed using a Nikon Eclipse E200 microscope (Tokyo, Japan) and Image View 4 software.

2.5. Intestinal Biomarkers

A 0.5 g sample of ileal tissue was placed in a freezing tube, and 4.5 mL of phosphate-buffered saline was added to prepare a 1:9 tissue homogenate. The homogenate was centrifuged at 3500 rpm for 15 min. After centrifugation, the supernatant was collected and stored for further analysis; the remaining steps were carried out strictly according to the instructions. The total protein concentration in the supernatant was measured using a commercial BCA assay kit (Nanjing Jiancheng Bioengineering Institute, Nanjing, China). This was done with the Bicinchoninic Acid Assay according to the manufacturer's guidelines. Each protein concentration measurement was performed in three technical replicates to ensure precision with intra-assay CV < 5%, inter-assay CV < 8%. Assays for chicken tissue immunoglobulins (IgG, JYM0001Ch; IgM, JYM0060Ch; sIgA, JYM0036Ch) and cytokines (TNF-α, JYM0033Ch; IFN-γ, JYM0007Ch; IL-1β, JYM0041Ch; IL-6, JYM0028Ch; IL-10, JYM0040Ch) were performed using ELISA kits (Genome Biotechnology Co., Ltd., Wuhan, China). For ELISA assays, samples were analyzed in duplicate per the manufacturer's recommendations, with intra-assay CV < 10%, inter-assay CV < 12%.

2.6. Real-Time qPCR Assay

RNA isolation was performed according to the manufacturer's guidelines (Takara, Dalian, China). Briefly, 50 mg of tissue samples were mechanically homogenized in 1 mL of Trizol lysate using an FA-25D homogenizer (FLUKO, Shanghai, China). mRNA extraction was conducted following the method described by Ji et al. [32]. RNA concentration and purity were measured using an Implen P330 spectrophotometer (Implen, Munich, Germany).

For cDNA synthesis, the *Evo M-MLV* RT MIX KIT (Accurate Bio-technology Co., Ltd., Changsha, China) was used. The reaction mixture consisted of 2 μL 500 μg/μL RNA, 2 μL gDNA Clean Reaction Mix Ver.2, and 6 μL enzyme-free water, which was incubated in a PCR instrument (Labcycler, Senso Quest, Germany) for 2 min. Subsequently, 4 μL 5 × *Evo M-MLV* RT Reaction Mix Ver.2 and 6 μL enzyme-free water were added. The reaction conditions were as follows: 37 °C for 15 min, 85 °C for 5 s, and 4 °C for cooling.

The qPCR assay was conducted using the YBR Green Pro Taq HS qPCR Kit (Rox Plus; Accurate Biotechnology Co., Ltd., Changsha, China). The PCR reaction system included 2 μL of cDNA, 0.5 μL of F-strand primer, 0.5 μL of R-strand primer, 10 μL of 2 × SYBR Green Pro Taq HS Premix, and 7.2 μL of enzyme-free water. RT-qPCR was performed on a LightCycler 480 II instrument (Roche, Basel, Switzerland) with the following program: 95 °C for 30 s (1 cycle), followed by 40 cycles of 95 °C for 5 s and 60 °C for 30 s.

β-actin was used as an endogenous normalization control to standardize the experimental data. The relative expression levels of the following target genes were analyzed: *Occludin (OCLN), zonula occludin-1 (TJP1), junctional adhesion molecule-2 (JAM2), mucin-2 (MUC2), interleukin-1beta (IL1B), interleukin-6 (IL6), interleukin-10 (IL10), tumor necrosis factor-alpha (TNFA), interferon-γ (IFNG), myeloid differentiation factor 88 (MyD88), Toll-like receptor 4 (TLR4)*, and *nuclear transcription factor-κBp50 (NFKB1)*. The relative mRNA expression of the target gene was calculated using the $2^{-\Delta\Delta Ct}$ method [33]. Primer sequences are provided in Table 2.

Table 2. Primers for RT–qPCR assessment of tissue RNA.

Gene Name	Primer Sequence (5′→3′)	Accession No.
OCLN	F: AGGTCTGCAACAGCATCACA R: ATGCCTTCCCAAAAAGCCCT	NM_205128.1
TJP1	F: TATGCACAAGGAGGTCAGCC R: TTGGCCGAAGCATTCCATCT	XM_040680630.2
JAM2	F: AGCCTCAAATGGGATTGGATT R: CATCAACTTGCATTCGCTTCA	XM_046907880.1
MUC2	F: AAATGTATCTGTCGCCCCTCA R: TGTCGCCATCCTTTATTGTTG	XM_046942297.1
IL1B	F: CAGCCTCAGCGAAGAGACCT R: ACTGTGGTGTGCTCAGAATCC	NM_204524.2
IL6	F: AAATCCCTCCTCGCCAATCT R: CCCTCACGGTCTTCTCCATAAA	NM_204628.2
IL10	F: AGGAGACGTTCGAGAAGATGGA R: TCAGCAGGTACTCCTCGATGT	NM_001004414.4
TNFA	F: CTGTGGGGCGTGCAGTG R: GGCACAAAAGAGCTGATGGC	NM_204267.2
IFNG	F: AGCTCCCGATGAACGACTTG R: TCCTCTGAGACTGGCTCCTT	NM_205149.2
MyD88	F: CCTGGCTGTGCCTTCGGA R: TCACCAAGTGCTGGATGCTA	NM_001030962.5
TLR4	F: TTCAGAACGGACTCTTGAGTGG R: CAACCGAATAGTGGTGACGTTG	NM_001030693.2
NFKB1	F: GAAGGAATCGTACCGGGAACA R: CTCAGAGGGCCTTGTGACAGTAA	NM_001396396.1
β-actin	F: GCCAACAGAGAGAAGATGACAC R: GTAACACCATCACCAGAGTCCA	NM_205518.2

Abbreviations: F = forward primer, R = reverse primer. Occludin (*OCLN*), zonula occludin-1 (*TJP1*), junctional adhesion molecule-2 (*JAM2*), mucin-2 (*MUC2*), interleukin-1beta (*IL1B*), interleukin-6 (*IL6*), interleukin-10 (*IL10*), tumor necrosis factor-alpha (*TNFA*), interferon-γ (*IFNG*), myeloid differentiation factor 88 (*MyD88*), toll-like receptor 4 (*TLR4*), and nuclear transcription factor-κBp50 (*NFKB1*).

2.7. Ileal Microbiota Analysis

Samples were transported on dry ice to Shanghai Meiji Biomedicals Co. (Shanghai, China) to ensure sample integrity. Total DNA was extracted from the cecal contents of 56-day-old yellow-feathered broilers using the Fast DNA™ Spin Kit (MP Biomedicals, Santa Ana, CA, USA). The V3–V4 variable region of the bacterial 16S ribosomal RNA was amplified using the following primer sequences: forward primer 341F (5′-ACTCCTACGGGAGGCAGCAG-3′) and reverse primer 806R (5′-GGACTA-CHVGGGTWTCTAAT-3′). After quantitative purification, the amplified sequences were sequenced on the Illumina Nextseq 2000 PE300 platform. Raw sequences were processed using QIIME software (version 1.9.1) and clustered into amplicon sequence variants (ASV) [34]. All raw sequencing datasets have been deposited in the NCBI Sequence Read Archive under BioProject PRJNA1199219.

2.8. Statistical Analysis

All data were analyzed using two-way ANOVA to evaluate the main effects of RYC and BS, as well as their interaction effects, with the replicates serving as the experimental unit. Mean comparisons were performed using Duncan's multiple range test. Statistical computations were conducted using SPSS version 21 (IBM Corp., Armonk, NY, USA) [35]. For cecal microbial data, analyses were performed on the online Majorbio cloud platform (www.majorbio.com, accessed on 10 December 2024). These analyses included alpha diversity analysis, beta diversity analysis, bacterial abundance analysis, and linear discriminant analysis effect size (LEfSe). The data analysis process involved three main steps: filtering raw data, applying the UPARSE algorithm for clustering, and annotating species. A $p < 0.05$ was considered statistically significant, and a p-value between 0.05 and 0.10 was considered indicative of a trend.

3. Results

3.1. Effects of RYC and BS Feeding on the Growth Parameters of Yellow-Feathered Broilers

As shown in Table 3, during the 56-day experimental period, body weight, ADFI, ADG, and F/G were not significantly affected by the main effects of RYC, BS, and their interaction in yellow-feathered broilers ($p > 0.05$).

Table 3. Effects of feeding RYC and BS on the growth performance of yellow-feathered broilers from 1 to 56 days of age.

Items	CON		RYC		SEM	p-Value		
	CON	BS	CON	RYC + BS		RYC	BS	RYC × BS
ADFI (g/day)	96.05	95.05	94.80	95.12	3.013	0.543	0.929	0.928
ADG (g/day)	36.65	35.55	35.43	36.49	0.572	0.893	0.985	0.307
F/G	2.60	2.63	2.64	2.61	0.05	0.949	0.920	0.724
BW (g)	2122.14	2046.10	2055.90	2098.10	25.691	0.876	0.748	0.246

Abbreviations: Average daily feed intake (ADFI), average daily gain (ADG), feed conversion ratio (F/G), body weight (BW). Control group (CON), basal diet; *Bacillus subtilis* group (BS), basal diet + 5 × 10^9 CFU/kg BS; *Rhodotorula* yeast culture group (RYC), basal diet + 5000 mg/kg RYC; BS + RYC group (BS + RYC), basal diet + 5 × 10^9 CFU/kg BS + 5000 mg/kg RYC. SEM, standard error of the mean ($n = 6$).

3.2. Effects of Feeding RYC and BS on the Intestinal Morphology of Yellow-Feathered Broilers

In all experimental groups, epithelial cells were observed to be well-preserved, and normal mucosal architecture was identified in the duodenum, jejunum, and ileum, including intact mucosa, submucosa, and muscular layers (Figure 1). Notably, the intervillus space (the distance between adjacent intestinal villi) was analyzed and the following trends were observed: The narrowest intervillus spacing (indicating denser villus packing) was exhibited by the BS group, followed by the BS + RYC group, then the RYC group, with the widest spacing being recorded in the CON group. This trend was consistently observed in the duodenum, jejunum, and ileum.

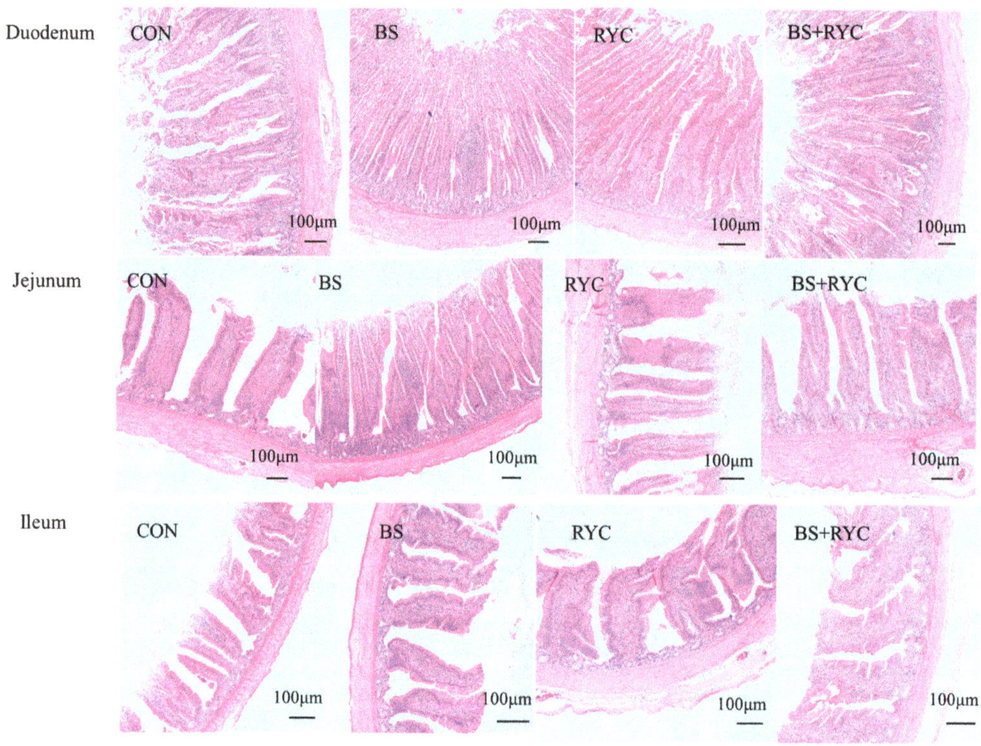

Figure 1. Optical microstructure of the intestine of yellow-feathered broilers supplemented with RYC and BS. Hematoxylin and eosin (H&E) staining results are presented for the duodenal, jejunal, and ileal intestinal segments. Control group (CON), basal diet: *Bacillus subtilis* group (BS), basal diet + 5 × 10^9 CFU/kg BS; *Rhodotorula* yeast culture group (RYC), basal diet + 5000 mg/kg RYC; BS + RYC group (BS + RYC), basal diet + 5 × 10^9 CFU/kg BS + 5000 mg/kg RYC. Figure 1 was examined at 40× magnification and a scale bar = 100 μm.

As shown in Table 4, in the jejunum, the CD was significantly affected by the RYC × BS interaction; however, further analysis using Duncan's multiple range test did not detect a significant effect ($p < 0.05$). In the jejunum, the VH/CD ratio was significantly affected by the RYC × BS interaction ($p < 0.05$). Further analysis using Duncan's multiple range test revealed that the VH/CD ratio in the BS + RYC group was significantly higher than that in the other three groups ($p < 0.05$). The VH or VH/CD ratio in the duodenum, VH and CD in the jejunum, or VH, CD, and VH/CD ratio in the ileum were not significantly affected by the main effects of RYC, BS, and their interaction ($p > 0.05$).

3.3. Effects of Feeding RYC and BS on the Intestinal Physical and Chemical Barrier of Yellow-Feathered Broilers

As shown in Table 5, the mRNA expression of *OCLN*, *JAM2*, and *TJP1* was significantly affected by the RYC × BS interaction ($p < 0.05$). Further analysis using Duncan's multiple range test on *OCLN* and *JAM2* revealed that the BS group was significantly higher than that in the other three groups ($p < 0.05$). Further analysis using Duncan's multiple range test on *TJP1* revealed that the BS group was significantly higher than that in the CON and RYC groups ($p < 0.05$), and the BS + RYC group was significantly higher than that in the CON group ($p < 0.05$).

Table 4. Effects of feeding RYC and BS on the intestinal morphology of yellow-feathered broilers.

Items		CON		RYC		SEM	p-Value		
		CON	BS	CON	RYC + BS		RYC	BS	RYC × BS
Duodenum	VH (μm)	1188.06	1247.31	1366.59	1248.56	35.89	0.239	0.245	0.688
	CD (μm)	73.13	66.22	76.91	69.67	1.58	0.175	0.947	0.020
	VH/CD	16.34	18.85	17.81	18.16	0.71	0.806	0.505	0.382
Jejunum	VH (μm)	954.70	1065.32	1011.16	1108.06	63.35	0.963	0.737	0.488
	CD (μm)	77.58	79.66	76.07	69.82	4.14	0.668	0.561	0.829
	VH/CD	12.31 [b]	13.30 [b]	13.23 [b]	15.77 [a]	0.45	0.200	0.016	0.013
Ileum	VH (μm)	707.28	789.69	711.66	726.35	48.62	0.769	0.798	0.675
	CD (μm)	71.43	65.10	62.66	58.99	2.19	0.748	0.101	0.249
	VH/CD	9.87	12.04	11.32	12.13	0.60	0.605	0.560	0.274

Abbreviations: VH (villus height); CD (crypt depth); the VH to CD ratio (VH/CD); control group (CON), basal diet; *Bacillus subtilis* group (BS), basal diet + 5×10^9 CFU/kg BS; *Rhodotorula* yeast culture group (RYC), basal diet + 5000 mg/kg RYC; BS + RYC group (BS + RYC), basal diet + 5×10^9 CFU/kg BS + 5000 mg/kg RYC. SEM, standard error of the mean ($n = 6$). [a,b] Means within a row with different superscripts differ significantly ($p < 0.05$).

Table 5. Effects of supplementation on RYC and BS on the expression regulation of tight junction-related proteins and *mucin2* genes in the ileum of yellow-feathered broilers.

Items	CON		RYC		SEM	p-Value		
	CON	BS	CON	RYC + BS		RYC	BS	RYC × BS
OCLN	1.00 [b]	1.32 [a]	1.01 [b]	0.77 [b]	0.06	0.012	0.696	0.009
JAM2	1.00 [c]	2.72 [a]	1.65 [bc]	1.96 [b]	0.17	0.816	<0.001	0.006
TJP1	1.00 [c]	2.26 [a]	1.38 [bc]	1.79 [ab]	0.13	0.809	<0.001	0.026
MUC2	1.00	1.87	1.42	2.22	0.11	0.009	<0.001	0.809

Abbreviations: Control group (CON), basal diet; *Bacillus subtilis* group (BS), basal diet + 5×10^9 CFU/kg BS; *Rhodotorula* yeast culture group (RYC), basal diet + 5000 mg/kg RYC; BS + RYC group (BS + RYC), basal diet + 5×10^9 CFU/kg BS + 5000 mg/kg RYC. SEM, standard error of the mean ($n = 6$). [a,b,c] Means within a row with different superscripts differ significantly ($p < 0.05$).

The mRNA expression of *MUC2* was not significantly affected by the RYC × BS interaction ($p > 0.05$). However, supplementation with either RYC or BS alone significantly increased the mRNA expression of *MUC2* ($p < 0.05$).

3.4. Effects of Feeding RYC and BS on the Intestinal Immune Barrier of Yellow-Feathered Broilers

3.4.1. Effects of Feeding RYC and BS on the Ileal Cytokine and Immunoglobulin Contents of Yellow-Feathered Broilers

As shown in Table 6, IgM was significantly affected by the RYC × BS interaction ($p < 0.05$). Specifically, IgM levels were not affected by BS supplementation alone, were significantly increased by RYC supplementation alone, and were significantly decreased when both RYC and BS were supplemented together. Further analysis using Duncan's multiple range test revealed that the RYC group had significantly higher IgM levels compared to the other three groups ($p < 0.05$). For IgG, BS supplementation significantly increased its content ($p < 0.05$), and there was no significant interaction between BS and RYC ($p > 0.05$). For IFN-γ, RYC supplementation significantly reduced its content ($p < 0.05$). For IL-1β, the RYC × BS interaction showed a trend toward significance ($p = 0.052$). BS supplementation significantly increased IL-1β concentration ($p < 0.05$), while RYC supplementation significantly decreased it ($p < 0.05$).

Table 6. Effects of feeding RYC and BS on the immune function of the ileum of yellow-feathered broilers.

Items	CON		RYC		SEM	p-Value		
	CON	BS	CON	BS + RYC		RYC	BS	RYC × BS
IgM (ng/mg prot.)	48.12 [b]	48.88 [b]	81.83 [a]	47.80 [b]	3.74	0.002	0.002	0.001
IgG (μg/mg prot.)	2.79	3.74	2.57	4.37	0.22	0.565	<0.001	0.235
sIgA (pg/mg prot.)	134.34	180.53	166.91	181.71	9.47	0.370	0.113	0.404
TNF-α (pg/mg prot.)	8.75	10.43	10.21	10.97	0.47	0.299	0.209	0.633
IFN-γ (pg/mg prot.)	22.25	21.97	15.34	20.87	1.03	0.041	0.166	0.128
IL-1β (pg/mg prot.)	11.55	23.94	11.31	17.82	1.27	0.037	<0.001	0.052
IL-6 (pg/mg prot.)	12.63	12.92	11.92	13.02	0.49	0.773	0.508	0.699
IL-10 (pg/mg prot.)	11.93	12.43	11.10	12.59	0.47	0.737	0.323	0.621

Abbreviations: Control group (CON), basal diet; *Bacillus subtilis* group (BS), basal diet + 5 × 10^9 CFU/kg BS; *Rhodotorula* yeast culture group (RYC), basal diet + 5000 mg/kg RYC; BS + RYC group (BS + RYC), basal diet + 5 × 10^9 CFU/kg BS + 5000 mg/kg RYC. SEM, standard error of the mean ($n = 6$). [a,b] Means within a row with different superscripts differ significantly ($p < 0.05$). Prot. indicates the total protein concentration as determined by the BCA method.

3.4.2. Effects of Feeding RYC and BS on the Expression of Ileum Immunity-Related Genes of Yellow-Feathered Broilers

As shown in Table 7, the mRNA expression of *TNFA*, *IL1B*, and *NFKB1* was significantly increased by BS supplementation ($p < 0.05$). For *IL10* and *MyD88*, RYC supplementation significantly increased their mRNA expression ($p < 0.05$). Additionally, the mRNA expression of *IL6* was significantly influenced by the RYC × BS interaction ($p < 0.05$). However, further analysis using Duncan's multiple range test revealed no significant differences among the four groups ($p = 0.117$).

Table 7. Effects of RYC and BS supplementation on ileal immunity-related gene expression in yellow-feathered broilers.

Items	CON		RYC		SEM	p-Value		
	CON	BS	CON	BS + RYC		RYC	BS	RYC × BS
TNFA	1.00	1.96	1.70	2.36	0.17	0.074	0.011	0.608
IFNG	1.00	1.11	1.09	1.35	0.06	0.154	0.112	0.513
IL1B	1.00	1.69	0.86	1.61	0.06	0.791	<0.001	0.785
IL6	1.00	1.99	1.45	1.86	0.16	0.601	0.341	0.031
IL10	1.00	1.30	2.24	2.58	0.22	0.004	0.417	0.962
TLR4	1.00	0.71	0.97	0.87	0.10	0.761	0.358	0.639
MyD88	1.00	1.18	1.23	1.68	0.09	0.038	0.067	0.421
NFKB1	1.00	1.53	1.23	1.41	0.06	0.776	0.001	0.379

Abbreviations: Control group (CON), basal diet; *Bacillus subtilis* group (BS), basal diet + 5 × 10^9 CFU/kg BS; *Rhodotorula* yeast culture group (RYC), basal diet + 5000 mg/kg RYC; BS + RYC group (BS + RYC), basal diet + 5 × 10^9 CFU/kg BS + 5000 mg/kg RYC. SEM, standard error of the mean ($n = 6$).

3.5. *Effects of Feeding RYC and BS on the Microbiota of the Cecum of Yellow-Feathered Broilers*

A total of 38,510 clean reads were obtained from the cecum chyme samples. The rarefaction curves were observed to flatten out (Figure 2A), suggesting that a sufficient number of individual samples were collected from all four groups. Since the rarefaction curves were used to estimate the completeness of microbial community sampling, this result confirmed that the data were adequate for microbial community analysis.

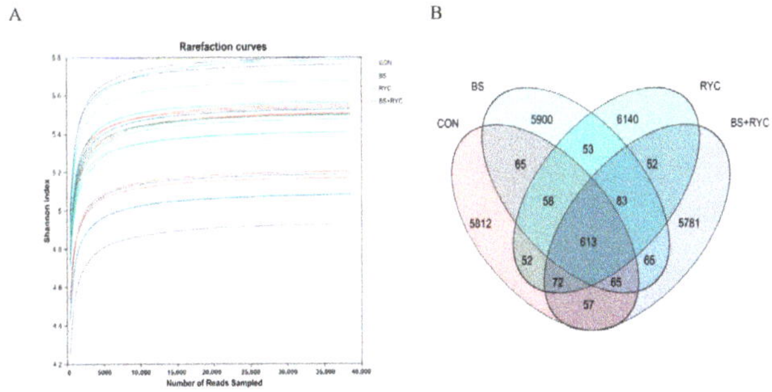

Figure 2. Effects of single or combined BS and RYC supplementation on microbial diversity in the cecum of yellow-feathered broilers. Rarefaction curves (**A**). Venn diagrams (ASV) (**B**). Control group (CON), basal diet; *Bacillus subtilis* group (BS), basal diet + 5 × 10^9 CFU/kg BS; *Rhodotorula* yeast culture group (RYC), basal diet + 5000 mg/kg RYC; BS + RYC group (BS + RYC), basal diet + 5 × 10^9 CFU/kg BS + 5000 mg/kg RYC. (n = 6).

To evaluate the effects of single or combined supplementation with BS and RYC, the count of shared and exclusive ASVs in the bacterial community of the cecum was analyzed using Venn diagrams. The results (Figure 2B) revealed that the four groups of cecum content samples collectively produced a total of 24,868 ASVs. Among these, 613 ASVs were identified as common across all four groups. The proportion of unique ASVs in each group was found to be 85.55% for the CON group, 85.48% for the BS group, 86.20% for the RYC group, and 85.16% for the BS + RYC group, respectively.

When assessing alpha diversity, the Chao index (Figure 3A), Ace index (Figure 3B), Simpson index (Figure 3C), or Shannon index (Figure 3D) was not significantly affected by the main effects of RYC, BS, and their interaction in yellow-feathered broilers ($p > 0.05$).

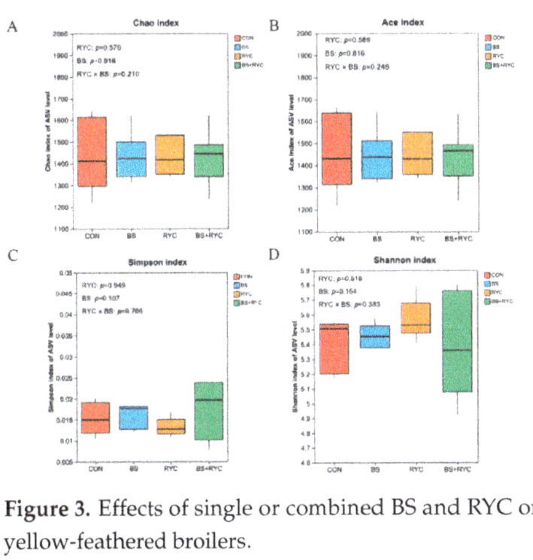

Figure 3. Effects of single or combined BS and RYC on the alpha diversity indices in the cecum of yellow-feathered broilers.

Chao index at the ASV level (**A**); Ace index at the ASV level (**B**); Simpson index at the ASV level (**C**); Shannon index at the ASV level (**D**). Control group (CON), basal diet; *Bacillus subtilis* group (BS), basal diet + 5 × 10^9 CFU/kg BS; *Rhodotorula* yeast culture group (RYC), basal diet + 5000 mg/kg RYC; BS + RYC group (BS + RYC), basal diet + 5 × 10^9 CFU/kg BS + 5000 mg/kg RYC. All values are presented as mean ± SD (error bars), with the sample size indicated as n = 6. no statistically significant differences (p > 0.05) were detected between groups for these indices.

Beta diversity indices, analyzed through PCoA (Figure 4A) and NMDS (Figure 4B), revealed significant shifts in microbial communities at the ASV level (R = 0.236, p = 0.002) across the four groups.

A B

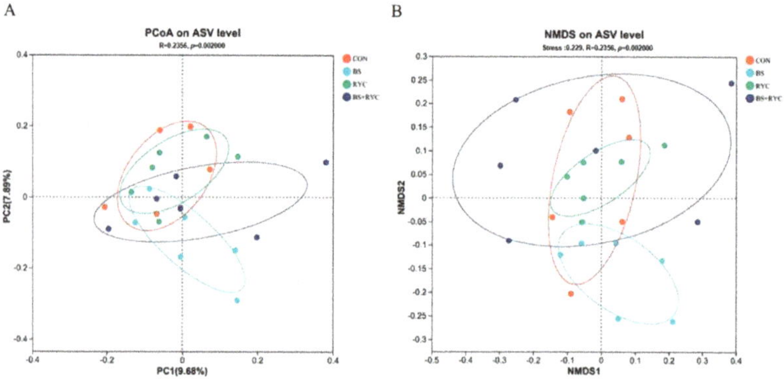

Figure 4. Effects of single or combined BS and RYC on the beta diversity indices in the cecum of yellow-feathered broilers. The assessment of cecal microbial communities was conducted using principal coordinate analysis (PCoA) (**A**) and nonmetric multidimensional scaling (NMDS) (**B**), employing the binary Curtis metric. Control group (CON), basal diet; *Bacillus subtilis* group (BS), basal diet + 5 × 10^9 CFU/kg BS; *Rhodotorula* yeast culture group (RYC), basal diet + 5000 mg/kg RYC; BS + RYC group (BS + RYC), basal diet + 5 × 10^9 CFU/kg BS + 5000 mg/kg RYC. (n = 6).

At the phylum level (Figure 5A), the cecal microbiota predominantly comprised *Firmicutes, Bacteroidota,* and *Actinobacteriota*, with *Firmicutes* accounting for over 90% of the total composition. At the genus level (Figure 5B), the dominant genera included *norank_f__norank_o__Clostridia_UCG-014, unclassified_f__Lachnospiraceae,* and *k_f__norank_o_ _Clostridia_vadinBB60_group*.

As shown in Figure 6, LEfSe analysis (LDA score ≥ 3) revealed distinct microbial signatures across the treatment groups. In the CON group, the abundance of *g_Turicibacter* and *g_Tyzzerella* was significantly higher. The BS group exhibited an increased abundance of *g_Marvinbryantia, o_Bacillales, f_Bacillaceae,* and *g_Bacillus*. In the RYC group, the abundance of *f_norank_o__Clostridia_UCG-014, o_Clostridia_UCG-014, g_norank_f__norank_o__Clostridia_UCG-014, o_Clostridia_vadinBB60_group, g_norank_f __norank_o__Clostridia_vadinBB60_group,* and *f_norank_o__Clostridia_vadinBB60_group* was elevated. The BS + RYC group showed an increased abundance of *g_Ruminococcus_torques _group, g_norank_f__norank_o__norank_c_Clostridia, f_norank_o__norank_c_Clostridia, p_ Desulfobacterota, c_Desulfovibrionia, f_Desulfovibrionaceae, o_Desulfovibrionales,* and *g_Bilophila*.

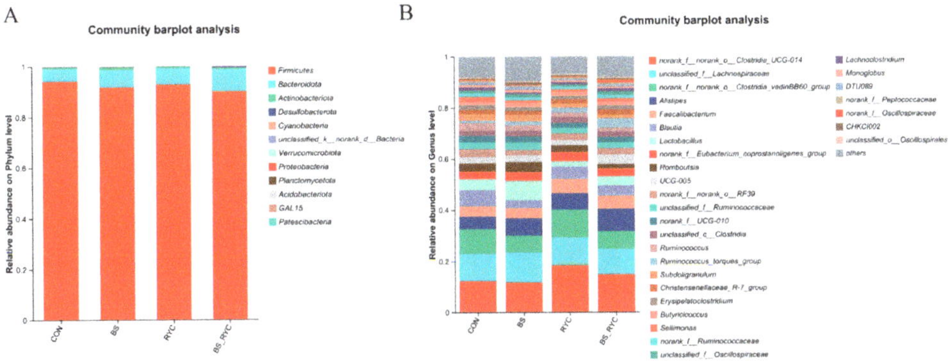

Figure 5. Effects of single or combined BS and RYC on the cecal microbiota of yellow-feathered broilers. The community composition of cecum microbiota at both the phylum and genus levels is represented as a percentage of abundance (**A**,**B**). Control group (CON), basal diet; *Bacillus subtilis* group (BS), basal diet + 5 × 10^9 CFU/kg BS; *Rhodotorula* yeast culture group (RYC), basal diet + 5000 mg/kg RYC; BS + RYC group (BS + RYC), basal diet + 5 × 10^9 CFU/kg BS + 5000 mg/kg RYC. (*n* = 6).

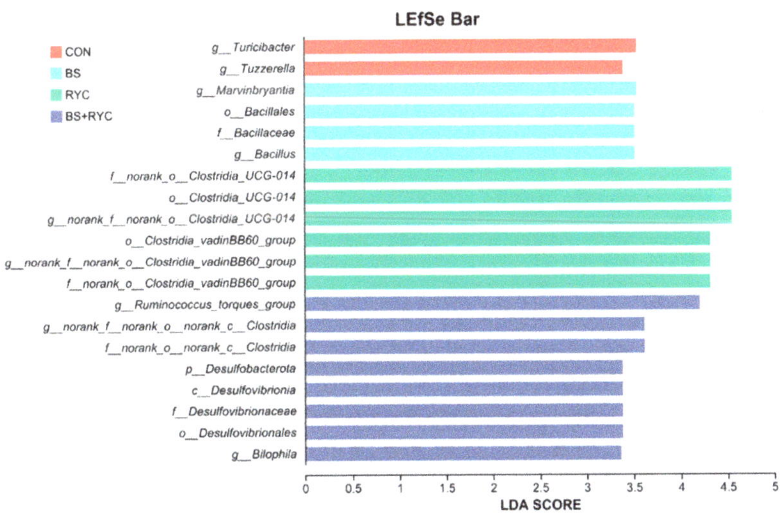

Figure 6. Effects of single or combined BS and RYC on cecal microbiota of yellow-feathered broilers. Linear discriminant analysis (LDA) discriminant histograms. Control group (CON), basal diet, *Bacillus subtilis* group (BS), basal diet + 5 × 10^9 CFU/kg BS, *Rhodotorula* yeast culture group (RYC), basal diet + 5000 mg/kg RYC, BS + RYC group (BS + RYC), basal diet + 5 × 10^9 CFU/kg BS + 5000 mg/kg RYC (*n* = 6).

4. Discussion

Throughout the feeding trial in this study, broilers across all experimental groups were observed to exhibit comparable feed intake levels and similar daily weight gain trends. These observations suggest that the primary effects of RYC and BS supplementation are not driven by enhanced nutrient utilization for growth but rather by their regulatory roles in intestinal function. Specifically, RYC and BS were found to enhance intestinal barrier integrity and stimulate immune activity, which may incur physiological costs such as elevated immune resource allocation and increased intestinal mucus production. Consequently, these metabolic demands are likely to divert energy away from direct body protein deposition [12]. The improvements in intestinal health observed in yellow-feathered broilers following RYC and BS primarily involved enhanced villus morphology, upregulation of the *MUC2* gene, and optimized cecal microbiota composition. These findings are consistent with and corroborated by previous studies [22,27,36].

The structural integrity of intestinal tissue and intercellular junction complexes is fundamental to maintaining gut barrier homeostasis. In broilers, the development of the small intestinal mucosa directly influences digestion and absorption efficiency, thereby impacting overall health and performance [37]. The VH/CD ratio, a key indicator of intestinal development and absorptive capacity [38], was significantly increased in yellow-feathered broilers receiving combined RYC and BS supplementation. Findings from this research further verify the interactive effects of co-administration of RYC and BS. Notably, cosupplementation markedly elevated the jejunal VH/CD ratio, indicating synergism. Supporting our results, Xie et al. [39] demonstrated that fermented feed containing BS and yeast significantly improved the VH/CD ratio in Xuefeng black-bone chickens. The underlying mechanism may involve bacterial proteins and small peptides derived from RYC, which facilitate BS colonization in the jejunum while simultaneously stimulating villus growth and crypt cell proliferation [14,31]. Notably, dietary supplementation with BS alone did not significantly enhance intestinal VH and CD in the present study, consistent with the findings of Aliakbarpour et al. [40], who observed no significant improvement in jejunal villus architecture in 42-day-old broilers. This lack of effect may be attributed to age-dependent microbiota maturation, wherein established gut microbial communities competitively exclude exogenous BS colonization, thereby diminishing its promotive effects on intestinal villi [41].

TJPs regulate the function of the intestinal epithelial barrier and ensure its integrity by controlling permeability [42]. Interestingly, our results indicate that singular RYC administration downregulates ileal *OCLN* mRNA expression, demonstrating antagonism when coadministered with BS. In contrast, Kyoung et al. [14] found that the yeast cell wall components had no significant effect on ileal *OCLN* gene expression in broilers. These divergent outcomes may be attributed to multiple factors, including distinct yeast strains employed, disparities in supplementation levels, and differences in the breed and age of the broilers utilized. We speculate that RYC likely targets anti-inflammatory mechanisms as a primary mode of action, reallocating cellular resources to strengthen gut anti-inflammatory competence rather than fortifying the intestinal barrier via increased TJP mRNA expression. The mechanism underlying this effect warrants further investigation. Studies by Bilal et al. [43] revealed that dietary supplementation with *Bacillus pumilus* and *Bacillus subtilis* markedly elevated ileal *JAM2* and *TJP1* mRNA levels in broilers at 42 days of age. Consistent with prior findings, BS administration in this study yielded concordant *JAM2* and *TJP1* mRNA expression trends in ileal tissues.

MUC2, the primary structural component of intestinal mucus, plays a pivotal role in gut barrier defense by physically preventing pathogen adhesion and toxin penetration [44]. Findings from this study reveal that both RYC and BS significantly upregulate ileal *MUC2* mRNA expression yet exhibit no synergistic effect. This enhancement likely involves BS promoting goblet cell proliferation [40], while RYC provides the essential nutrients for epithelial cell metabolism and mucus production [45]. By strengthening the intestinal epithelial barrier, these mechanisms collectively enhance the integrity of the mucosal microstructure.

Immunoglobulin levels serve as critical biomarkers for evaluating humoral immune competence in poultry. The present study demonstrated that RYC supplementation significantly increased IgM concentrations in ileum tissue, consistent with Wang et al. [46], who report that yeast-derived products enhance jejunal immunoglobulins, and Kyoung et al. [14], who found that yeast cell walls increase the number of goblet cells in the duodenum of broilers. The underlying mechanism may involve β-glucans in RYC activating TLR2-mediated signaling in intestinal epithelial and antigen-presenting cells, subsequently promoting B lymphocyte differentiation into antibody-producing cells [47]. Findings from this study also reveal that dietary BS administration markedly elevated IgG levels in the ileum of yellow-feathered broilers, consistent with Dong et al.'s [48] report on serum IgG concentration in broilers fed diets containing *Bacillus subtilis* strain BYS2. Surprisingly, the combined supplementation of RYC and BS demonstrated an antagonistic effect on IgM in the ileal tissues of yellow-feathered broilers. We postulate that this observed antagonism may be attributable to organism-level dynamic immunoglobulin homeostasis. This phenomenon may reflect (i) the earlier emergence of IgM in primary immune responses, followed by (ii) IgG class switching during secondary responses [49]. Such sequential activation patterns could optimize humoral immunity across different stages of pathogen challenge, with IgM providing rapid first-line defense while IgG mediates sustained protection.

Cytokines play a crucial role in cell-mediated immunity, and their expression patterns reflect immune regulation. IFN-γ is a key Th1 cytokine, and its downregulation in our study suggests that RYC may help maintain immune homeostasis by promoting T regulatory cell differentiation. This aligns with Zhou et al. [50], who found that yeast cell wall polysaccharides suppress *IFNG* and *IL1B* upregulation in the ileum during *E. coli* challenge. The β-glucans in RYC likely contributes to this effect by enhancing *IL10* secretion, which inhibits IFN-γ production—consistent with Wang et al. [51], who observed similar immunomodulation with yeast hydrolysates in broilers. In addition, RYC contains yeast cell wall components (β-glucans and mannan oligosaccharides), which can reduce intestinal inflammation by adsorbing harmful bacteria [15]. Meanwhile, BS supplementation significantly increased ileal IL-1β levels, indicating its immune-stimulating properties, possibly due to transient inflammatory responses triggered by microbial colonization [18]. Khan et al. [52] reported similar findings, with BS elevating *IL1B* expression in the spleen and kidney. The combination of RYC and BS appears to establish a balanced regulatory model. While BS maintains baseline IL-1β (indicating immune activation), RYC counteracts excessive inflammation by scavenging reactive oxygen species via carotenoids [45,53].

Interestingly, while RYC alone reduced IL-1β protein levels, the BS + RYC group showed elevated *IL1B* mRNA expression. This discrepancy may reflect post-transcriptional regulation or compensatory feedback mechanisms. RYC could suppress IL-1β protein synthesis via NF-κB inhibition, whereas BS colonization transiently stimulates IL1B transcription in immune cells, consistent with Rajput et al. [54], who found that Saccharomyces boulardii and BS B10 upregulated IL1B in avian dendritic cells. Overall, BS appears to dominantly regulate IL1B mRNA levels in broilers, while RYC modulates downstream inflammatory responses. An interactive effect was observed for RYC and BS regarding IL-6 at the transcriptional level, while IL-6 cytokine levels in the ileum of yellow-feathered broilers remained unaffected. The underlying mechanisms merit future in-depth exploration.

The TLR4/NF-κB pathway is crucial for the regulation of the intestinal immune system [55]. Research conducted by Rajput et al. [56] demonstrates that *Saccharomyces boulardii* and *Bacillus subtilis* strain B10 modulate broiler gut immune responses through upregulation of TLR, MyD88, and NF-κB mRNA expression. This activation induces NF-κB nuclear translocation via the MyD88-dependent pathway, resulting in the upregulation of *NFKB1* mRNA expression in the ileum tissue, which subsequently leads to the upregulation of *TNFA* and *IL1B* mRNA expression [52]. These results align with the previous findings reported by Yang et al. [57]. In contrast, RYC treatment specifically upregulated *IL10* mRNA expression, suggesting its potential role in mitigating intestinal inflammation through enhanced anti-inflammatory responses and mucosal protection mechanisms [23]. While BS activated the NF-κB signaling pathway and enhanced inflammatory factors IL-1β, the β-glucan component of RYC simultaneously enhanced IL-10 expression through T cell-mediated pathways. Studies conducted by Bai et al. [58] demonstrate that yeast culture supplementation in the diet of pseudobagrus ussuriensis mitigates intestinal inflammation through regulation of the TLR2-MyD88-NF-κB signaling cascade. Together, these complementary actions establish a sophisticated immune balance regulatory network. This mechanism creates a balance between immune system activation and inflammation control, which helps prevent excessive immune damage while strengthening the intestinal mucosa's defense [59].

The intestinal microbiota plays a pivotal role in modulating local immune responses and overall immune function in poultry [60]. In the present study, *Firmicutes* emerged as the dominant phylum in the cecal microbiota of yellow-feathered broilers, aligning with findings from Guo et al. [61]. Dietary supplementation with BS probiotics significantly increased the abundance of *Bacillus* spp. in the intestinal microbiota. This observation is consistent with previous research by Liu et al. [31], demonstrating that *Bacillus*-based probiotics in poultry feed yield similar microbial shifts. BS spores exhibit remarkable resilience, surviving gastric acid and bile salt stress before proliferating extensively in the posterior intestinal tract under favorable conditions [29].

RYC supplementation influenced the cecal microbial community structure, particularly enriching anaerobic *Clostridia* species such as *norank_f__norank_o__Clostridia_UCG-014* and *g_norank_f__norank_o__Clostridia_vadinBB60_group* [62]. The mannan oligosaccharides in RYC served as a selective carbon source for these *Clostridia*, promoting their colonization and metabolic activity [63]. These bacteria are critical for fiber degradation and butyrate synthesis, with elevated butyrate levels enhancing intestinal barrier integrity [64,65]. RYC jointly optimized the cecal bacterial community composition of yellow-feathered broilers by providing exclusive carbon sources, combining BS-suppressed pathogenic or competing bacteria, and the construction of an anaerobic microenvironment [25,66].

The BS-supplemented group exhibited significant enrichment of taxa classified under the *Bacillus* order, consistent with findings by Liu et al. [31], who reported that BS HC6 supplementation in white-feathered broilers enhanced cecal microbiota pathways linked to energy metabolism. In contrast, RYC supplementation prominently enriched the *Clostridia UCG-014* and *Clostridia vadinBB60* branches, both core members of the *Clostridia* class. Notably, the presence of unclassified microbial groups suggests that RYC may stimulate the proliferation of metabolically active but poorly characterized microbiota [36,67]. The specific interaction effect of RYC and BS co-addition showed significant enrichment of *p_Desulfobacterota, g_Ruminococcus_torques_group*, and *g_Bilophila*. The different experimental groups formed a characteristic flora structure. The BS group developed a *Bacillus*-dominated colony structure, while the BS + RYC group developed a *Desulfovibrio*-dominated colony structure. The CON group demonstrated distinct functional differentiation characteristics of the *Firmicutes* phylum, whereas the RYC group demonstrated those of the genus *Clostridia*. These structural differences likely modulate intestinal barrier function through multiple metabolic pathways, including short-chain fatty acid production, bile acid transformation, and sulfur cycling [68].

5. Conclusions

These findings highlight that dietary supplementation with RYC and BS primarily influences intestinal health and immune function rather than growth performance in yellow-feathered broilers. The combined use of RYC and BS may offer synergistic benefits for improving intestinal morphology. The combined use of RYC and BS revealed an antagonistic interaction, where the effects of one supplement were counteracted by the other in physical and chemical barriers and immune response. These findings highlight the complex interplay between RYC and BS in modulating gut health and immune responses, providing valuable insights for optimizing broiler nutrition and health management strategies in yellow-feathered broilers.

Author Contributions: Conceptualization, X.S., A.G. and Y.X.; methodology, X.S. and K.W.; validation, X.S., K.W. and Y.L.; formal analysis, X.S.; investigation, X.S., X.L., M.C., J.D., G.Z. and G.Y.; writing-original draft preparation, X.S.; supervision, A.G. and Y.X.; project administration, A.G.; funding acquisition, A.G. All authors have read and agreed to the published version of the manuscript.

Funding: This research was funded by the Inner Mongolia Natural Science Foundation Project (2023LHMS03064) and the High-level Achievement Cultivation Project of Animal Science College, Inner Mongolia Agricultural University (CG202436).

Institutional Review Board Statement: The experimental methodologies were assessed and granted approval by the Animal Care and Use Committee at Inner Mongolia Agricultural University, located in Hohhot, China (reference number NND2024124).

Informed Consent Statement: Not applicable.

Data Availability Statement: Data are available on request owing to restrictions. 16S rRNA sequencing data are available under NCBI BioProject PRJNA1199219 (accessible after 1 September 2025). All other datasets, including growth performance and so on, are available from the corresponding author upon reasonable request.

Acknowledgments: The authors express sincere gratitude to Donglin Wu, Yuanyuan Xing, and Ming Li for their valuable contributions during the manuscript revision stages.

Conflicts of Interest: The authors declare no conflicts of interest.

References

1. Liu, Y.; Tang, Y.; Mei, H.; Liu, Z.; Li, Z.; Ma, X.; Luo, Z.; Huang, W.; Li, Y.; Yu, M. Feeding citrus pomace fermented with combined probiotics improves growth performance, meat quality, fatty acid profile, and antioxidant capacity in yellow-feathered broilers. *Front. Vet. Sci.* **2024**, *11*, 1469947. [CrossRef] [PubMed]
2. Fang, X.; Ye, H.; Zhang, S.; Guo, L.; Xu, Y.; Zhang, D.; Nie, Q. Investigation of potential genetic factors for growth traits in yellow-feather broilers using weighted single-step genome-wide association study. *Poult. Sci.* **2023**, *102*, 103034. [CrossRef] [PubMed]
3. Gao, L.; Liu, C.; Wu, J.; Cui, Y.; Zhang, M.; Bi, C.; Shan, A.; Dou, X. EGCG improve meat quality, restore lipid metabolism disorder and regulate intestinal flora in high-fat fed broilers. *Poult. Sci.* **2025**, *104*, 104875. [CrossRef]
4. Salvo Romero, E.; Alonso Cotoner, C.; Pardo Camacho, C.; Casado Bedmar, M.; Vicario, M. The intestinal barrier function and its involvement in digestive disease. *Rev. Esp. Enferm. Dig.* **2015**, *107*, 686–696. [CrossRef] [PubMed]
5. Song, M.; Ren, C.; Liu, Y.; Ye, X.; Wang, Y.; Xie, J.; Zhao, F. Comparison of the characteristics of small intestinal fluid from white-feathered and yellow-feathered broilers. *Poult. Sci.* **2024**, *103*, 104417. [CrossRef]
6. Flemming, S.; Luissint, A.C.; Kusters, D.H.M.; Raya-Sandino, A.; Fan, S.; Zhou, D.W.; Hasegawa, M.; Garcia-Hernandez, V.; García, A.J.; Parkos, C.A.; et al. Desmocollin-2 promotes intestinal mucosal repair by controlling integrin-dependent cell adhesion and migration. *Mol. Biol. Cell* **2020**, *31*, 407–418. [CrossRef]
7. Yang, L.; Liu, G.; Lian, K.; Qiao, Y.; Zhang, B.; Zhu, X.; Luo, Y.; Shang, Y.; Gu, X.L. Dietary leonurine hydrochloride supplementation attenuates lipopolysaccharide challenge-induced intestinal inflammation and barrier dysfunction by inhibiting the NF-κB/MAPK signaling pathway in broilers. *J. Anim. Sci.* **2019**, *97*, 1679–1692. [CrossRef]

8. Latue, P.E.; Ariyadi, B.; Kurniawati, A.; Al Anas, M. Positive effect of fermented sorghum on productivity, jejunal histomorphology, and tight junction gene expression in broiler chickens. *Poult. Sci.* **2025**, *104*, 104548. [CrossRef]
9. Jha, R.; Das, R.; Oak, S.; Mishra, P. Probiotics (Direct-Fed Microbials) in Poultry Nutrition and Their Effects on Nutrient Utilization, Growth and Laying Performance, and Gut Health: A Systematic Review. *Animals* **2020**, *10*, 1863. [CrossRef]
10. Sjofjan, O.; Adli, D.N.; Harahap, R.P.; Jayanegara, A.; Utama, D.T.; Seruni, A.P. The effects of lactic acid bacteria and yeasts as probiotics on the growth performance, relative organ weight, blood parameters, and immune responses of broiler: A meta-analysis. *F1000Research* **2021**, *10*, 183. [CrossRef]
11. Bilal, R.M.; Hassan, F.U.; Saeed, M.; Rafeeq, M.; Zahra, N.; Fraz, A.; Saeed, S.; Khan, M.A.; Mahgoub, H.A.M.; Farag, M.R.; et al. Role of Yeast and Yeast-Derived Products as Feed Additives in Broiler Nutrition. *Anim. Biotechnol.* **2023**, *34*, 392–401. [CrossRef] [PubMed]
12. Wu, C.; Yang, Z.; Song, C.; Liang, C.; Li, H.; Chen, W.; Lin, W.; Xie, Q. Effects of dietary yeast nucleotides supplementation on intestinal barrier function, intestinal microbiota, and humoral immunity in specific pathogen-free chickens. *Poult. Sci.* **2018**, *97*, 3837–3846. [CrossRef] [PubMed]
13. Johnson, C.N.; Hashim, M.M.; Bailey, C.A.; Byrd, J.A.; Kogut, M.H.; Arsenault, R.J. Feeding of yeast cell wall extracts during a necrotic enteritis challenge enhances cell growth, survival and immune signaling in the jejunum of broiler chickens. *Poult. Sci.* **2020**, *99*, 2955–2966. [CrossRef] [PubMed]
14. Kyoung, H.; Kim, E.; Cho, J.H.; Lee, H.; Kim, Y.; Park, K.I.; Kim, H.B.; Song, M. Dietary yeast cell wall enhanced intestinal health of broiler chickens by modulating intestinal integrity, immune responses, and microbiota. *Poult. Sci.* **2023**, *102*, 102660. [CrossRef] [PubMed]
15. Tapingkae, W.; Srinual, O.; Lumsangkul, C.; Doan, H.V.; Chiang, H.I.; Manowattana, A.; Boonchuay, P.; Chaiyaso, T. Industrial-Scale Production of Mycotoxin Binder from the Red Yeast *Sporidiobolus pararoseus* KM281507. *J. Fungi* **2022**, *8*, 353. [CrossRef]
16. Li, Z.; Li, C.; Cheng, P.; Yu, G. *Rhodotorula mucilaginosa*-alternative sources of natural carotenoids, lipids, and enzymes for industrial use. *Heliyon* **2022**, *8*, e11505. [CrossRef]
17. Wickramasuriya, S.S.; Park, I.; Lee, Y.; Kim, W.H.; Przybyszewski, C.; Gay, C.G.; van Oosterwijk, J.G.; Lillehoj, H.S. Oral Deliv-ery of *Bacillus subtilis* Expressing Chicken NK-2 Peptide Protects Against Eimeria acervulina Infection in Broiler Chickens. *Front. Vet. Sci.* **2021**, *8*, 684818. [CrossRef]
18. Yang, J.; Huang, K.; Wang, J.; Wu, D.; Liu, Z.; Yu, P.; Wei, Z.; Chen, F. Combined Use of *Bacillus subtilis* yb-114,246 and *Bacillus licheniformis* yb-214,245 Improves Body Growth Performance of Chinese Huainan Partridge Shank Chickens by Enhancing Intestinal Digestive Profiles. *Probiotics Antimicrob. Proteins* **2021**, *13*, 327–342. [CrossRef]
19. Park, I.; Lee, Y.; Goo, D.; Zimmerman, N.P.; Smith, A.H.; Rehberger, T.; Lillehoj, H.S. The effects of dietary *Bacillus subtilis* supplementation, as an alternative to antibiotics, on growth performance, intestinal immunity, and epithelial barrier integrity in broiler chickens infected with Eimeria maxima. *Poult. Sci.* **2020**, *99*, 725–733. [CrossRef]
20. Qiu, K.; Li, C.L.; Wang, J.; Qi, G.H.; Gao, J.; Zhang, H.J.; Wu, S.G. Effects of Dietary Supplementation with *Bacillus subtilis*, as an Alternative to Antibiotics, on Growth Performance, Serum Immunity, and Intestinal Health in Broiler Chickens. *Front. Nutr.* **2021**, *8*, 786878. [CrossRef]
21. Ningsih, N.; Respati, A.N.; Astuti, D.; Triswanto, T.; Purnamayanti, L.; Yano, A.A.; Putra, R.P.; Jayanegara, A.; Ratriyanto, A.; Irawan, A. Efficacy of *Bacillus subtilis* to replace in-feed antibiotics of broiler chickens under necrotic enteritis-challenged experiments: A systematic review and meta-analysis. *Poult. Sci.* **2023**, *102*, 102923. [CrossRef]
22. Yang, J.; Wang, J.; Huang, K.; Liu, Q.; Guo, L.; Xu, X.; Zhang, H.; Zhu, M. Selenium-enriched *Bacillus subtilis* yb-114246 improved growth and immunity of broiler chickens through modified ileal bacterial composition. *Sci. Rep.* **2021**, *11*, 21690. [CrossRef]
23. Fazelnia, K.; Fakhraei, J.; Yarahmadi, H.M.; Amini, K. Dietary Supplementation of Potential Probiotics *Bacillus subtilis*, *Bacillus licheniformis*, and *Saccharomyces cerevisiae* and *Synbiotic* Improves Growth Performance and Immune Responses by Modulation in Intestinal System in Broiler Chicks Challenged with Salmonella Typhimurium. *Probiotics Antimicrob. Proteins* **2021**, *13*, 1081–1092. [CrossRef] [PubMed]
24. Cheng, Y.H.; Zhang, N.; Han, J.C.; Chang, C.W.; Hsiao, F.S.; Yu, Y.H. Optimization of surfactin production from *Bacillus subtilis* in fermentation and its effects on *Clostridium perfringens*-induced necrotic enteritis and growth performance in broilers. *J. Anim. Physiol. Anim. Nutr.* **2018**, *102*, 1232–1244. [CrossRef]
25. Zhao, Y.; Guo, L.; Xia, Y.; Zhuang, X.; Chu, W. Isolation, Identification of Carotenoid-Producing *Rhodotorula* sp. from Marine Environment and Optimization for Carotenoid Production. *Mar. Drugs* **2019**, *17*, 161. [CrossRef]
26. Maina, A.N.; Schulze, H.; Kiarie, E.G. Response of broiler breeder pullets when fed hydrolyzed whole yeast from placement to 22 wk of age. *Poult. Sci.* **2024**, *103*, 103383. [CrossRef]

27. Selionova, M.I.; Trukhachev, V.I.; Zagarin, A.Y.; Kulikov, E.I.; Belyaeva, N.P. Effects of Dietary Supplementation Using Phytobiotics with Different Functional Properties on Expression of Immunity Genes, Intestinal Histology, Growth, and Meat Productivity of Broiler Chickens. *Vet. Sci.* **2025**, *12*, 302. [CrossRef]
28. Sun, Z.; Wang, T.; Demelash, N.; Zheng, S.; Zhao, W.; Chen, X.; Zhen, Y.; Qin, G. Effect of Yeast Culture (*Saccharomyces cerevisiae*) on Broilers: A Preliminary Study on the Effective Components of Yeast Culture. *Animals* **2019**, *10*, 68. [CrossRef]
29. Wang, J.; Ishfaq, M.; Miao, Y.; Liu, Z.; Hao, M.; Wang, C.; Wang, J.; Chen, X. Dietary administration of *Bacillus subtilis* KC1 improves growth performance, immune response, heat stress tolerance, and disease resistance of broiler chickens. *Poult. Sci.* **2022**, *101*, 101693. [CrossRef]
30. Ministry of Agriculture and Rural Affairs of China. *Nutrient Requirements of Yellow-Feathered Broilers (NY/T 3645-2020)*; China Agriculture Press: Beijing, China, 2020.
31. Liu, S.; Xiao, G.; Wang, Q.; Zhang, Q.; Tian, J.; Li, W.; Gong, L. Effects of Dietary *Bacillus subtilis* HC6 on Growth Performance, Antioxidant Capacity, Immunity, and Intestinal Health in Broilers. *Animals* **2023**, *13*, 2915. [CrossRef]
32. Ji, Y.; Liu, X.; Lv, H.; Guo, Y.; Nie, W. Effects of *Lonicerae flos* and Turmeric extracts on growth performance and intestinal health of yellow-feathered broilers. *Poult. Sci.* **2024**, *103*, 103488. [CrossRef] [PubMed]
33. Livak, K.J.; Schmittgen, T.D. Analysis of relative gene expression data using real-time quantitative PCR and the 2(-Delta Delta C(T)) Method. *Methods* **2001**, *25*, 402–408. [CrossRef] [PubMed]
34. Bolyen, E.; Rideout, J.R.; Dillon, M.R.; Bokulich, N.A.; Abnet, C.C.; Al-Ghalith, G.A.; Alexander, H.; Alm, E.J.; Arumugam, M.; Asnicar, F.; et al. Asnicar F Reproducible, interactive, scalable and extensible microbiome data science using QIIME 2. *Nat. Biotechnol.* **2019**, *37*, 852–857. [CrossRef]
35. IBM Corp. *IBM SPSS Statistics for Windows (Version 21.0) [Computer Software]*; IBM Corp.: Armonk, NY, USA, 2012; Available online: https://www.ibm.com/analytics/spss-statistics-software (accessed on 10 December 2024).
36. Sun, Z.; Wang, T.; Aschalew, N.D.; Zhao, W.; Chen, X.; Zhang, X.F.; Zhen, Y.G.; Qin, G.X. Effects of yeast cultures with different fermentation times on the growth performance, caecal microbial community and metabolite profile of broilers. *J. Anim. Physiol. Anim. Nutr.* **2020**, *104*, 212–223. [CrossRef] [PubMed]
37. Peterson, L.W.; Artis, D. Intestinal epithelial cells: Regulators of barrier function and immune homeostasis. *Nat. Rev. Immunol.* **2014**, *14*, 141–153. [CrossRef]
38. Marquis, V.; Schulthess, J.; Molist, F.; Santos, R.R. Effect of a Yeast β-Glucan on the Performance, Intestinal Integrity, and Liver Function of Broiler Chickens Fed a Diet Naturally Contaminated with *Fusarium* Mycotoxins. *Toxins* **2025**, *17*, 51. [CrossRef]
39. Xie, Y.; Liu, J.; Wang, H.; Luo, J.; Chen, T.; Xi, Q.; Zhang, Y.; Sun, J. Effects of fermented feeds and ginseng polysaccharides on the intestinal morphology and microbiota composition of Xuefeng black-bone chicken. *PLoS ONE* **2020**, *15*, e0237357. [CrossRef]
40. Aliakbarpour, H.R.; Chamani, M.; Rahimi, G.; Sadeghi, A.A.; Qujeq, D. The *Bacillus subtilis* and Lactic Acid Bacteria Probiotics Influences Intestinal Mucin Gene Expression, Histomorphology and Growth Performance in Broilers. *Asian-Australas. J. Anim. Sci.* **2012**, *25*, 1285–1293. [CrossRef]
41. Yang, J.; Zhan, K.; Zhang, M. Effects of the Use of a Combination of Two Bacillus Species on Performance, Egg Quality, Small Intestinal Mucosal Morphology, and Cecal Microbiota Profile in Aging Laying Hens. *Probiotics Antimicrob. Proteins* **2020**, *12*, 204–213. [CrossRef]
42. Awad, W.A.; Hess, C.; Hess, M. Enteric Pathogens and Their Toxin-Induced Disruption of the Intestinal Barrier Through Alteration of Tight Junctions in Chickens. *Toxins* **2017**, *9*, 60. [CrossRef]
43. Bilal, M.; Si, W.; Barbe, F.; Chevaux, E.; Sienkiewicz, O.; Zhao, X. Effects of novel probiotic strains of *Bacillus pumilus* and *Bacillus subtilis* on production, gut health, and immunity of broiler chickens raised under suboptimal conditions. *Poult. Sci.* **2021**, *100*, 100871. [CrossRef]
44. Qiao, Y.; Liu, C.; Guo, Y.; Zhang, W.; Guo, W.; Oleksandr, K.; Wang, Z. Polysaccharides derived from *Astragalus membranaceus* and *Glycyrrhiza uralensis* improve growth performance of broilers by enhancing intestinal health and modulating gut microbiota. *Poult. Sci.* **2022**, *101*, 101905. [CrossRef] [PubMed]
45. Hui, J.; Li, L.; Li, R.; Wu, M.; Yang, Y.; Wang, J.; Fan, Y.; Zheng, X. Effects of supplementation with β-carotene on the growth performance and intestinal mucosal barriers in layer-type cockerels. *Anim. Sci. J.* **2020**, *91*, e13344. [CrossRef] [PubMed]
46. Wang, T.; Cheng, K.; Yu, C.Y.; Li, Q.M.; Tong, Y.C.; Wang, C.; Yang, Z.B.; Wang, T. Effects of a yeast-derived product on growth performance, antioxidant capacity, and immune function of broilers. *Poult. Sci.* **2021**, *100*, 101343. [CrossRef]
47. Martín, D.; Ordás, M.C.; Morel, E.; Nuñez-Ortiz, N.; Díaz-Rosales, P.; Vicente-Gil, S.; Zarza, C.; Jensen, L.; Tafalla, C. Effect of β-glucans on rainbow trout (*Oncorhynchus mykiss*) IgM+ B cells. *Fish Shellfish Immunol.* **2024**, *151*, 109740. [CrossRef] [PubMed]
48. Dong, Y.; Li, R.; Liu, Y.; Ma, L.; Zha, J.; Qiao, X.; Chai, T.; Wu, B. Benefit of Dietary Supplementation with *Bacillus subtilis* BYS2 on Growth Performance, Immune Response, and Disease Resistance of Broilers. *Probiotics Antimicrob. Proteins* **2020**, *12*, 1385–1397. [CrossRef] [PubMed]

49. Yang, T.C.; Li, H.; Huang, G.N.; Wang, S.Y. Detection of IgM and IgG complexes provides new insight into immune regulation of patients with malignancies: A randomized controlled trial. *Int. Immunopharmacol.* **2007**, *7*, 1433–1441. [CrossRef]
50. Zhou, J.; Fu, Y.; Qi, G.; Dai, J.; Zhang, H.; Wang, J.; Wu, S. Yeast cell-wall polysaccharides improve immunity and attenuate inflammatory response via modulating gut microbiota in LPS-challenged laying hens. *Int. J. Biol. Macromol.* **2023**, *224*, 407–421. [CrossRef]
51. Wang, T.; Cheng, K.; Li, Q.; Wang, T. Effects of yeast hydrolysate supplementation on intestinal morphology, barrier, and anti-inflammatory functions of broilers. *Anim. Biosci.* **2022**, *35*, 858–868. [CrossRef]
52. Khan, S.; Khalid, A.; Yang, R.; Khalid, F.; Zahid, M.H.; Liu, H.; Zhang, Y.; Wang, Z. Effect of *Bacillus subtilis* Supplemented Diet on Broiler's Intestinal Microbiota and TLRs Gene Expression. *Probiotics Antimicrob. Proteins* **2024**, *16*, 2251–2268. [CrossRef]
53. Wang, M.Y.; Zhang, Y.; Tong, Y.X.; Guo, P.T.; Zhang, J.; Wang, C.K.; Gao, Y.Y. Effects of lutein on jejunal mucosal barrier function and inflammatory responses in lipopolysaccharide-challenged yellow-feather broilers. *Poult. Sci.* **2022**, *101*, 102191. [CrossRef]
54. Rajput, I.R.; Hussain, A.; Li, Y.L.; Zhang, X.; Xu, X.; Long, M.Y.; You, D.Y.; Li, W.F. Saccharomyces boulardii and *Bacillus subtilis* B10 modulate TLRs mediated signaling to induce immunity by chicken BMDCs. *J. Cell. Biochem.* **2014**, *115*, 189–198. [CrossRef] [PubMed]
55. Bi, Y.; Wei, H.; Chai, Y.; Wang, H.; Xue, Q.; Li, J. Intermittent mild cold acclimation ameliorates intestinal inflammation and immune dysfunction in acute cold-stressed broilers by regulating the TLR4/MyD88/NF-κB pathway. *Poult. Sci.* **2024**, *103*, 103637. [CrossRef] [PubMed]
56. Rajput, I.R.; Ying, H.; Yajing, S.; Arain, M.A.; Weifen, L.; Ping, L.; Bloch, D.M.; Wenhua, L. *Saccharomyces boulardii* and *Bacillus subtilis* B10 modulate TLRs and cytokines expression patterns in jejunum and ileum of broilers. *PLoS ONE* **2017**, *12*, e0173917. [CrossRef]
57. Yang, J.; Wang, J.; Huang, K.; Zhu, M.; Liu, Q.; Liu, G.; Chen, F.; Zhang, H.; Qin, S. Selenium enriched *Bacillus subtilis yb-1114246* activated the TLR2-NF-κB1 signaling pathway to regulate chicken intestinal β-defensin 1 expression. *Food Funct.* **2021**, *12*, 5913–5926. [CrossRef]
58. Bai, Y.; Liao, Y.; Song, Y.; Wang, J.; Deng, X.; Luan, L.; An, N.; Zhou, W.; Liang, T.; Yang, Y.; et al. Dietary yeast culture alleviates intestinal-hepatic damage related to TLR2-MyD88-NF-κB signaling pathway and antioxidant capability in *Pseudobagrus ussuriensis*. *Fish Shellfish Immunol.* **2022**, *130*, 428–435. [CrossRef]
59. Tarasiuk-Zawadzka, A.; Fichna, J. Interaction between nutritional factors and the enteric nervous system in inflammatory bowel diseases. *J. Nutr. Biochem.* **2025**, *10*, 109959. [CrossRef]
60. Yue, Y.; Luasiri, P.; Li, J.; Laosam, P.; Sangsawad, P. Research advancements on the diversity and host interaction of gut microbiota in chickens. *Front. Vet. Sci.* **2024**, *11*, 1492545. [CrossRef]
61. Guo, P.; Lin, S.; Lin, Q.; Wei, S.; Ye, D.; Liu, J. The digestive tract histology and geographical distribution of gastrointestinal microbiota in yellow-feather broilers. *Poult. Sci.* **2023**, *102*, 102844. [CrossRef]
62. Rodrigues, D.R.; Briggs, W.; Duff, A.; Chasser, K.; Murugesan, R.; Pender, C.; Ramirez, S.; Valenzuela, L.; Bielke, L.R. Comparative effectiveness of probiotic-based formulations on cecal microbiota modulation in broilers. *PLoS ONE* **2020**, *15*, e0225871. [CrossRef]
63. Oba, S.; Washida, K.; Shimada, Y.; Sunagawa, T.; Tanihiro, R.; Sugiyama, H.; Nakamura, Y. Yeast mannan increases *Bacteroides thetaiotaomicron* abundance and suppresses putrefactive compound production in in vitro fecal microbiota fermentation. *Biosci. Biotechnol. Biochem.* **2020**, *84*, 2174–2178. [CrossRef] [PubMed]
64. Liu, L.; Li, Q.; Yang, Y.; Guo, A. Biological Function of Short-Chain Fatty Acids and Its Regulation on Intestinal Health of Poultry. *Front. Vet. Sci.* **2021**, *8*, 736739. [CrossRef] [PubMed]
65. Melaku, M.; Zhong, R.; Han, H.; Wan, F.; Yi, B.; Zhang, H. Butyric and Citric Acids and Their Salts in Poultry Nutrition: Effects on Gut Health and Intestinal Microbiota. *Int. J. Mol. Sci.* **2021**, *22*, 10392. [CrossRef] [PubMed]
66. Oketch, E.O.; Yu, M.; Hong, J.S.; Chaturanga, N.C.; Seo, E.; Lee, H.; Hermes, R.G.; Smeets, N.; Taechavasonyoo, A.; Kirwan, S.; et al. Laying hen responses to multi-strain *Bacillus*-based probiotic supplementation from 25 to 37 weeks of age. *Anim. Biosci.* **2024**, *37*, 1418–1427. [CrossRef]
67. Wang, J.; Zheng, Z.; Yang, H.; Chen, J.; Xiao, Y.; Ji, X.; Zhang, Z.; He, H.; Ding, B.; Tang, B. Effect of β-1,3/1,6-glucan on gut microbiota of yellow-feathered broilers. *AMB Express* **2022**, *12*, 115. [CrossRef]
68. Qiu, M.; Hu, J.; Peng, H.; Li, B.; Xu, J.; Song, X.; Yu, C.; Zhang, Z.; Du, X.; Bu, G.; et al. Research Note: The gut microbiota varies with dietary fiber levels in broilers. *Poult. Sci.* **2022**, *101*, 101922. [CrossRef]

veterinary sciences

MDPI

Article

Ssc-miR-130b Enhances Cell Proliferation and Represses Adipogenesis of Primary Cultured Intramuscular Preadipocytes in Pigs

Yunqiu Yang [1,†], Yongfang Chen [1,†], Lijun Wang [1], Min Du [2], Rui Zhang [3], Yao Lu [4] and Shifeng Pan [1,2,5,*]

1 College of Veterinary Medicine, Yangzhou University, Yangzhou 225009, China; yyq13375299380@163.com (Y.Y.); chenyongfang77@163.com (Y.C.); wlj6379@163.com (L.W.)
2 Department of Animal Sciences, Washington State University, Pullman, WA 99163, USA; min.du@wsu.edu
3 Meat Processing Key Laboratory of Sichuan Province, Chengdu University, Chengdu 610106, China; zhangrui@cdu.edu.cn
4 Clinical Medical College, Yangzhou University, Yangzhou 225009, China; lubber0916@163.com
5 Jiangsu Co-Innovation Center for Prevention and Control of Important Animal Infectious Diseases and Zoonoses, Yangzhou University, Yangzhou 225009, China
* Correspondence: sfpan@yzu.edu.cn; Tel.: +86-51487979274; Fax: +86-51487972218
† These authors contributed equally to this work.

Simple Summary: The goal of increasing intramuscular fat (IMF) content and thus improving the meat quality of livestock and poultry involves some increasingly dynamic and challenging concerns. MicroRNAs have proven to be critical regulators of lipid accumulation, among which miR-130b has been demonstrated to inhibit adipogenesis in porcine subcutaneous preadipocytes. In this study, the role and underlying mechanism of miR-130b in regulating cell proliferation and adipogenesis of primary cultured porcine intramuscular preadipocytes (PIMPA) are further studied. Results showed that overexpression of miR-130b accelerated PIMPA proliferation by obviously promoting cell cycle progression, while inhibiting adipogenic differentiation by dramatically reducing PPAR-γ and its downstream target gene expression. Our results indicate that miR-130b plays an inhibitory role in IMF deposition and may serve as a potential candidate for improving meat quality.

Abstract: In the efforts towards germplasm innovation of livestock and poultry, strategies to improve meat quality have faced some increasingly challenging and dynamic concerns. Intramuscular fat (IMF) content and backfat thickness are two important traits contributing to meat quality. MicroRNAs (miRNAs)—a class of endogenous noncoding RNAs maintaining cell homeostasis by inhibiting target gene expression—have been proven as critical regulators of body fat deposition, thus affecting farm animal production. Our previous in vitro and in vivo models of pigs have clarified that miR-130b overexpression can obviously suppress adipogenesis of subcutaneous preadipocytes and lower backfat thickness. However, the way miR-130b regulates proliferation and adipogenesis of primary cultured porcine intramuscular preadipocytes (PIMPA) and the underlying mechanism are still unknown. PIMPA derived from longissimus dorsi muscle were employed to examine the role of miR-130b in proliferation and adipogenesis and to further elucidate its underlying mechanism. Lipid deposition in cytoplasm was evaluated by TG quantification and ORO-staining, and EDU-staining was employed to measure cell proliferation. Adipogenic and proliferation-related gene expression were conducted by qPCR and Western blot. MiR-130b overexpression markedly stimulated proliferation of PIMPA by increasing cell cycle-related gene expression. Furthermore, overexpression of miR-130b significantly inhibited adipogenic differentiation of PIMPA, mainly by inhibiting expression of adipogenic differentiation marker genes PPAR-γ and SREBP1. In addition, we proved that miR-130b significantly inhibited expression of PPAR-γ downstream target genes and ultimately repressed adipogenesis. Ssc-miR-130b accelerated proliferation but inhibited adipogenic differentiation of PIMPA, contributing to an enhanced knowledge of the function of ssc-miR-130b in lipid deposition, and providing potential implications for enhancing pork quality.

Keywords: ssc-miR-130b; adipogenic differentiation; proliferation; primary cultured intramuscular preadipocytes; pig

1. Introduction

In domestic animals, adipose tissue growth is a dominant factor of meat quality and production performance, among which intramuscular fat (IMF) content and backfat thickness are two critical indexes [1]. Especially, the content of IMF significantly influences essential attributes such as tenderness and juiciness [2]. Accordingly, enhancing IMF content remains a key focus in animal science research. IMF deposition is a highly intricate biological process governed by multiple factors [3]. At a cellular level, IMF content is primarily determined by proliferation and adipogenesis of preadipocytes, a population of cells widely distributed within IMF. The preadipocyte number reflects their proliferative capacity, while their differentiation into lipid-laden cells influences lipid accumulation. Together, these processes contribute to the growth and development of adipose tissue [4]. At the molecular level, preadipocyte activity is a tightly regulated and highly intricate process. It is primarily governed by the coordinated interaction of genes related to lipid metabolism, key transcription factors, and epigenetic regulators, all of which play an essential role in orchestrating adipogenesis and fat deposition [5]. Under such circumstances, to clarify the potential mechanism underlying proliferation and adipogenic differentiation of the preadipocyte is of great economic and academic value for enhancing meat quality in both livestock and poultry production.

Accumulating evidence suggest that proliferation and differentiation of preadipocytes involves a highly orchestrated series of intricate biological processes, among which microRNAs (miRNAs) are one group of the most important regulatory factors [6]. As endogenous non-coding small RNAs, MiRNAs inhibit target gene expression [7]. A growing number of studies have previously suggested that miRNAs play an important role in regulating proliferation, differentiation, development, and apoptosis [8]. Moreover, it is proven that miRNAs direct adipocyte fate decisions that are important for proliferation and adipogenic differentiation, which greatly contributes to adipose tissue growth [9]. Recent computational and experimental studies have also shown that miRNAs serve as key regulators in various biological processes related to lipid deposition. Furthermore, more and more studies have highlighted that as crucial epigenetic regulators, miRNAs play a vital role in IMF deposition [10]. In Erhualian pigs, it has been shown that miR-32-5p is a key contributor to IMF accumulation by inhibiting KLF3 [11], whereas miR-381-3p has been shown to suppress IMF deposition via repressing FABP3 [12]. However, the role of miR-130b in controlling proliferation and adipogenesis of porcine intramuscular preadipocytes (PIMPA) and the underlying mechanism remain largely unclear.

It was demonstrated previously that miR-130b expression in plasma of obese mouse models and patients exhibited a strong positive correlation with body mass index (BMI), and the miR-130b secretion in adipocytes significantly increased during fat formation, showing that blood miR-130b was a more reliable indicator predicting metabolic syndrome compared to TG level [13]. Moreover, miR-130b has also been considered as an obesity-related miRNA [14]. Also, it was proven that miR-130 served as an inhibitor of adipogenesis of human preadipocytes. Overexpression of miR-130 significantly reduced lipid accumulation, while inhibition of miR-130 obviously increased lipid accumulation by directly targeting PPAR-γ [15,16]. Furthermore, miR-130b duplex was able to directly suppress KLF3 expression, thereby inhibiting the adipogenic differentiation, fatty acid synthesis, and lipid deposition of goat intramuscular adipocytes [17]. In addition, miR-130a/b overexpression significantly inhibited lipid deposition in bovine intramuscular preadipocyte by targeting PPAR-γ and CYP2U1 [18]. In our previous studies, both in vivo and in vitro experiments in pigs revealed that miR-130b downregulates PPAR-γ expression and suppresses backfat thickness [19]. Our subsequent study revealed that lipid accumulation was significantly decreased by MV-shuttled miR-130b via suppressing PPAR-γ expression in primary cultured porcine adipocytes [20], and this function was further confirmed in HFD-induced obese mice [21]. The aforementioned findings collectively suggest that miR-130b is a potential inhibitor of subcutaneous fat deposition. However, backfat thickness and IMF are two important indexes of pig meat quality; although the inhibition of miR-130b in pig subcutaneous lipid deposition has been largely clarified, the actual role of miR-130b in proliferation and adipogenic differentiation of PIMPA is still largely unknown, and the underlying molecular mechanisms need to be further clarified.

The study utilized PIMPA to further investigate the impact of miR-130b on IMF deposition by regulating proliferation and adipogenic differentiation. MiR-130b mimic and miR-130b inhibitor were used to examine its biological function in the process of adipogenesis and provided evidence that overexpression of miR-130b reduced lipid accumulation during PIMPA differentiation. Suppression of endogenous miR-130b using an inhibitor enhanced adipogenic differentiation. The above results demonstrated firstly that ssc-miR-130b promoted proliferation and inhibited adipogenic differentiation of PIMPA, which provides new understanding to further unravel the role of miR-130b. In summary, our findings have provided a meaningful point of view on miR-130b in porcine IMF deposition, which has potential significance for improving meat quality.

2. Material and Methods

2.1. Ethics Statements

Experiments involving animals were conducted in compliance with the guidelines for the care and use of laboratory animals established by the Ministry of Science and Technology of the People's Republic of China (approval no. SYXK-SU-2007-0005) and were approved by Yangzhou University.

2.2. Isolation and Culture of PIMPA

In the study, the postnatal Erhualian piglets at 3–7 d of age were anesthetized with injection intraperitoneally of 50 mg/kg body weight pentobarbital sodium followed by exsanguination. The longissimus dorsi muscle samples were meticulously prepared by removing visible blood vessels and connective tissues before being finely chopped into 0.5–1.0 mm^3 pieces. The primary PIMPA were derived from longissimus dorsi muscle of the piglets by digestion of collagenase, and purification by percoll density gradient centrifugation method. The purified PIMPA were then plated with a 3.0×10^4 cells/well density in 24-well or 6-well plates and maintained at 37 °C in a humidified atmosphere with 5% CO_2. Before adipogenic differentiation, cells were cultured in growth medium consisting of DMEM/F-12 (Hyclone, Logan, UT, USA) supplemented with 10% fetal bovine serum (FBS) and 1% antibiotic–antimycotic mixture. Culture medium was then changed every 2 days.

2.3. Experimental Groups and Adipogenic Differentiation

When 85–90% confluence was reached, the primary cultured PIMPA were randomly divided into four groups: (1) mNC group with miR-130b mimic control transfection; (2) mimic group with miR-130b mimic transfection; (3) iNC group with miR-130b inhibitor control transfection; (4) and inhibitor group with miR-130b inhibitor transfection. To induce adipogenic differentiation, PIMPA were first cultured in differentiation medium (DM) following confluence for two days. The DM was composed of 10% FBS, 1 μM dexamethasone (DEX), 1.7 μM insulin, and 0.5 mM IBMX. Then, the DM was replaced subsequently with DMEM containing 10% FBS as well as 1.7 μM insulin for another two days. Following this, the regular medium was changed every other day until the end of 7 days.

2.4. MiR-130b Mimic and Inhibitor Transfection

Primary cultured PIMPA were subjected to serum starvation for 4 h before miR-130b mimic and inhibitor (Suzhou GenePharma Co., Ltd., Suzhou, China) transfection. miR-130b mimic, miR-130b inhibitor, or the corresponding negative controls (NCs) were transfected into PIMPA using ribo Fect™ CP Transfection Kit (Ribobio, Guangzhou, China) at a concentration of 100 nM/well, as suggested by the manufacturer's instructions. The cells were harvested at 48 h (early differentiation stage) and 7 d (late differentiation stage) after transfection and adipogenic differentiation for further analysis.

2.5. Oil Red O (ORO) Staining and Quantification

For ORO staining, mature adipocytes at day 7 of induced differentiation were washed with PBS and fixed in 4% paraformaldehyde for 30 min. Following fixation, the adipocytes were rinsed three times with PBS and stained with a 1% filtered ORO solution (Solarbio, Beijing, China) for 30 min. Excess stain was removed after washing the cells with PBS three times, after which they were examined under an IX53 inverted fluorescence microscope (Olympus, Tokyo, Japan) to identify ORO positively stained adipocytes, which appeared red. For quantification, three images were randomly captured per observation field to determine the ORO-stained area. Additionally, ORO retained in the cells was extracted using 100% isopropanol (v/v) and quantified by detecting OD510 values.

2.6. Total RNA Extraction and qRT-PCR

Total RNA was extracted and reverse transcribed into cDNA. The Applied Biosystems 7500 Real-Time PCR System (Thermo Fisher Scientific, Inc., Shanghai, China.) was used to perform qPCR experiment. The PCR amplification protocol: initial denaturation at 94 °C for 5 s, followed by 40 cycles of annealing at 62 °C for 40 s and extension. All experiments were conducted independently in triplicate. mRNA expression levels were normalized to peptidylprolyl isomerase A (PPIA). Primers used for gene detection were synthesized by Genewiz, Inc. (Suzhou, China), with primer sequences shown in Table 1.

Table 1. Primer sequence of qPCR reaction.

Gene Name	GenBank		Primer Sequence (5′-3′)	Annealing Temperature (°C)	PCR Product (bp)
Porcine					
P27	NM_214316.1	F R	GTCCCTTTCAGTGAGAACCGATAC TTGCTGCCACATAACGGAATCAT	61	134
PCNA	NM_001291925.1	F R	GTGATTCCACCACCATGTTC TGAGACGACTCCATGCTCTG	57.9	145
Cyclinb1	NM_001170768.1	F R	AACTGCTCTTGGAGACATCGGT TGGTTCAGGCTCCAGTTCAGG	61.8	196
FAS	NM_001099930.1	F R	GTCCTGCTGAAGCCTAACTC TCCTTGGAACCGTCTGTG	57.2	206
GR	NM_001008481.1	F R	CCAAACTCTGCCTTGTGTGTTC TGTGCTGTCCTTCCACTGCT	61	108

Table 1. *Cont.*

Gene Name	GenBank		Primer Sequence (5′-3′)	Annealing Temperature (°C)	PCR Product (bp)
FTO	NM_001112692.1	F R	GGAGAAAGCCAATATCGACACC TCTGCTCTTCCTGTCCACCTC	60.1	109
Perilipin	NM_001038638.1	F R	GCCTGACTTTGCTGGATGG CTTGGTGCTGGTGTAGGTCTTCT	60.5	119
11β-HSD1	NM_214248.3	F R	CCATGCTGAAGCAGAGCAAC AAGAACCCGTCCAGAGCAAA	59.6	115
SCD1	NM_213781.1	F R	CCCAGCCGTCAAAGAGAA CGATGGCGTAACGAAGAAA	56.5	200
PPAR-α	NM_001044526.1	F R	GAGCCTGAGGAAACCCTTCT GGTCTCCGCACCAAATGA	58.2	128
PPAR-γ	NM_214379.1	F R	GCCCTTCACCACTGTTGATT GAGTTGGAAGGCTCTTCGTG	58.5	210
Caspase-3	NM_214131.1	F R	TGCTGCAAATCTCAGGGAGACCT GTGCCTCGGCAGGCCTGAAT	64.6	288
PPIA	NM_214353.1	F R	AGGATTTATGTGCCAGGGTG ATGGACAAGATGCCAGGAC	50.9	126

2.7. Western Blot

For total protein extraction, protein concentration determination, and electrophoresis, refer to our published studies [19–21]. Membranes were then incubated overnight at 4 °C with primary antibodies against PPAR-γ and SREBP-1 diluted in TBST. Following incubation, membranes were incubated with goat anti-rabbit IgG peroxidase-conjugated secondary antibodies (Bioworld Technology, St. Louis Park, MN, USA) at a 1:10,000 dilution. Protein bands were detected by an enhanced chemiluminescence (ECL) system with the LumiGlo substrate (SuperSignal West Pico Trial Kit, Pierce, Rockford, IL, USA). The bands were visualized and captured using the VersaDoc 4000MP imaging system (Bio-Rad, Hercules, CA, USA), and band intensity was automatically quantified using Quantity One software (Bio-Rad), Version 4.3.0. β-actin was served as the internal reference protein, and protein expression levels were shown as relative fold changes compared to mNC or iNC cells, with Table 2 presenting the details of the antibodies used in the study.

Table 2. Antibody information in the study.

Antibody Name	Species Source	Number	Company
PPAR-γ	Rabbit polyclonal antibody	BS6442	Bioworld (Shanghai, China)
β-actin	Rabbit polyclonal antibody	AP0060	Bioworld (Shanghai, China)
SREBP1	Rabbit polyclonal antibody	14088-1-AP	Proteintech (Wuhan, China)
Caspase-3	Rabbit polyclonal antibody	19677-1-AP	Proteintech (Wuhan, China)
Bcl-2	Rabbit polyclonal antibody	26593-1-AP	Proteintech (Wuhan, China)

2.8. Real-Time PCR Quantification of miR-130b

Specific primers used for miR-130b expression quantification were shown in Table 3. RT-qPCR was conducted using an Mx3000P real-time PCR system (Stratagene, Santa Clara, CA, USA). Firstly, genomic DNA contamination of total RNA was removed by treating with RNase-free DNase Kit (Promega, Madison, WI, USA). Subsequently, the purified total RNA (4 μg) was treated with the Poly(A) Tailing Kit (AM1350, Ambion, TX, USA) polyadenylation following the manufacturer instructions.

The total volume of 20 μL polyadenylation reaction consisted of 4 μg RNA (1 μg/μL), 2 μL of 25 mM $MnCl_2$, 4 μL of 5× E-PAP buffer, 0.8 μL of E-PAP enzyme, 2 μL of 10 mM ATP, 0.2 pmol each of external controls (E5, E2, and E1), and RNase-free water. After extraction of phenol-chloroform and precipitation of ethanol, the poly(A)-tailed RNA was diluted in diethylpyrocarbonate (DEPC)-treated water.

Table 3. miR-130b related sequences.

Name	Sequences (5′-3′)
miR-130b mimic	CAGTGCAATGATGAAAGGGCAT GCCCTTTCATCATTGCACTGTT
mimic NC	TTCTCCGAACGTGTCACGTTT ACGTGACACGTTCGGAGAATT
miR-130b inhibitor	ATGCCCTTTCATCATTGCACTG
inhibitor NC	CAGTACTTTTGTGTAGTACAA
Universal primer	TAGAGTGAGTGTAGCGAGCA
External reference	GTGACCCACGATGTGTATTCGC
Poly (T) adapter	TAGAGTGAGTGTAGCGAGCACAGAATTAATACG ACTCACTATAGGTTTTTTTTTTTTTTTTVN

For cDNA synthesis, 1 μg/μL oligo(dT)-adapter primers were used. The RT1 reaction contained a 10 μL volume of poly(A)-tailed RNA (2 μg), 1 μL oligo(dT)-adapter primer, and DEPC-treated water, and was incubated for 5 min at 70 °C. The subsequent RT2 reaction (25 μL total volume) included RT1 reaction product (10 μL), 10 mM dNTPs (1.25 μL), M-MLV 5× buffer (5 μL), M-MLV reverse transcriptase (1 μL, 200 U/μL), RNase inhibitor (0.5 μL, 40 U/μL), and DEPC water. The cDNA synthesis reaction was conducted by incubation at 42 °C for 1 h. The PCR reaction (25 μL in total volume) consisted of 2 μL RT2 product, 2 μL primers (10 μM), 12.5 μL SYBR Green PCR Master Mix (TaKaRa, Tokyo, Japan), and 8.5 μL sterile deionized water. E5 small nuclear RNA expression was used for normalize miR-130b expression. Data are presented as fold changes relative to mNC or iNC cells, using the $2^{-\Delta\Delta Ct}$ method.

2.9. The Effect of miR-130b on PIMPA Proliferation

When PIMPA reached 85–90% confluence, cells were transfected with miR-130b inhibitor, miR-130b mimic, or their respective negative controls (iNC and mNC) using Lipo2000 transfection reagent (Invitrogen Life Technologies, Carlsbad, CA, USA). Two days post-transfection, the CCK-8 Cell Counting Kit (Vazyme, Nanjing, China) was used for PIMPA proliferation, and Varioskan LUX Multimode Reader (Thermo Fisher Scientific, Waltham, MA, USA) was used for detection OD450 values.

To evaluate the proliferation-related gene expression, p27, cyclin B1 (CCNB1), and proliferating cell nuclear antigen (PCNA) mRNA levels were quantified by RT-qPCR in triplicate.

Additionally, DNA synthesis and cell proliferation were analyzed by the EdU incorporation assay. After 2 days of transfection, PIMPA were incubated with 10 μM EdU reagent A (Beyotime, Beijing, China) in cell medium at 37 °C for 2 h. After being fixed with 4% paraformaldehyde, PIMPA were then incubated for 30 min with Click Reaction Solution (Beyotime, Beijing, China). Hoechst 33342 was used to counterstain nuclei for 30 min, and fluorescence signals were observed by an IX53 microscope, and at least three visual fields were randomly taken from each sample to determine the total number of PIMPA and EdU-positive PIMPA for statistical analysis.

2.10. The Effect of miR-130b on the Differentiation of PIMPA

To investigate the role of miR-130b in regulating PIMPA adipogenesis, RNA was extracted from adipocytes induced to differentiate into adipocytes for 7 days. The expression of miR-130b, and PPAR-γ, a known miR-130b target gene, as well as key markers associated with differentiation and lipid metabolism—including fatty acid synthase (FAS), glucocorticoid receptor (GR), fat mass and obesity-associated protein (FTO), Perilipin,

11β-hydroxysteroid dehydrogenase type 1 (11β-HSD1), stearoyl-CoA desaturase 1 (SCD-1), and PPAR-α—were evaluated by RT-qPCR.

2.11. Statistical Analysis

Statistical analyses were performed by SPSS 26.0 software, and all data are shown as mean ± SEM. Differences between two groups were assessed using Student's *t*-test (unpaired, parametric). For multiple group comparisons, one-way ANOVA was conducted, followed by Duncan's test and Bonferroni correction. A *p*-value lower than 0.05 represents statistically significant differences.

3. Results

3.1. Isolation and Identification of PIMPA

The longissimus dorsi muscle tissue was aseptically separated, which was then digested by collagenase type I to isolate PIMPA, and then the PIMPA was purified by differential adhesion method. The PIMPA was plated into the cell culture plates at a density of 3×10^4 cells/cm^2; after growing 4 days, the density of PIMPA reached 85%, showing short fusiform or irregular small triangles, similar to fibroblasts, with no lipid drops in the cytoplasm (Figure 1A). When PIMPA reached 85–90% confusion, "cocktail method" (0.5 mM 3-IBMX, 1 μM DEX, and 1.7 μM insulin) was used to induce adipogenic differentiation. After adipogenic induction of DEX, IBMX, and insulin for 7 days, most of the PIMPA had differentiated into mature adipocytes, and a large number of lipid droplets could be seen under the microscope (Figure 1B). ORO staining method was also used to identify the cellular lipid accumulation, and results showed that a mass of ORO-stained cells were observed, showing that PIMPA underwent differentiation into mature adipocytes following treatment with differentiation-inducing solution (Figure 1C). The above results suggested that PIMPA was successfully isolated and could be used for further analysis.

A B C

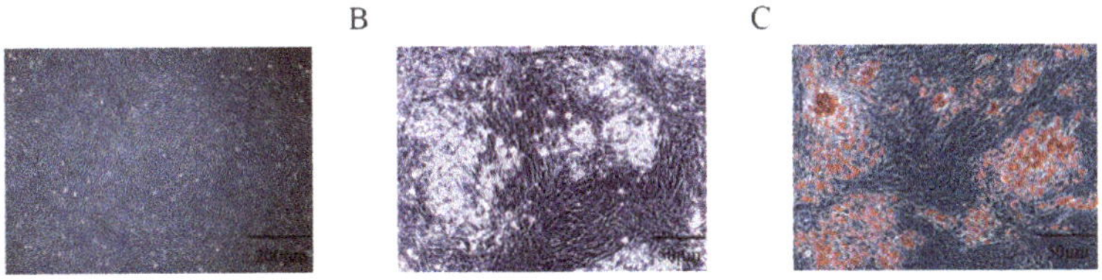

Figure 1. Isolation and identification of PIMPA. (**A**). PIMPA proliferation reached 80% fusion (200 μm). (**B**). Intracellular lipid droplets under microscope after adipogenic induction (scale bar = 50 μm); (**C**). ORO positively stained cells to identify lipid (scale bar = 50 μm).

3.2. MiR-130b Overexpression Significantly Promoted the Proliferation of PIMPA

The RT-qPCR analysis indicated that, after transfected with miR-130b mimic, miR-130b expression was significantly elevated in PIMPA ($p < 0.01$) (Figure 2A). The EdU staining result revealed a significant increase in both the number and fluorescence intensity of proliferating green fluorescent cells in the miR-130b mimic group compared to control group. This finding suggests that miR-130b mimic markedly enhanced the proliferation of EdU-labeled preadipocytes. (Figure 2B,C). On the other hand, transfection with the miR-130b inhibitor resulted in reduced miR-130b expression compared to iNC group ($p < 0.05$) (Figure 2D). These results suggest that miR-130b mimic and miR-130b inhibitor were successfully transfected into PIMPA. The proliferation activity assay results indicated that miR-130b mimic significantly enhanced the viability of PIMPA. However, the EdU staining result indicated that miR-130b inhibitor showed no significant difference compared with iNC group (Figure 2E,F).

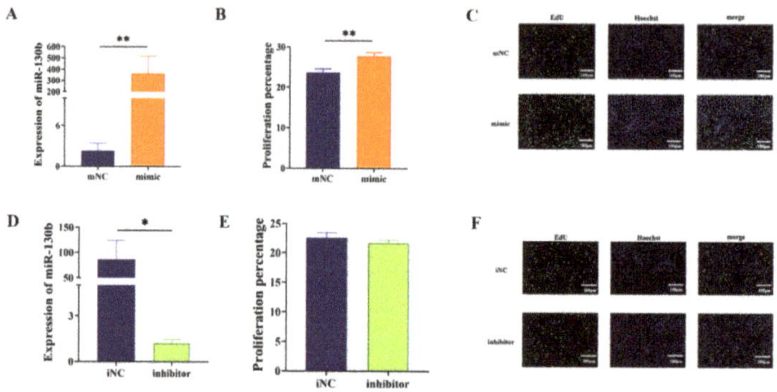

Figure 2. Effect of miR-130b mimic and inhibition on PIMPA proliferation. (**A**). miR-130b expression. (**B**). Cell proliferation rate by Image J (V1.8.0.112) method. (**C**). Proliferative activity by EdU method (scale bar = 100 μm). (**D**). miR-130b expression. (**E**). Cell proliferation rate by Image J method. (**F**). Proliferative activity by EdU method (scale bar = 100 μm). Values are presented as the mean ± SEM. ** Represents $p < 0.01$ mNC vs. miR-130b mimic; * Represents $p < 0.05$ iNC vs. miR-130b inhibitor.

3.3. Overexpression of miR-130b Promoted the Expression of PCNA and cyclinb1 and Inhibited the Expression of P27 in PIMPA

The RT-qPCR results showed that compared with mNC group, miR-130b mimic significantly upregulated the expression of proliferation marker genes proliferating cell nuclear antigen (*PCNA*) and *cyclinb1*, while downregulating the *P27* mRNA expression at day 2 during adipogenic differentiation of PIMPA ($p < 0.01$) (Figure 3A). On the other hand, miR-130b inhibitor showed no significant difference in *PCNA*, *cyclinb1*, and *P27* compared with iNC group (Figure 3B). These findings indicated that miR-130b overexpression markedly increased the PIMPA proliferation through the cell cycle and DNA replication pathways.

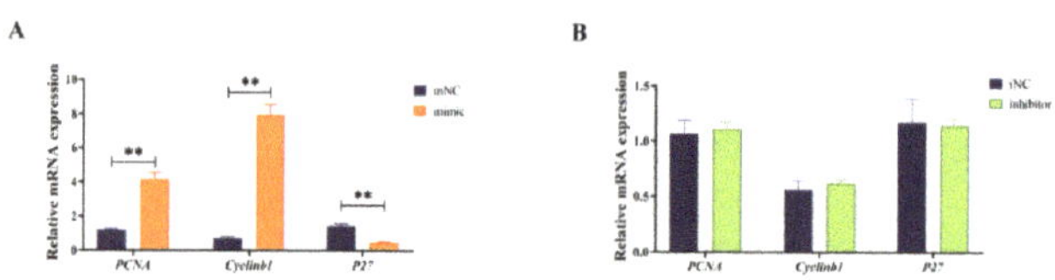

Figure 3. Effect of miR-130b mimic and inhibition on cell cycle-related gene expression of PIMPA. (**A**). Effect of miR-130b mimic on expression of cell cycle-related gene PCNA, cyclinb1, and P27. (**B**). Effect of miR-130b inhibition on expression of PCNA, cyclinb1, and P27. Values are presented as the mean ± SEM. ** Represents $p < 0.01$ mNC vs. miR-130b mimic.

3.4. Overexpression of miR-130b Significantly Inhibited the Adipogenic Differentiation of PIMPA

Both ORO staining and TG quantitative detection kit were used to detect the effect of miR-130b on the adipogenic differentiation of PIMPA. The results depicted in Figure 4 confirm the successful transfection of miR-130b mimic, miR-130b inhibitor, mNC, and iNC into PIMPA by day 7. Additionally, ORO staining revealed that miR-130b mimic markedly reduced lipid droplet accumulation and intracellular TG content compared to the mNC group (Figure 4A). Conversely, transfection with miR-130b inhibitor led to a significant increase in both lipid droplet accumulation and intracellular TG content (Figure 4B). Quantification of the ORO staining area further validated these findings, demonstrating that the miR-130b mimic significantly reduced the stained area compared to mNC group (Figure 4A). In contrast, miR-130b inhibitor increased the area of ORO staining (Figure 4B). In addition, the results of quantitative detection of intracellular TG content were consistent with those of ORO staining results, showing that miR-130b mimic significantly reduced the intracellular TG content by 50% (Figure 4A), while miR-130b inhibitor significantly increased the intracellular TG content by 80% (Figure 4B). These findings indicate that miR-130b acts as a negative regulator of adipogenic differentiation in PIMPA.

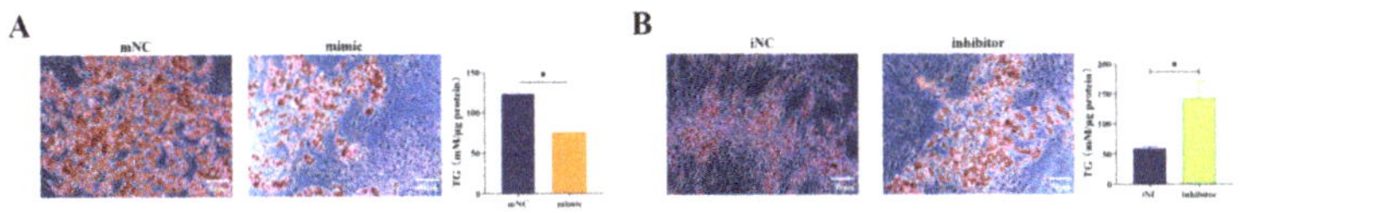

Figure 4. Effect of miR-130b mimic and inhibition on adipogenic differentiation of PIMPA. (**A**). Effect of miR-130b mimic on lipid accumulation and intracellular TG content by ORO staining and TG quantitative detection kit. (**B**). Effect of miR-130b inhibition on lipid accumulation and intracellular TG content by ORO staining and TG quantitative detection kit. Scale bar = 50 µm. Values are presented as the mean ± SEM. * Represents $p < 0.05$ mNC vs. miR-130b mimic.

3.5. Overexpression of miR-130b Significantly Inhibited the Expression of Key Genes for Adipogenesis

On day 7 after transfection with miR-130b mimic, qPCR and Western blot were employed to analyze the expression levels of essential genes related to lipid metabolism. The results showed that compared with mNC, the mRNA expression of adipogenic differentiation marker genes PPAR-γ, FTO, and GR; lipogenic genes FAS and SCD-1; Perilipin (a critical regulator of fat storage and breakdown); and lipolytic gene PPAR-α were all significantly decreased in miR-130b mimic group ($p < 0.05$) (Figure 5A). In addition, Western blot results also showed that the protein expression of PPAR-γ and SREBP1 were markedly reduced ($p < 0.01$) (Figure 5B,C). The above results showed that miR-130b mimic significantly inhibited the adipogenic differentiation of PIMPA through reducing PPAR-γ and its downstream de novo fatty acid synthesis.

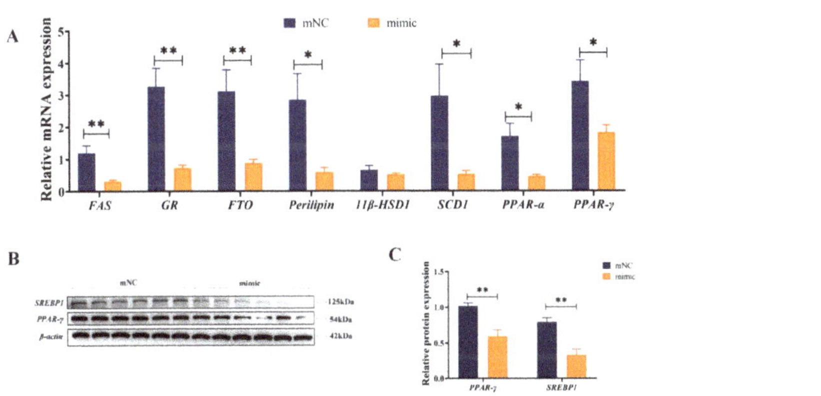

Figure 5. Effect of miR-130b mimic on the expression of key adipogenic genes of PIMPA. (**A**). Relative mRNA expression of FAS, GR, FTO, Perilipin, 11β-HSD1, SCD-1, PPAR-α, and PPAR-γ. (**B**,**C**). Relative protein expression of SREBP-1 and PPAR-γ. Values are presented as the mean ± SEM. * Represents $p < 0.05$; ** Represents $p < 0.01$ mNC vs. miR-130b mimic. The original images of the Western blot are published as Supplementary Materials—Figure S1–S3.

3.6. miR-130b Inhibition Significantly Promoted the Expression of Key Genes for Adipogenesis

On the contrary, on day 7 after transfection with miR-130b inhibitor, qPCR and Western blot results showed that compared to iNC group, the mRNA expression of adipogenic differentiation marker genes PPAR-γ, FTO, GR, FAS, SCD-1, Perilipin, and PPAR-α were all significantly increased in miR-130b inhibitor group ($p < 0.05$) (Figure 6A). However, protein expression of PPAR-γ and SREBP1 showed no significant difference between the two groups (Figure 6B,C). These results demonstrated that miR-130b inhibition significantly up-regulated mRNA levels of lipid metabolism and thus promoted adipogenic differentiation of PIMPA.

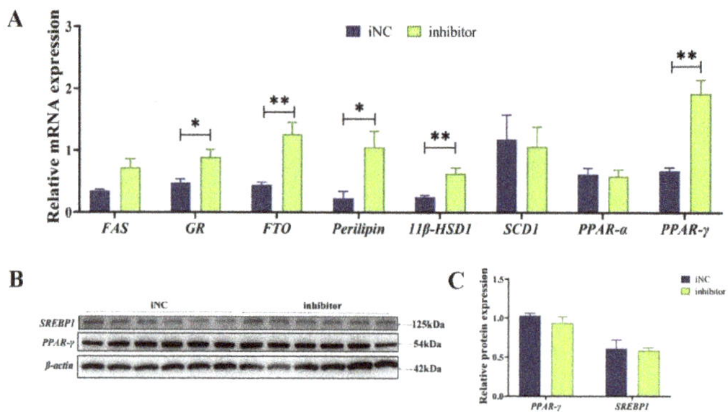

Figure 6. Effect of miR-130b inhibition on the expression of key adipogenic genes of PIMPA. (**A**). Relative mRNA expression of FAS, GR, FTO, Perilipin, 11β-HSD1, SCD-1, PPAR-α, and PPAR-γ. (**B**,**C**). Relative protein expression of PPAR-γ and SREBP-1. Values are presented as the mean ± SEM. * Represents $p < 0.05$; ** Represents $p < 0.01$ iNC vs. miR-130b inhibitor. The original images of the Western blot are published as Supplementary Materials, Figure S1–S3.

4. Discussion

Genetic selection of livestock and poultry breeds has mainly focused on increasing growth rate and lean meat percentage in the past few decades, and a large number of candidate genes had received strong selection to improve livestock performance. However, the rapid growth is inevitably accompanied by the reduction of IMF deposition, which eventually leads to the deterioration of livestock and poultry meat quality. Previous studies have proven that in longissimus dorsi muscle of Laiwu pigs with significantly different IMF content, miRNAs were differentially expressed, which were mainly concentrated in lipid metabolism, lipid storage, Wnt, mTOR, and PPAR signaling pathways by GO and KEGG analysis, suggesting that differentially expressed miRNAs played a vital role in regulating pig IMF content [22]. Moreover, a growing number of reports showed that miRNAs play important roles in controlling preadipocyte proliferation and adipogenesis in livestock animals. However, this is the first study to systematically report the effect of miR-130b on proliferation and differentiation of porcine preadipocytes, suggesting that miR-130b can serve as a candidate gene to regulate porcine intramuscular fat deposition and enhance meat quality.

Previous study has shown that miR-130b participates in the inhibition of mouse muscle cell line C2C12 proliferation and promotes myogenic differentiation by targeting Sp1 [23]. Moreover, miR-130b was involved in oxidative metabolism of muscle lipid via targeting PGC-1α [13]. The above results have fully demonstrated the vital role of miR-130b in the development and metabolism of muscle tissue. A previous study in goat has demonstrated that miR-130b-3p/miR-130b-5p duplex reduced adipogenesis by directly inhibiting KLF3 expression, suggesting that KLF3 is a candidate gene of miR-130b [17]. Therefore, the expression of KLF3 after miR-130b mimic or inhibitor treatment should be detected in our future study. However, in the study, we found that miR-130b mimic significantly reduced the mRNA and protein expression of PPAR-γ, one of the most important transcription factors regulating the differentiation of all adipocytes; furthermore, PPAR-γ downstream factors, FAS, GR, FTO, Perilipin and SREBP-1, were all downregulated. Moreover, in our previous studies, we have verified that PPAR-γ is a target of miR-130b [19–21]. Moreover, miR-301b~miR-130b-PPARγ axis has been proved to play an important role in MSCs adipogenesis [24]. Therefore, in the study, we could conclude that miR-130b inhibited adipogenic differentiation of PIMPA by blocking PPAR-γ expression, thus dominating IMF deposition in pigs.

Numerous previous studies have demonstrated that miR-130b functions in regulating the proliferation process of various cells [25]. However, we provided the first evidence demonstrating that miR-130b enhanced the proliferation of PIMPA, characterized by increased cell viability and EdU-labeled cell density of PIMPA, as well as increased expression of proliferation marker genes Cyclinb1 and PCNA, and inhibited expression of P27. These results were in accordance with previous study findings that miR-429 mimics could, respectively, promote the proliferation of porcine preadipocytes [26]. It is known that reduced cell proliferation is related to arrested cell cycle-dependent mechanisms, which is strictly controlled by an array of cyclins and cyclin-dependent kinases (CDKs), among which p27 is one of the broad-spectrum CDKs inhibitors, negatively regulating the transition from G1 to S phase of cell cycle. In the study, we found that miR-130b mimic markedly downregulated P27 expression, suggesting that miR-130b promoted proliferation of PIMPA by blocking the G1/S phase transition and thus negatively arrested the cell cycle. Furthermore, Cyclinb1 has been shown to accelerate the proliferation and promote cell cycle of various cell types by moving from G1 to S phase. PCNA has also been shown to play a vital role in cell proliferation, and has been considered as a potential target of cancer therapy [27]. Our current results indicated for the first time that miR-130b promoted proliferation of PIMPA by regulating expression of PCNA, cyclinb1, and P27, which was consistent with previous findings in many types of cells, showing enhanced effect of miR-130b on proliferation. However, which target gene was involved in miR-130b regulating the cell cycle needs to be further studied. Moreover, some other studies have shown diametrically opposite results concerning the effect of miR-130b on proliferation, based on the inhibition role of miR-130b in proliferation. The above results suggested that different role of miR-130b in regulating cell proliferation was closely related to cell types.

Numerous studies have demonstrated that PPAR-γ plays a vital role in lipid accumulation. In order to know whether miR-130b reduced lipid accumulation is related to changes of PPAR-γ signaling pathway, we further conducted the expression of PPAR-γ and its downstream target genes in differentiated PIMPA, and the results showed that miR-130b mimic decreased the accumulation of lipid droplets and expression of PPAR-γ, FTO, GR, FAS, and SCD-1, suggesting that miR-130b might reduce lipid accumulation of PIMPA by repressing the expression of PPAR-γ and its downstream target genes. This was further validated by ORO staining and RT-qPCR experiments following the overexpression or inhibition of miR-130b in PIMPA. These results are consistent with our previous study, which demonstrated that miR-130b overexpression inhibited lipid accumulation and differentiation in primary cultured preadipocytes derived from subcutaneous adipose tissue. miR-130b overexpression inhibited lipid accumulation and differentiation of primary cultured subcutaneous adipose tissue-derived preadipocytes [20]. It is known that PPAR-γ, FTO, GR, FAS, and SCD-1 expression were positively associated with the extent of adipocyte differentiation and have been identified as essential genes for lipogenesis. For example, PPAR-γ is a marker gene of adipogenic differentiation, acting as a molecular switch in the process of adipocyte differentiation [28]. Numerous studies have demonstrated that PPAR-γ overexpression significantly induced lipid transport and TG synthesis in adipocytes, and ultimately resulted in the hypertrophy of adipose tissue [28]. In the study, miR-130b mimic decreased the expression of PPAR-γ as well as its downstream target genes related to adipogenesis and lipogenesis, including FAS and GR. It was demonstrated that PPAR-γ was able to accelerate the activation of adipocyte-specific genes and activate the expression of lipid-metabolizing enzymes, such as aP2, LPL, and FAS [29]. Furthermore, the expression of SREBP-1c, also known as adipocyte differentiation-dependent factor 1 (ADD1), was also inhibited in miR-130b mimic cells. Previous studies have shown that as a transcription factor, SREBP-1c was able to promote PPAR-γ expression and the production of endogenous PPAR-γ ligands [30]. Furthermore, SREBP-1c can also directly increase the expression of lipogenic genes, such as ACC, FAS, and SCD-1 [31]. Therefore, we can sum up that decreased PPAR-γ and SREBP-1c resulted in repressed expression of FAS and GR. Consequently, our current data demonstrated that miR-130b significantly suppressed adipocyte differentiation and lipid accumulation in PIMPA through inhibiting adipogenesis-related genes in mature adipocytes.

It has been well reported that lipid droplet-associated proteins, locating on the surface of lipid droplets, are correlated with lipid accumulation and participate in regulating lipid metabolism, leading to trigger activation or inhibition signaling of lipolysis [32]. Lipolysis and perilipin-1 are correlated with lipid accumulation. Perilipin-1, the predominant lipid droplet-associated protein, serves as a key regulatory gatekeeper in controlling intracellular lipolysis in adipocytes under basal conditions [33]. Perilipin-1 interacts with lipid droplets, forming a protective barrier that restricts lipase access, thereby reducing lipase-mediated TG hydrolysis and promoting TG storage [34]. In the study, both PPAR-γ and Perilipin-1 were significantly inhibited in miR-130b mimic cells compared with mNC group. Numerous studies have demonstrated that a PPAR-responsive element has been identified within the Perilipin-1 gene. Therefore, PPAR-γ regulates Perilipin-1 protein expression by binding to the Perilipin-1 gene promoter, thereby increasing Perilipin-1 levels. Further studies showed that two different PPAR-γ activators (darglitazone and rosiglitazone) could induce the expression of Perilipin-1 in adipocytes [35]. Therefore, our results suggested that miR-130b mimic resulted in inhibition of PPAR-γ, which thus led to decreased Perilipin-1 expression and activated lipolysis.

Numerous studies have demonstrated that FTO is a key regulator of adipogenesis. APrevious research showed that FTO promoted adipogenesis through inhibiting Wnt/β-catenin signaling pathway in PIMPA [36], which is consistent with our result that miR-130b significantly inhibited FTO expression and thus inhibited adipogenic differentiation of PIMPA. In addition, FTO overexpression in differentiating PIMPA also significantly increased the mRNA expression of PPAR-γ and FAS [37]. Our findings provided the first functional evidence demonstrating the role of miR-130b in PIMPA differentiation by regulating FTO expression. Moreover, a genome-wide study identified the FTO gene on porcine chromosome 6, a region potentially harboring genes associated with meat quality traits, including IMF content [38]. Moreover, the porcine FTO (pFTO) gene has been extensively reported to be linked with fat-related traits, such as IMF deposition, IMF content, and the total lipid percentage in muscle [39]. These above data suggest a novel role of miR-130b in regulating both PIMPA differentiation and proliferation, with PPAR-γ identified as a potential target of miR-130b in these processes. Furthermore, due to the physiological and anatomical similarities between pigs and humans, as well as the resemblance in adipocyte lipogenic patterns, miR-130b may also play a crucial role in human obesity treatment. Therefore, further investigations into the function of miR-130b in human preadipocytes are warranted.

5. Conclusions

In summary, miR-130b overexpression accelerated the proliferation of PIMPA by obviously promoting cell cycle progression, while inhibiting adipogenic differentiation by dramatically reducing expression of PPAR-γ and its downstream target genes. This study contributes to understanding the roles of miR-130b in the proliferation and differentiation of PIMPA, and it also provides a theoretical platform for improving fat content by manipulating the expression of miR-130b in pigs.

Supplementary Materials: The following supporting information can be downloaded at: https://www.mdpi.com/article/10.3390/vetsci12040375/s1, Figures S1–S3: The original image of the Western blot.

Author Contributions: Y.Y., Y.C. and L.W. performed experiments and wrote the manuscript. R.Z. and Y.L. helped to analyze data. M.D. and S.P. designed experiments and edited the manuscript. All authors have read and agreed to the published version of the manuscript.

Funding: The study was supported by the National Natural Science Foundation of China (No. 32072809, 31501923), the Natural Science Foundation of Jiangsu Province (BK20211119), the National Foreign Experts Project (High-end Foreign Experts Introduction Programs of Ministry of Science and Technology, G2023014067L), the Open Fund of Meat Processing Key Laboratory of Sichuan Province (No. 22-R-17), and the Priority Academic Program Development of Jiangsu Higher Education Institutions (PAPD).

Institutional Review Board Statement: Ethics approval (approval no. SYXK-SU-2007-0005) was obtained from the Institute of Animal Care and Use Committee of Yangzhou University (Yangzhou, China).

Informed Consent Statement: Not applicable.

Data Availability Statement: The datasets analyzed during the current study are available in this paper and Supplementary Materials.

Acknowledgments: We thank Min Du of Washington State University for the critical reading of the manuscript.

Conflicts of Interest: The authors declare no conflicts of interest.

Abbreviations

ACC	acetyl-CoA carboxylase
DEPC	diethylpyrocarbonate
FAS	fatty acid synthesis
FTO	fat mass and obesity-associated protein
GR	glucocorticoid receptor
IMF	intramuscular fat
miRNAs	microRNAs
ORO	oil red o
PCNA	proliferating cell nuclear antigen
PIMPA	porcine primary intramuscular preadipocytes
PPAR-α	peroxisome proliferator activated receptor-α
PPAR-γ	peroxisome proliferator-activated receptor-gamma
PPIA	peptidylprolyl isomerase A
RT-qPCR	reverse transcription-quantitative polymerase chain reaction
SCD-1	stearyl-CoA-desaturase-1
SREBP-1	sterol regulatory element binding protein-1
TNF	tumor necrosis factor
11β-HSD1	11 beta-hydroxysteroid dehydrogenase type 1

References

1. Wood, J.D.; Enser, M.; Fisher, A.V.; Nute, G.R.; Sheard, P.R.; Richardson, R.I.; Hughes, S.I.; Whittington, F.M. Fat deposition, fatty acid composition and meat quality: A review. *Meat Sci.* **2008**, *78*, 343–358. [CrossRef] [PubMed]
2. Yu, T.Y.; Tian, X.K.; Li, D.; He, Y.L.; Yang, P.Y.; Cheng, Y.; Zhao, X.; Sun, J.C.; Yang, G.S. Transcriptome, proteome and metabolome analysis provide insights on fat deposition and meat quality in pig. *Food Res. Int.* **2023**, *166*, 112550. [CrossRef]
3. Qimuge, N.; He, Z.Z.; Qin, J.; Sun, Y.M.; Wang, X.; Yu, T.Y.; Dong, W.Z.; Yang, G.S.; Pang, W.J. Overexpression of DNMT3A promotes proliferation and inhibits differentiation of porcine intramuscular preadipocytes by methylating p21 and PPARg promoters. *Gene* **2019**, *696*, 54–62. [CrossRef]
4. Yi, X.D.; He, Z.Z.; Tian, T.T.; Kou, Z.Y.; Pang, W.J. LncIMF2 promotes adipogenesis in porcine intramuscular preadipocyte through sponging MiR-217. *Anim. Biotechnol.* **2023**, *34*, 268–279. [CrossRef] [PubMed]
5. Li, M.X.; Sun, X.M.; Cai, H.F.; Sun, Y.J.; Plath, M.; Li, C.; Lan, X.; Lei, C.J.; Lan, X.Y.; Lei, C.Z.; et al. Long non-coding RNA ADNCR suppresses adipogenic differentiation by targeting miR-204. *Biochim. Biophys. Acta* **2016**, *1859*, 871–882. [CrossRef]
6. Aryal, B.; Singh, A.K.; Rotllan, N.; Price, N.; Fernández-Hernando, C. MicroRNAs and lipid metabolism. *Curr. Opin. Lipidol.* **2017**, *28*, 273–280. [CrossRef] [PubMed]
7. Bartel, D.P. MicroRNAs: Genomics, biogenesis, mechanism, and function. *Cell* **2004**, *116*, 281–297. [CrossRef]
8. Rottiers, V.; Näär, A.M. MicroRNAs in metabolism and metabolic disorders. *Nat. Rev. Mol. Cell Biol.* **2012**, *13*, 239–250. [CrossRef]
9. Rodríguez-Barrueco, R.; Latorre, J.; Devis-Jáuregui, L.; Lluch, A.; Bonifaci, N.; Llobet, F.J.; Olivan, M.; Coll-Iglesias, L.; Gassner, K.; Davis, M.L.; et al. A microRNA Cluster Controls Fat Cell Differentiation and Adipose Tissue Expansion By Regulating SNCG. *Adv. Sci.* **2022**, *9*, e2104759. [CrossRef]
10. Liu, Y.K.; Wei, Y.L.; Dou, Y.Q.; Li, C.L.; Song, C.L.; Zhang, Z.; Qi, K.L.; Li, X.J.; Qiao, R.M.; Wang, K.J.; et al. Effect of miR-149-5p on intramuscular fat deposition in pigs based on metabolomics and transcriptomics. *BMC Genom.* **2023**, *24*, 293. [CrossRef]
11. Liu, H.C.; Wei, W.; Lin, W.M.; Yu, W.S.; Luo, W.; Niu, Y.F.; Zhang, L.F.; Chen, J. miR-32-5p Regulates Lipid Accumulation in Intramuscular Fat of Erhualian Pigs by Suppressing KLF3. *Lipids* **2021**, *56*, 279–287. [CrossRef] [PubMed]
12. Jiang, Y.; Liu, J.L.; Liu, H.T.; Zhang, W.; Li, X.J.; Liu, L.Q.; Zhou, M.; Wang, J.R.; Su, S.G.; Ding, X.D.; et al. miR-381-3p Inhibits Intramuscular Fat Deposition through Targeting FABP3 by ceRNA Regulatory Network. *Biology* **2022**, *11*, 1497. [CrossRef]
13. Wang, Y.C.; Li, Y.; Wang, X.Y.; Zhang, D.; Zhang, H.H.; Wu, Q.; He, Y.Q.; Wang, J.Y.; Zhang, L.; Xia, H.F.; et al. Circulating miR-130b mediates metabolic crosstalk between fat and muscle in overweight/obesity. *Diabetologia* **2013**, *56*, 2275–2285. [CrossRef] [PubMed]
14. Tekcan, E.; Kara, N.; Aydın, H.M.; Abur, Ü.; Abbaszadeh, M. Evaluation of the promoter methylation status of hypoxia factor 3A and interleukin-6 genes and expression levels of mir-130b and mir-146b in childhood obesity. *Rev. Assoc. Med. Bras.* **2022**, *68*, 1276–1281. [CrossRef] [PubMed]
15. Lee, E.K.; Lee, M.J.; Abdelmohsen, K.; Kim, W.; Kim, M.M.; Srikantan, S.; Martindale, J.L.; Hutchison, E.R.; Kim, H.H.; Marasa, B.S.; et al. miR-130 suppresses adipogenesis by inhibiting peroxisome proliferator-activated receptor gamma expression. *Mol. Cell Biol.* **2011**, *31*, 626–638. [CrossRef]
16. Alonso-Villa, E.; Mangas, A.; Bonet, F.; Campuzano, Ó.; Quezada-Feijoo, M.; Ramos, M.; García-Padilla, C.; Franco, D.; Toro, R. The Protective Role of miR-130b-3p Against Palmitate-Induced Lipotoxicity in Cardiomyocytes Through PPARγ Pathway. *Int. J. Mol. Sci.* **2024**, *25*, 12161. [CrossRef]
17. Li, Y.Y.; He, C.S.; Ran, L.; Wang, Y.; Xiong, Y.; Wang, Y.L.; Zhu, J.J.; Lin, Y.Q. miR-130b duplex (miR-130b-3p/miR-130b-5p) negatively regulates goat intramuscular preadipocyte lipid droplets accumulation by inhibiting Krüppel-like factor 3 expression. *J. Anim. Sci.* **2023**, *101*, skad184. [CrossRef]
18. Ma, X.Y.; Wei, D.W.; Cheng, G.; Li, S.J.; Wang, L.; Wang, Y.N.; Wang, X.Y.; Zhang, S.; Wang, H.; Zan, L. Bta-miR-130a/b regulates preadipocyte differentiation by targeting PPARG and CYP2U1 in beef cattle. *Mol. Cell. Probes* **2018**, *42*, 10–17. [CrossRef]
19. Pan, S.F.; Zheng, Y.T.; Zhao, R.Q.; Yang, X.J. MicroRNA-130b and microRNA-374b mediate the effect of maternal dietary protein on offspring lipid metabolism in Meishan pigs. *Br. J. Nutr.* **2013**, *109*, 1731–1738. [CrossRef]
20. Pan, S.F.; Yang, X.J.; Jia, Y.M.; Li, R.S.; Zhao, R.Q. Microvesicle-shuttled miR-130b reduces fat deposition in recipient primary cultured porcine adipocytes by inhibiting PPAR-g expression. *J. Cell Physiol.* **2014**, *229*, 631–639. [CrossRef]
21. Pan, S.F.; Yang, X.J.; Jia, Y.; Li, Y.M.; Chen, R.R.; Wang, M.; Cai, D.M.; Zhao, R.Q. Intravenous injection of microvesicle-delivery miR-130b alleviates high-fat diet-induced obesity in C57BL/6 mice through translational repression of PPAR-γ. *J. Biomed. Sci.* **2015**, *22*, 86. [CrossRef] [PubMed]
22. Wang, L.X.; Xie, Y.H.; Chen, W.; Zhang, Y.; Zeng, Y.Q. miR-34a Regulates Lipid Droplet Deposition in 3T3-L1 and C2C12 Cells by Targeting LEF1. *Cells* **2022**, *12*, 167. [CrossRef]
23. Wang, Y.C.; Yao, X.; Ma, M.; Zhang, H.; Wang, H.; Zhao, L.; Liu, S.N.; Sun, C.; Li, P.; Wu, Y.T.; et al. miR-130b inhibits proliferation and promotes differentiation in myocytes via targeting Sp1. *J. Mol. Cell Biol.* **2021**, *13*, 422–432. [CrossRef] [PubMed]

24. Liu, L.L.; Liu, H.H.; Chen, M.T.; Ren, S.S.; Cheng, P.P.; Zhang, H. miR-301b~miR-130b-PPARγ axis underlies the adipogenic capacity of mesenchymal stem cells with different tissue origins. *Sci. Rep.* **2017**, *7*, 1160. [CrossRef]
25. Chen, W.K.; Zhang, H.J.; Zou, M.X.; Wang, C.; Yan, Y.G.; Zhan, X.L.; Li, X.L.; Wang, W.J. LncRNA HOTAIR influences cell proliferation via miR-130b/PTEN/AKT axis in IDD. *Cell Cycle* **2022**, *21*, 323–339. [CrossRef]
26. Peng, Y.; Chen, F.F.; Ge, J.; Zhu, J.Y.; Shi, X.E.; Li, X.; Yu, T.Y.; Chu, G.Y.; Yang, G.S. miR-429 Inhibits Differentiation and Promotes Proliferation in Porcine Preadipocytes. *Int. J. Mol. Sci.* **2016**, *17*, 2047. [CrossRef] [PubMed]
27. Horsfall, A.J.; Abell, A.D.; Bruning, J.B. Targeting PCNA with Peptide Mimetics for Therapeutic Purposes. *Chembiochem* **2020**, *21*, 442–450. [CrossRef]
28. Samulin, J.; Berget, I.; Grindflek, E.; Lien, S.; Sundvold, H. Changes in lipid metabolism associated gene transcripts during porcine adipogenesis. *Comp. Biochem. Physiol. B Biochem. Mol. Biol.* **2009**, *153*, 8–17. [CrossRef]
29. Moseti, D.; Regassa, A.; Kim, W.K. Molecular Regulation of Adipogenesis and Potential Anti-Adipogenic Bioactive Molecules. *Int. J. Mol. Sci.* **2016**, *17*, 124. [CrossRef]
30. Payne, V.A.; Au, W.S.; Lowe, C.E.; Rahman, S.M.; Friedman, J.E.; O'Rahilly, S.; Rochford, J.J. C/EBP transcription factors regulate SREBP1c gene expression during adipogenesis. *Biochem. J.* **2009**, *425*, 215–223. [CrossRef]
31. Czech, M.P.; Tencerova, M.; Pedersen, D.J.; Aouadi, M. Insulin signalling mechanisms for triacylglycerol storage. *Diabetologia* **2013**, *56*, 949–964. [CrossRef]
32. Granneman, J.G.; Moore, H.P.; Granneman, R.L.; Greenberg, A.S.; Obin, M.S.; Zhu, Z.X. Analysis of lipolytic protein trafficking and interactions in adipocytes. *J. Biol. Chem.* **2007**, *282*, 5726–5735. [CrossRef] [PubMed]
33. Sztalryd, C.; Brasaemle, D.L. The perilipin family of lipid droplet proteins: Gatekeepers of intracellular lipolysis. *Biochim. Biophys. Acta Mol. Cell Biol. Lipids* **2017**, *1862 Pt B*, 1221–1232. [CrossRef]
34. Brasaemle, D.L.; Rubin, B.; Harten, I.A.; Gruia-Gray, J.; Kimmel, A.R.; Londos, C. Perilipin A increases triacylglycerol storage by decreasing the rate of triacylglycerol hydrolysis. *J. Biol. Chem.* **2000**, *275*, 38486–38493. [CrossRef]
35. Dalen, K.T.; Schoonjans, K.; Ulven, S.M.; Weedon-Fekjaer, M.S.; Bentzen, T.G.; Koutnikova, H.; Auwerx, J.; Nebb, H.I. Adipose tissue expression of the lipid droplet-associating proteins S3-12 and perilipin is controlled by peroxisome proliferator-activated receptor-gamma. *Diabetes* **2004**, *53*, 1243–1252. [CrossRef] [PubMed]
36. Chen, X.L.; Luo, Y.L.; Jia, G.; Liu, G.M.; Zhao, H.; Huang, Z.Q. FTO Promotes Adipogenesis through Inhibition of the Wnt/β-catenin Signaling Pathway in Porcine Intramuscular Preadipocytes. *Anim. Biotechnol.* **2017**, *28*, 268–274. [CrossRef] [PubMed]
37. Chen, X.L.; Zhou, B.; Luo, Y.L.; Huang, Z.Q.; Jia, G.; Liu, G.M.; Zhao, H. Tissue Distribution of Porcine FTO and Its Effect on Porcine Intramuscular Preadipocytes Proliferation and Differentiation. *PLoS ONE* **2016**, *11*, e0151056. [CrossRef]
38. Ovilo, C.; Pérez-Enciso, M.; Barragán, C.; Clop, A.; Rodríquez, C.; Oliver, M.A.; Toro, M.A.; Noruera, J.L. A QTL for intramuscular fat and backfat thickness is located on porcine chromosome 6. *Mamm. Genome* **2000**, *11*, 344–346. [CrossRef]
39. Fan, B.; Du, Z.Q.; Rothschild, M.F. The fat mass and obesity-associated (FTO) gene is associated with intramuscular fat content and growth rate in the pig. *Anim. Biotechnol.* **2009**, *20*, 58–70. [CrossRef]

veterinary sciences

MDPI

Article

Angelicae Pubescentis Radix Remitted Intestine Damage in Mice Induced by *Escherichia coli* via Mediating Antioxidant Defense, Inflammatory Mediators, and Restoring Gut Microbiota

Kehong Deng [1], Chang Xu [2], Qing He [2], Muhammad Safdar [3], Mudassar Nazar [4], Xiaocong Li [1,*] and Kun Li [2,*]

1 Department of Pharmacy, Hubei Three Gorges Vocational and Technical College, Yichang 443000, China; yc-dkh@163.com

2 College of Veterinary Medicine, Nanjing Agricultural University, Nanjing 210095, China; 9201710122@stu.njau.edu.cn (C.X.); 9201710121@stu.njau.edu.cn (Q.H.)

3 Department of Breeding and Genetics, Cholistan University of Veterinary and Animal Sciences, Bahawalpur 63100, Pakistan; msafdar@cuvas.edu.pk

4 Department of Clinical Medicine and Surgery, Faculty of Veterinary Sciences, Constituent College Burewala, University of Agriculture Faisalabad, Burewala 61010, Pakistan; mudassar786.rana@gmail.com

* Correspondence: lxcctgu@126.com (X.L.); lk3005@njau.edu.cn (K.L.)

Simple Summary: This study investigated the therapeutic potential of *Angelicae Pubescentis* Radix (APR) in mice infected with *Escherichia coli* (*E. coli*) and found that APR treatment significantly reduced both bacterial load and intestinal injury. Serum analysis indicated that APR treatment also alleviated the inflammation and oxidative stress induced by *E. coli* infection. Intestinal microbiota sequencing further showed that APR treatment increased the abundance of intestinal probiotics such as *Ligilactobacillus*, *Paludicola*, and *Blautia_A_1417806* while also enhancing the enrichment of functional pathways associated with antioxidant defense.

Abstract: This study aims to explore the therapeutic potential of *Angelicae Pubescentis* Radix (APR), a traditional Chinese medicine that is widely known for its anti-inflammatory, anti-oxidative, and anti-microbial properties, using a mouse model. In this study, 30 mice were selected and divided into three groups: control group (CD), infection group (ED), and treatment group (TD). Mice in the TD were gavaged with APR oil (0.15 mL/kg/day) for 20 days, while mice in the CD and ED received an equal volume of normal saline. On the 21st day, mice in the ED and TD were infected with multi-drug-resistant *E. coli* (1×10^7 CFU/mL) derived from diarrheal yak. Twenty-four hours later, all mice were euthanized, and blood, organs, and intestinal samples were collected for analysis. The results of intestinal sections and intestinal bacterial load revealed that APR treatment significantly reduced ($p < 0.05$) both bacterial load and intestinal injury. Serum analysis indicated that APR treatment also alleviated the inflammation and oxidative stress induced by *E. coli* infection. Intestinal microbiota sequencing further showed that APR treatment increased the abundance of intestinal probiotics such as *Ligilactobacillus*, *Paludicola*, and *Blautia_A_1417806* while also enhancing the enrichment of functional pathways associated with antioxidant defense. In conclusion, APR treatment effectively alleviates diseases caused by *E. coli* infection, promotes the growth of beneficial intestinal bacteria, and improves the antioxidant capacity in animals. Additionally, these findings confirm APR's role in addressing immediate effects rather than chronic adaptations. Future studies should investigate the prolonged effects of APR treatment beyond the acute phase.

 https://doi.org/10.3390/vetsci12040354

Keywords: *Angelicae Pubescentis* Radix; bacterial diarrhea; intestinal micro-ecosystem; antioxidant defense; gut microbiota

1. Introduction

Bacterial diarrhea, primarily caused by enteropathogenic bacteria such as *E. coli*, *Salmonella* [1], and *Shigella* [1], is a major public health issue worldwide. These pathogens typically enter the human body via contaminated food or water [2], disturbing the intestinal microecological balance [3], triggering an inflammatory response [4], and resulting in symptoms such as diarrhea, abdominal pain, fever, and others. Traditional treatments largely rely on antibiotics to rapidly control the infection by killing or inhibiting the growth of bacteria [5]. However, the overuse and misuse of antibiotics have led to the emergence of drug-resistant "superbugs" [6] and can disrupt the natural intestinal microbiota, leading to secondary infections or chronic intestinal diseases [7], such as *Clostridium difficile*-associated diarrhea (CDAD), which complicates treatment and increases risks [8]. In this regard, traditional Chinese medicine (TCM) has gained attention as a promising alternative due to its natural, safe, and relatively inexpensive characteristics [9–12]. Many studies have shown that compound Chinese herbal extracts can significantly improve immune functions [13], maintain the balance of intestinal flora [14], and enhance antioxidant defense, thereby reducing disease incidence [15,16]. However, the mechanisms of action of TCM and antibiotics differ, and with increasing bacterial resistance, blindly replacing antibiotics with Chinese medicine may not achieve the desired effects.

Angelicae Pubescentis Radix (APR), a member of the Umbelliferae family, is rich in volatile oils, coumarins, polysaccharides, and other bioactive compounds [17]. These compounds demonstrate significant immune-regulating, anti-inflammatory, and antioxidant effects [18]. Specifically, coumarin compounds, such as imperatorin, possess remarkable anti-inflammatory effects [19] by inhibiting the nuclear factor kappa B (NF-κB) signaling pathway, thereby alleviating intestinal inflammatory mediators and promoting intestinal mucosal repair [20]. Additionally, polysaccharides in APR can activate immune cells and enhance the body's ability to resist bacterial invasion [17,20]. Similarly, coumarins like columbianadin (CBN) and osthole also inhibit NF-κB activation, reducing the expression of pro-inflammatory cytokines like tumor necrosis factor-α (TNF-α), interleukin-6 (IL-6), and interleukin-1β (IL-1β) [17,20]. The suppression of Nucleotide-binding oligomerization domain-containing protein 1 (NOD1)/NF-κB signaling to block LPS-induced cytokine production in immune cells has been proved [20]. CBN inhibits Janus Kinase 2/Signal Transducer and Activator of Transcription 3 (JAK2/STAT3) phosphorylation, interrupting cytokine-derived inflammatory cascades [17]. Additionally, coumarins modulate Signal Transducer and Activator of Transcription (MAPK) pathways (e.g., p. 38), reducing oxidative stress and tissue damage [17,20]. Hence, coumarins like osthole and CBN enhance intestinal barrier function by reducing oxidative stress and inflammatory mediators' release [20]. Additionally, Bergapten and umbelliferone exhibit analgesic and anti-inflammatory effects, potentially stabilizing gut epithelial cells during immune challenges [20].

APR polysaccharides bind to Toll-like receptors (TLRs) and pattern recognition receptors (PRRs), activating macrophages and dendritic cells [21,22]. This triggers the production of cytokines, e.g., Granulocyte-Macrophage Colony-Stimulating Factor (GM-CSF), that promote hematopoiesis and immune cell differentiation, increased phagocytic activity, and antigen presentation, boosting adaptive immune response [21,22].

Angelica sinensis polysaccharides (ASP) activate the Wingless-related integration site/β-catenin pathway (Wnt/β-catenin), reducing oxidative injury in gut epithelial cells [23]. Similarly, ASP-Hb-AP (a polysaccharide variant) upregulates antioxidant enzyme expression, e.g., superoxide dismutase (SOD) and catalase (CAT), in intestinal cells, alleviating oxidative damage. Polysaccharides bind TLRs on epithelial cells, inducing anti-inflammatory cytokine release [21] and also stimulating mucin (MUC2) [23] production via gut microbiota, strengthening the mucosal layer [23].

The gut, as the largest digestive organ in the body, plays a crucial role in overall health. The intestinal microbiota is closely associated with various physiological processes, including metabolism, immunity, and nutrient absorption [24]. The intestinal flora, comprising the microbiota, its metabolites, and the environment in which they exist, is fundamental to maintaining body health [25]. Once disrupted, it can lead to a range of diseases [26]. At present, over 1000 species of intestinal microbial members have been identified in the human gut [27]. Under normal conditions, the intestinal microbiota remains in a stable state, but an abnormal physiological condition can trigger a variety of health issues [28]. Additionally, this intestinal microbiota provides essential nutrients and supports the normal functioning of the intestinal barrier, playing a key role in defending against pathogens [29–31]. The interaction between symbiotic bacteria, metabolites, intestinal epithelial cells, and the immune system helps to maintain the balance of the intestinal environment and serves as a natural barrier against pathogen invasion [32]. Therefore, in this study, *E. coli* was employed to induce bacterial diarrhea in model animals, and the therapeutic effects of APR were investigated from the perspectives of intestinal microecological changes, antioxidant stress, and inflammation, as well as the potential of replacing antibiotics with traditional Chinese medicine.

2. Materials and Methods

2.1. Oil Extraction

Angelicae Pubescentis Radix (1.0 kg) was purchased from a Chinese herb farm in Yichang, China. The main roots were carefully collected and shade-dried. Then, the velamina of the roots were removed, and the roots were soaked in hexyl hydride (5 L) for 10 h. The resulting extracts were concentrated at 45 °C and dried via vacuum to achieve oil with a concentration of (2 g/mL) as depicted in a previous study [33]. The key active ingredients in the APR oil extract include coumarins such as imperatorin, CBN, osthole, Bergapten, and umbelliferone, along with volatile oils and APR polysaccharides [17,20].

2.2. Animal Experiment

Thirty ICR mice (15 male and 15 female), four weeks old and weighing 24.16 ± 0.75 g, were obtained from the Experimental Animal Center at Yangzhou University. Mice were acclimatized for 3 days under controlled conditions (26 °C, 50–60% humidity, 12 h light/dark cycle). Feed and water were provided ad libitum, with environmental parameters monitored daily. Four-week-old ICR mice were selected to standardize immune and metabolic responses, as juvenile rodents are widely used in enteric infection models to ensure uniformity in pathogen susceptibility and therapeutic outcomes [3,15]. This age minimizes developmental variability while allowing acute observation of *E. coli*-induced intestinal damage and APR intervention. The feed used in the given study was purchased from Yizheng Anliu Biotechnology Co. Ltd. (Yizheng, China; standards: GB14924.1-2010, GB14924.2-2010, GB14924.3-2010) and mainly consisted of corn, wheat, soybean meal, bran, fish meal, various amino acids, and multivitamins. The feed provided was composed of ≤10% moisture, ≥19% protein, ≥4% crude protein, ≤5% crude fiber, ≤8% crude ash,

1.0–1.8% calcium, 0.6–1.2% phosphorus, with calcium to phosphorus ratio (Ca: P); 1.2:1–1.7:1, ≥0.92% lysine, and methionine + cystine ≥ 0.53%. The remaining ~48% was carbohydrate (corn) content. The feed was stored in dry, ventilated conditions at a temperature of 20–25 °C, with a relative humidity of 50–60%, in a sealed (air-tight) pack to avoid environmental exposure. The shelf life of feed was 3 months (2 months in the rainy season).

After three days of acclimatization, the mice were randomly divided into three groups: the control group (CD), the infection group (ED), and the treatment group (TD). ICR mice in the TD were gavaged with APR oil (0.15 mL/kg/day) for 20 days, while those in the CD and ED received an equal volume of normal saline. On the 21st day, the ED and TD were inoculated with multi-drug-resistant *E. coli* (clinical isolate PP859186, 1×10^7 CFU/mL) derived from diarrheal yak [3]. The strain's pathogenicity is attributed to its clinical origin and resistance profile, though serotype-specific markers (e.g., K88/K99) were not characterized. Meanwhile, mice in the CD were administered an equivalent amount of Luria–Bertani nutrient medium (Solarbio Science & Technology Co., Ltd., Beijing, China). Twenty-four hours post-inoculation, all mice were euthanized to collect blood, organs, and intestinal samples. The key parameters, such as body weight, diarrhea rate and mortality, organ index, organ bacterial loads, serum oxidation resistance, inflammation levels, and microbiota composition, were analyzed (Figure 1).

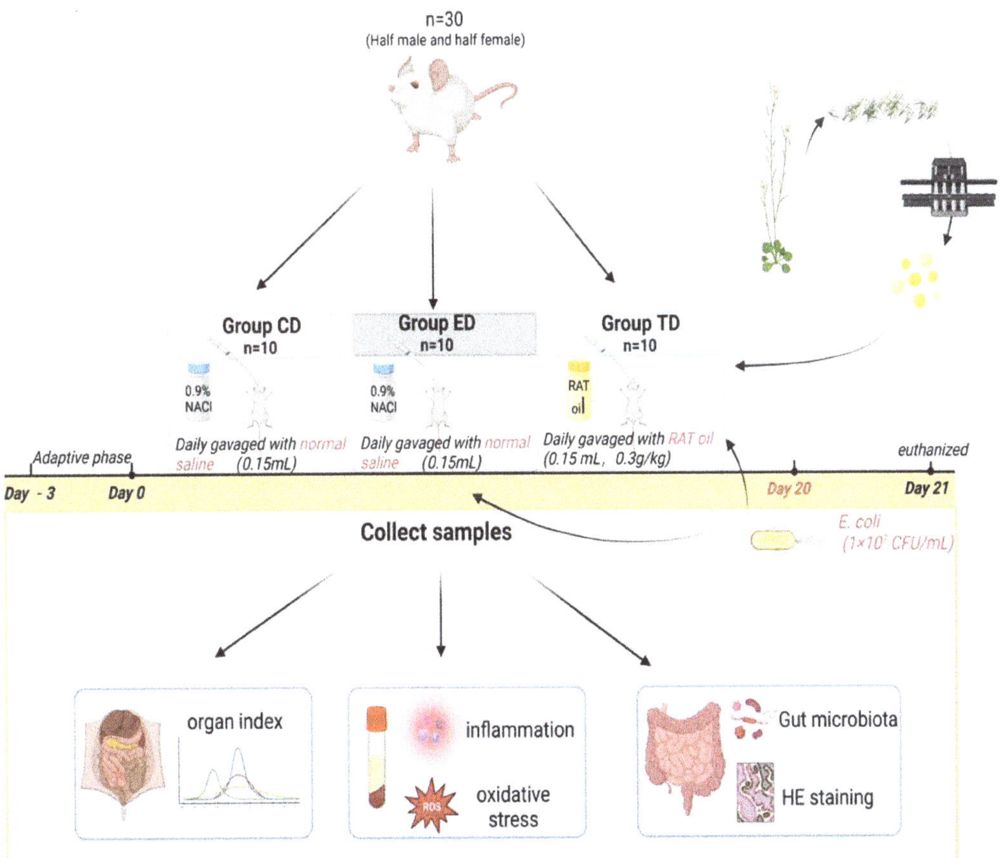

Figure 1. Experimental design of the mice model induced by *E. coli*.

2.3. Pathological Analysis

Jejunum tissues (0.5–1 cm/animal) from mice in CD, TD, and ED were collected and fixed in a 4% formaldehyde solution for 48 h. The fixed samples were then sent to Wuhan Pinuofei Biological Technology (China) for hematoxylin and eosin (H&E) staining. Pathological analysis of villus integrity and damage was conducted using an Olympus CX33 microscope (Olympus, Hachioji, Tokyo, Japan), as described in a previous study [15].

2.4. Examination of Antioxidant and Inflammatory Levels in Mice

Blood samples were collected and centrifuged at 3500× g for 20 min to separate the serum. The obtained serum samples were analyzed for antioxidant and inflammatory markers using commercial assay kits, including malondialdehyde (MDA), SOD, total antioxidant capacity (T-AOC), glutathione peroxidase (GSH-Px), IL-6, IL-10, TNF-α and IL-1β, sourced from Solarbio Science & Technology Co, Ltd. (Beijing, China) and Jiancheng Bioengineering Research Institute (Nanjing, China).

2.5. Microbiota Sequencing and Analysis

The total microbial DNA was extracted from rectal samples of mice in the CD (n = 5), TD (n = 5), and ED (n = 5) groups using the FastDNA Spin Kit for Feces (Beijing Think-Far Technology Co, Ltd., Beijing, China). The quality and quantity of DNA were checked via Ultramicrospectrophotometer YSNano-100 (Yeasen Biotechnology, Shanghai, China) and agarose gel electrophoresis (1.5%), following the protocol described in a previous study [34,35]. The V3-V4 region of the 16S rRNA gene was amplified using 338F-806R primer pairs (F: 5′-ACTCCTACGGGAGGCAGCAG-3′, R:5′-GGACTACHVGGGTWTCTAAT-3′) [36]. The amplified products were used to construct sequencing libraries via the Hieff NGS® OnePot II DNA Library Prep Kit for Illumina® (Yeasen, China). Microbiota sequencing was subsequently performed on the Illumina platform at Bioyi Biotechnology Co, Ltd. (Wuhan, China).

The raw sequencing data from the CD, TD, and ED underwent quality control and denoising via Divisive Amplicon Denoising Algorithm 2 (DADA2, Version: 1.26.0) [37]. The filtered data were used to produce amplicon sequence variants (ASVs) by blasting with the Greengenes database with ≥97% similarity and assigned taxonomy table via Quantitative Insights into Microbial Ecology (QIIME 2 Version: 2024.10) platform [38]. A Venn diagram was generated to visualize shared ASVs among the CD, TD, and ED [39]. The analysis of alpha and beta diversities of mice in different groups was conducted via the QIIME2 platform to explore the species diversity of samples and groups [40]. Species difference and biomarker analyses were performed via Linear discriminant effect Size (LefSe) analysis and analysis of variance (ANOVA) [41]. Finally, microbiota function prediction was performed using the Phylogenetic Investigation of Communities by Reconstruction of Unobserved States (PICRUSt2; Version: 2.6.) [42], and the abundance of metabolic pathways among different groups was analyzed by employing the Kyoto Encyclopedia of Genes and Genomes (KEGG) and MetaCyc Metabolic Pathway (MetaCyc) databases [43].

2.6. Statistical Analysis

Variances among different mice groups (CD, TD, and ED) were analyzed using ANOVA and Student's t-test via IBM SPSS (25.0). One-way ANOVA with Tukey's post hoc test was used for comparisons across all three groups (e.g., body weight changes, organ indices, bacterial loads across intestinal segments, antioxidant markers, inflammatory cytokines, alpha diversity indices, and taxonomic and functional pathway abundances), while Student's t-test was applied for pairwise group comparisons (e.g., weight loss in

ED vs. CD/TD, bacterial load reduction in TD vs. ED, recovery of antioxidant markers in TD vs. ED). Results are presented as means ± standard error of the mean (SEM), with statistical significance set at $p < 0.05$.

3. Results

3.1. Effect of APR Oil on Body Weight in E. coli-Induced Mice

Following *E. coli* induction, no significant differences ($p \geq 0.05$) in body weight were observed among the groups (Figure 2a). However, post-infection, the ED displayed a net weight loss (~3%), while the TD and CD experienced weight gain. By the experiment's conclusion, the CD exhibited the most substantial relative weight gain, averaging approximately 10%, while the TD showed a moderate increase of ~5%. In contrast, the ED demonstrated a significant ($p < 0.05$) average weight loss of ~3% (Figure 2b). These findings highlight distinct weight trends among the groups, reflecting the physiological impacts of infection and treatment. Additionally, the CD maintained a 100% survival rate throughout the experiment, while the TD had a slight reduction in survival, and the ED experienced a significant drop ($p < 0.05$) in survival rate by day 21 (Figure 2c).

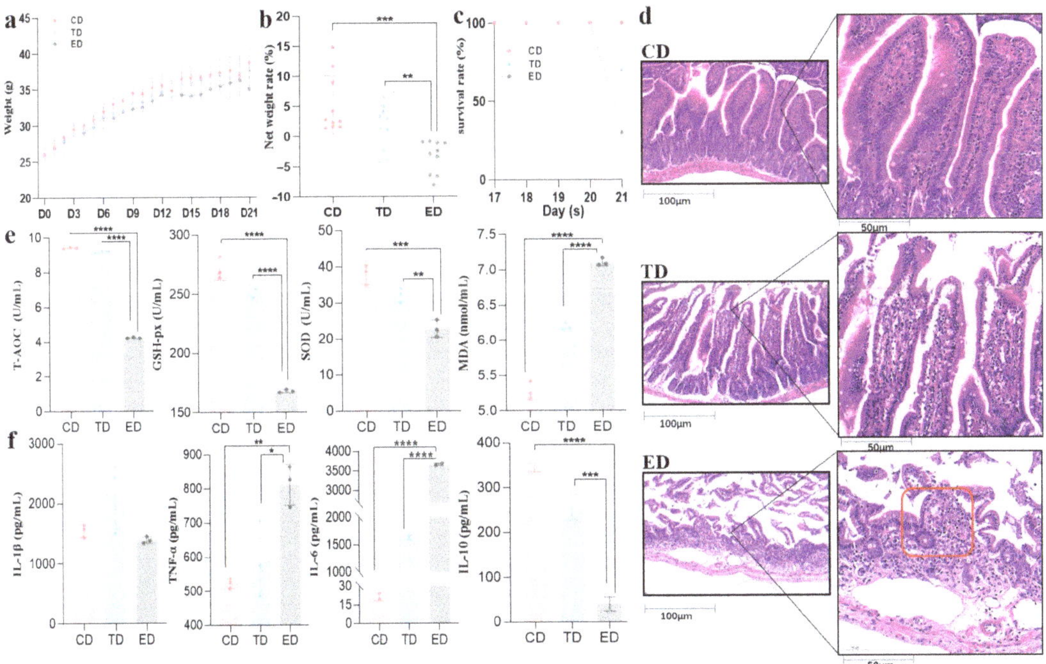

Figure 2. APR oil regulated body weight, intestinal integrity, inflammation, and antioxidant capacity in *E. coli*-induced mice. (**a**) Body weight, (**b**) Net weight change on day 21, (**c**) Survival rate, (**d**) H&E staining (red circled area in group ED shows severely damaged and atrophied villi, (**e**) Antioxidant capacity, (**f**) Inflammation levels. Scale bar = 50 μm. Statistical significance is indicated as * $p < 0.05$, ** $p < 0.01$, *** $p < 0.001$, **** $p < 0.0001$. Data are presented as mean ± SEM.

The pathological examination (Figure 2d) of jejunum tissues revealed clear differences in the intestinal tissue integrity. The CD exhibited intact intestinal mucosa, a well-organized tissue structure, and complete, neatly arranged villi. In contrast, the ED showed severe intestinal damage, including inflammatory cell infiltration, disrupted tissue structure, and villi characterized by obvious shortening, aggregation, and destruction. After treatment with APR oil, the infected mice maintained relatively complete intestinal mucosa and villi, although there were some defects in the upper mucosa and other phenomena. The integrity was markedly higher compared to the ED, and the structure was close to normal.

Serum oxidative stress markers also highlighted the protective effects of APR oil. T-AOC, GSH-px, and SOD levels, which were significantly decreased ($p < 0.05$) in the ED, showed recovery in the TD, approaching levels observed in the CD. Conversely, MDA levels, which were elevated significantly ($p < 0.05$) in the ED, were slightly alleviated in the APR oil treatment group (Figure 2e). Inflammatory cytokine analysis further supported these findings. IL-6 and TNF-α levels were significantly elevated ($p < 0.05$) in the serum of the ED but returned to normal in the TD. Additionally, IL-10, an anti-inflammatory cytokine, was substantially higher in the TD compared to the ED, which exhibited minimal levels (Figure 2f). Therefore, these results suggest that APR oil effectively mitigates weight loss, restores intestinal integrity, and alleviates inflammation and oxidative stress in *E. coli*-induced mice.

3.2. Effect of APR Oil on Organ Index and Bacterial Load in E. coli-Induced Mice

After blood collection from the eyeballs, about 0.1 g of heart, liver, spleen, lung, kidney, duodenum, jejunum, ileum, cecum, and colon were weighed and placed in a sterilized 2 mL centrifuge tube, and 1 mL of normal saline and two grinding beads were added. After grinding, the supernatant was taken and diluted 10 times its original volume. The diluted 20 μL bacterial solution was inoculated into the MacConkey agar and cultured in the incubator at 37 °C for 18 h. The colony number was recorded, and the bacterial load in each organ and intestine was calculated.

The organ indices of the spleen, liver, and lungs in the ED were significantly different ($p < 0.05$) from those in the control group, indicating notable organ changes due to *E. coli* infection. In contrast, the organ indices in the APR oil treatment group were similar to those observed in the CD, suggesting that APR oil may help mitigate these changes. The bacterial load in the intestinal tract also displayed interesting trends. In the ED, the bacterial load in the duodenum, jejunum, ileum, cecum, and colon were significantly increased ($p < 0.05$) compared to the CD and TD. Specifically, the bacterial load in the ED was approximately 100 times higher than that in the TD, indicating a substantial increase in intestinal bacteria due to *E. coli* infection (Figure 3a,b).

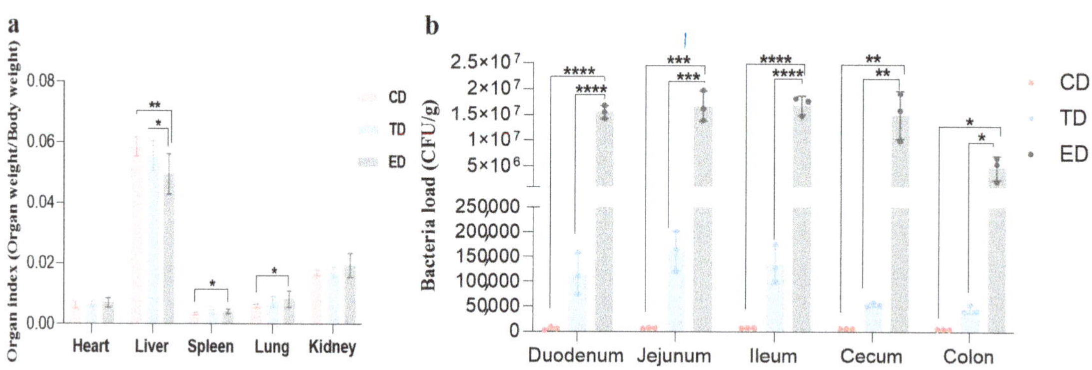

Figure 3. Organ index and bacterial loads in different intestinal segments of mice in different groups. (**a**) Organ index, (**b**) Bacterial loads. Statistical significance is indicated as * $p < 0.05$, ** $p < 0.01$, *** $p < 0.001$, **** $p < 0.0001$. Data are presented as mean ± SEM.

3.3. Effect of APR Oil on Intestinal Flora Composition in Mice Induced by E. coli

Sequencing data revealed more than 72,000, 64,000, and 62,000 unfiltered original sequences from the CD, ED, and TD, respectively. After denoising and filtering, at least 51,000, 42,000, and 41,000 high-quality sequences were obtained from these groups (Table 1).

Table 1. Analysis of sequencing data for different groups.

Samples	Input	Filtered	Denoised	Merged	Non-Chimeric	Non-Singleton
CD1	249,221	217,178	215,454	179,602	172,753	172,736
CD2	86,646	80,467	79,688	77,022	73,834	73,823
CD3	112,225	105,324	103,483	94,350	64,494	64,269
CD4	72,401	55,552	54,831	51,812	51,031	51,023
CD5	83,730	68,872	67,471	57,687	52,697	52,666
ED1	64,315	59,486	58,720	55,822	54,138	54,128
ED2	250,698	212,903	210,729	177,267	176,877	176,860
ED3	78,105	73,538	72,079	64,401	42,999	42,693
ED4	126,352	107,042	105,197	92,901	80,996	80,929
ED5	130,055	115,104	114,016	110,494	110,016	110,013
TD1	75,095	70,498	69,375	62,991	45,933	45,725
TD2	80,405	75,227	73,190	61,520	41,184	40,824
TD3	76,459	72,214	70,685	64,095	42,846	42,492
TD4	135,212	115,539	112,866	97,505	86,143	86,054
TD5	62,930	54,114	53,429	51,836	50,733	50,727

Alpha diversity analysis was employed to compare the microbial diversity within each group (Figure 4a). The Good's coverage index ranged from 99.43 to 99.96% across the three groups, indicating that the sequencing results could reflect the entire bacterial structure. The α-diversity correlation index, a measure of species richness, was highest in the TD (1139.61 $\pm$ 406.73), followed by the ED (814.89 $\pm$ 367.2), and lowest in the CD (611.41 $\pm$ 679.61).

The Shannon index, which reflects both richness and evenness, was also highest in the TD (6.34 $\pm$ 2.24), followed by the ED (5.21 $\pm$ 3.65) and the CD (4.64 $\pm$ 2.18). The Pielou_e index, which measures evenness, ranged from 0.278 to 0.678 in the CD, 0.177–0.714 in the ED, and 0.474–0.724 in the TD. The Observed_species index, representing the number of distinct species observed, ranged from 288.1 to 1202.5 in the CD, 421.7–1174.2 in the ED, and 403.2–1734 in the TD (Table 2). Additionally, both the rarefaction curve (Figure 4b) and the rank-abundance curve (Figure 4c) approached a plateau with an increasing number of sequences, suggesting that the sequencing depth was sufficient to reflect the complete microbiota composition and maximum species diversity has been captured. However, some curves (e.g., CD1 and ED1) appear to continue increasing slightly, suggesting that additional sequencing might uncover more diversity in these samples. Overall, sequencing depth seems sufficient for robust analyses, but further sequencing could be considered for samples where saturation is not fully achieved.

β-diversity analysis, including PCoA (Figure 4d) and NMDS (Figure 4e), revealed significant differences ($p < 0.05$) in microflora composition among the three groups, with dispersed sample distributions indicating group-specific microbial variations.

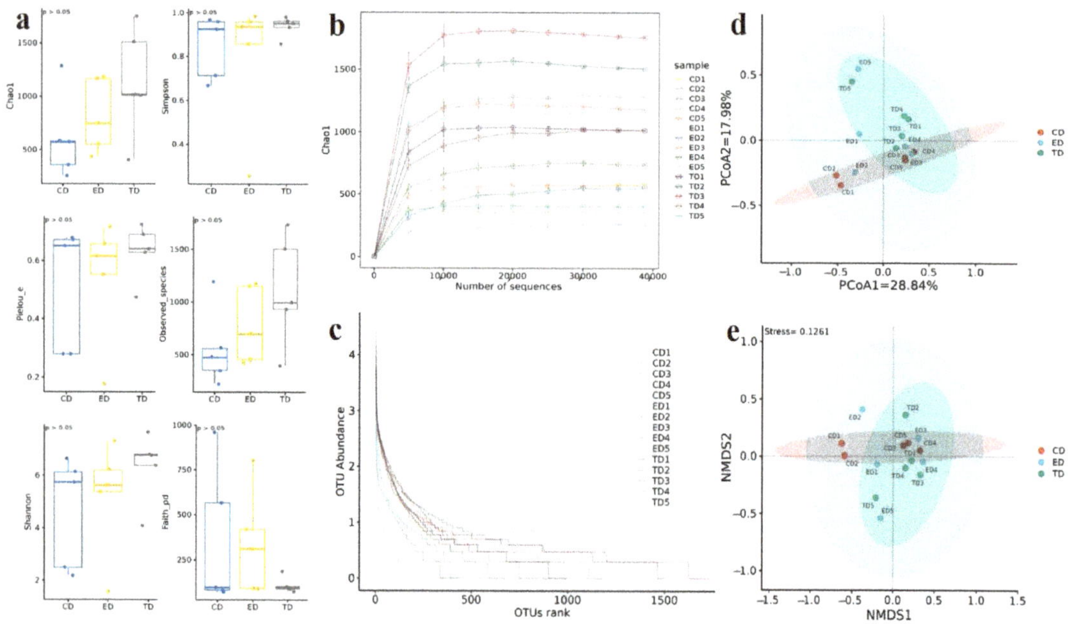

Figure 4. Alpha and beta diversity analysis. (**a**) Diversity indexes, (**b**) Rarefaction curve, (**c**) Rank abundance curve, (**d**) Principal coordinate analysis (PCoA), (**e**) Non-metric multidimensional scaling (NMDS).

Table 2. Analysis of alpha diversity indices.

Samples	Chao1	Faith_pd	Goods_Coverage	Observed_Species	Pielou_e	Shannon
CD1	582.0832203	958.4556284	0.996555103	476.3	0.27947093	2.485988474
CD2	255.5041792	567.6606101	0.998574081	228.1	0.279320731	2.188011412
CD3	1290.61498	95.08548944	0.994327265	1202.5	0.65048806	6.655672854
CD4	357.2362835	80.52636973	0.99972152	355	0.678150492	5.745070009
CD5	571.7104581	69.77193845	0.999022743	561.3	0.670765459	6.12584253
ED1	429.8264161	419.2213773	0.999316693	421.7	0.615342611	5.365819805
ED2	551.5114524	800.9651922	0.996910938	456.4	0.177518042	1.56818406
ED3	1182.085674	87.01808259	0.998344593	1174.2	0.71550874	7.296371476
ED4	746.050341	89.12529358	0.997117219	695	0.657047117	6.203063941
ED5	1164.977309	310.826434	0.999169718	1151.8	0.552819016	5.621986171
TD1	1009.620328	72.66430364	0.997883039	994.9	0.640729954	6.380648383
TD2	1510.369014	99.88844622	0.998630808	1506	0.724254771	7.645599344
TD3	1750.916658	87.09434071	0.996910938	1734	0.628727328	6.765034476
TD4	1020.439937	93.64806703	0.995603631	934.3	0.689977552	6.808484768
TD5	406.7303696	184.7245038	0.999623537	403.2	0.4741676	4.104083743

3.4. Effect of APR Oil on Intestinal Flora Composition in E. coli-Induced Mice

At the phylum level, 56 phyla were identified across the groups. Bacteroidota (CD: 28.47%, ED: 29.17%, TD: 47.29%) and Firmicutes_D (CD: 25.13%, ED: 32.93%, TD: 11.59%) were the most abundant. Campylobacterota was most abundant in the CD (32.97%), Proteobacteria dominated the ED (14.00%), and Firmicutes_A was prevalent in the TD

(19.21%). Other less abundant phyla included Deferribacterota (CD: 5.72%, ED: 1.21%, TD: 0.54%), Actinobacteriota (CD: 0.48%, ED: 3.08%, TD: 1.29%), and Desulfobacterota_I (CD: 0.18%, ED: 0.46%, TD: 1.65%).

At the class level, Bacteroidia (CD: 28.47%, ED: 29.12%, TD: 47.27%) and Bacilli (CD: 25.13%, ED: 32.93%, TD: 11.59%) were the most prevalent. Campylobacteria, abundant in the CD (32.97%), was reduced in the ED (5.28%) and TD (7.27%) groups. Gammaproteobacteria was most abundant in the ED (12.85%), with lower levels in CD (0.52%) and TD (9.65%). Clostridia_258483 (CD: 4.17%, ED: 7.98%, TD: 19.20%) was present in all groups. In total, 9176 ASVs were generated, with 1880 unique to the CD, 2872 unique to the ED, and 4424 unique to the TD.

At the genus level, *Ligilactobacillus* was most abundant in the CD and TD (CD: 15.05%, TD: 10.86%) but scarce in the ED (2.18%). *Helicobacter_C_479931* (29.34%) and *Paramuribaculum* (6.29%) were predominant in the CD, while *Paramuribaculum* (11.59%) and *Helicobacter_D* (8.74%) were more common in the TD. *Malacoplasma_A_271108* (22.97%) was the most abundant genus in the ED. Additionally, 260 ASVs were shared across all three groups (Figure 5a). Figure 5b–f shows the top ten phyla, classes, orders, families, and genera among all three groups.

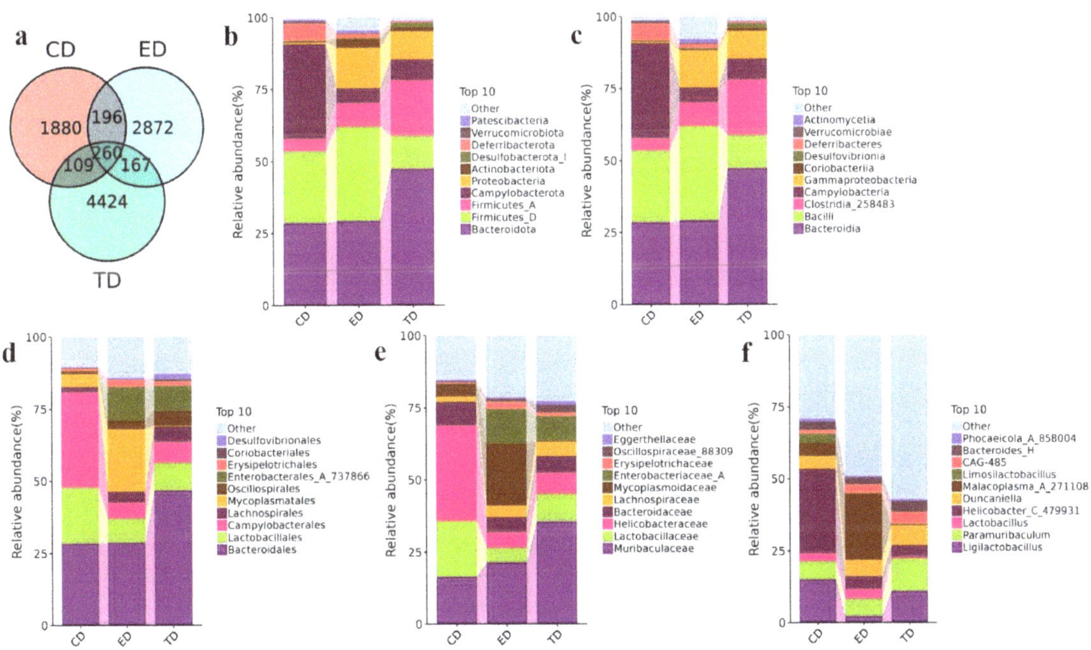

Figure 5. Venn diagram of ASVs and APR-mediated changes in the intestinal microbiota of *E. coli*-induced mice in different taxa. (**a**) Venn diagram of ASVs, (**b**) Phylum, (**c**) Class, (**d**) Order, (**e**) Family, (**f**) Genus.

Then, the re-analysis demonstrated significant differences ($p < 0.05$) in both phylum and genus composition. Firmicutes A was notably higher in the TD (Figure 6a), while *Ligilactobacillus* was more abundant in the CD and TD. The TD also showed increased levels of *Desulfovibrio R 446353*, *UBA644*, *Paludicola*, *Cupidesulfovibrio*, and *Blautia_A_141780*, while *CAG-95* and *CAG-510* were more abundant in the ED (Figure 6b).

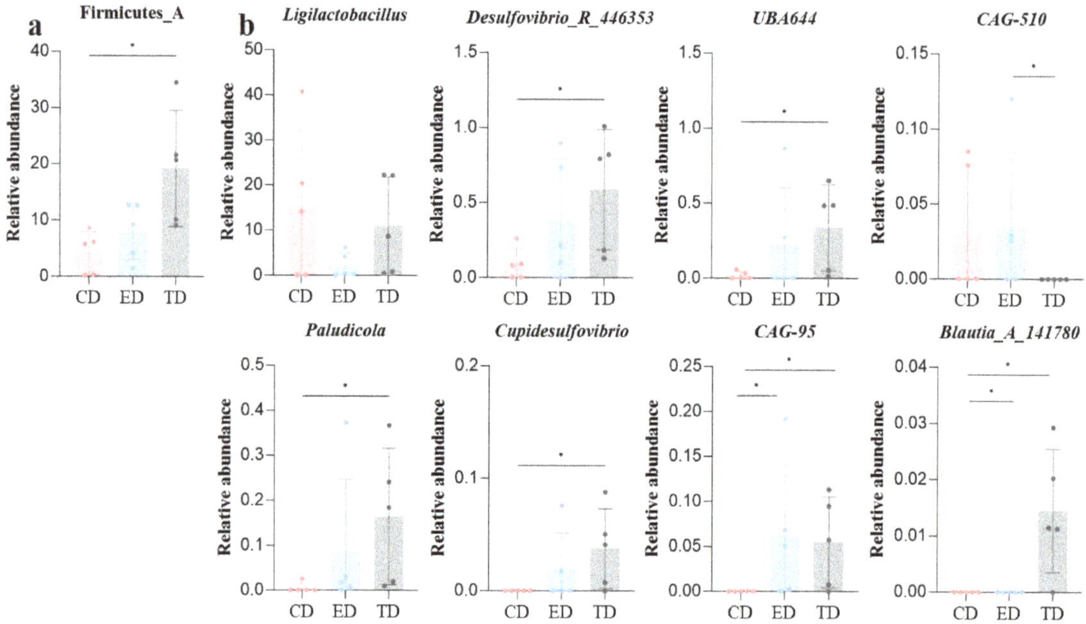

Figure 6. Microbiota differences among mice in different groups revealed by ANOVA analysis (**a**) Phyla and (**b**) Genus. Statistical significance is indicated as * $p < 0.05$, Data are presented as mean ± SEM (n = 5).

3.5. Analysis of Differential Metabolic Pathways Using KEGG and MetaCyc Databases

Differential metabolic pathways were analyzed using the KEGG and MetaCyc databases. The KEGG results revealed several pathways with LDA scores greater than 2 in the CD, including aminoacyl-tRNA biosynthesis, D-glutamine and D-glutamate metabolism, D-alanine metabolism, photosynthesis, lysine biosynthesis, ribosome production, protein export, mismatch repair, and peptidoglycan biosynthesis. In contrast, the ED showed higher LDA scores in pathways such as the biosynthesis of ansamycins, African trypanosomiasis, and the degradation of valine, leucine, and isoleucine.

At LDA SCORE (log 10) in the CD, D-glutamine and D-glutamate metabolism, D-alanine metabolism, aminoacyl-tRNA biosynthesis, lysine biosynthesis in group ED; biosynthesis of ansamycins, valine, and leucine and isoleucine degradation, in group TD; pentose and glucuronate interconversions, ascorbate and aldarate metabolism, tropane, piperidine, and pyridine alkaloid biosynthesis, beta-lactam resistance, etc., related pathways were found to be significantly ($p < 0.05$) upregulated (Figure 7a). Similarly, MetaCyc pathway analysis indicated that in group CD, PWY-7228, PWY0-162, PWY-5100, PWY-7197, PWY-7229, in group ED; PWY-7013, glucargalactsuper-PWY, galactardeg-PWY, in group TD; PWY0-1261, PWY-6629, sulfate-CYS-PWY, FAO-PWY and PWY-7242 pathways were found to be significant ($p < 0.05$) expressed (Figure 7b). These findings highlight distinct metabolic shifts across the groups in response to *E. coli* infection and APR oil treatment (Figure 7a,b).

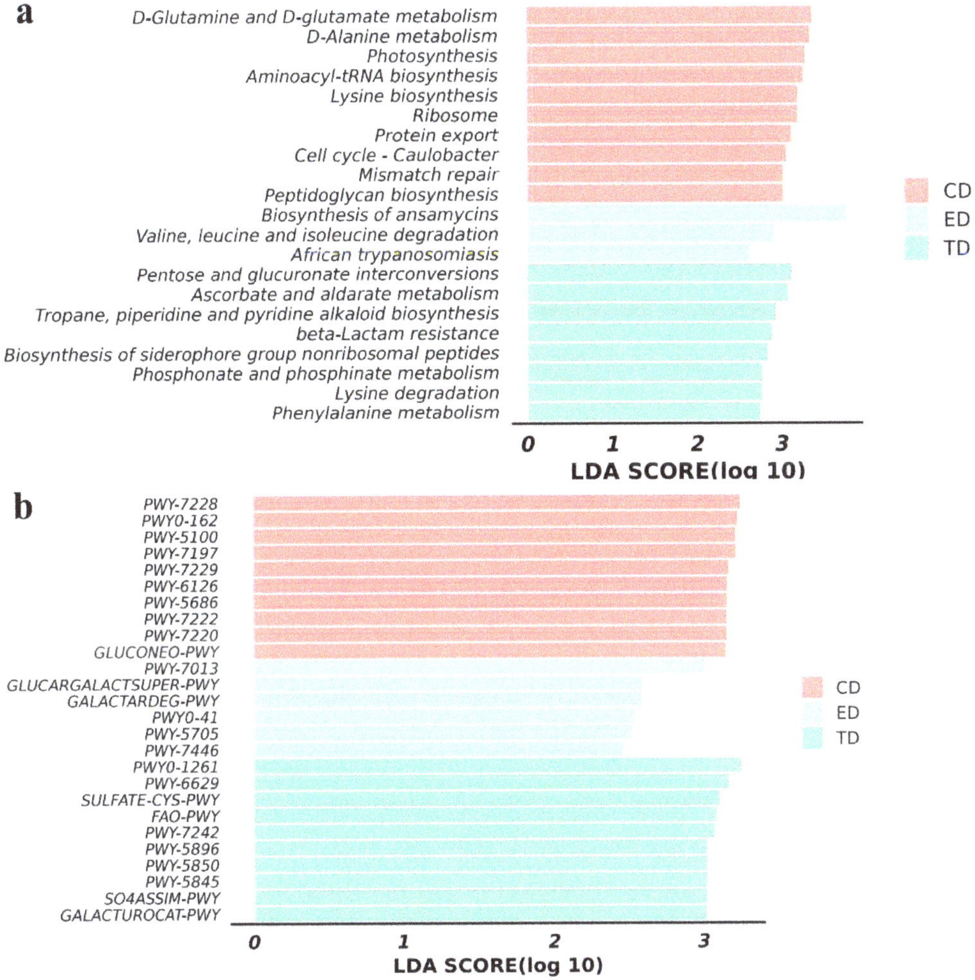

Figure 7. Functional analysis of mice microbiota (**a**) KEGG pathway analysis and (**b**) MetaCyc pathway analysis.

4. Discussion

Our findings provide strong evidence that APR exerts protective effects in the early stages of bacterial infection by reducing oxidative stress and modulating inflammatory responses. The significant reduction in bacterial load and improvement in antioxidant markers highlight the acute therapeutic potential of APR rather than suggesting prolonged microbiota shifts. While previous studies have explored the chronic benefits of herbal treatments, our study underscores the importance of early intervention [44]. During the experimental period, the body weight of mice in the CD showed a fluctuating upward trend. In contrast, the weight of mice in the TD and ED decreased initially after being exposed to *E. coli*, though the TD exhibited a slight recovery. This suggests that APR treatment may enhance resistance to bacterial infection, reducing symptoms such as diarrhea and weight loss [17].

Additionally, cytokine analysis revealed that IL-6 and TNF-α levels were elevated in the infection group but significantly reduced in the APR treatment group within 24 h, emphasizing the immediate anti-inflammatory effects of APR. This aligned with the study's objective of assessing acute responses rather than inferring long-term immune modulation. IL-10 is a key anti-inflammatory cytokine that blocks immune cell activation and cytokine production in innate immune cell types [45]. Conversely, pro-inflammatory cytokines such as IL-1β and TNF-α are secreted in significant quantities during pathological conditions [46,47]. The experimental results showed that APR treatment significantly reduced the inflammatory response in mice post-*E. coli* challenge. The literature indicated that other Chinese herbs, like Licorice root, have also shown anti-inflammatory and soothing effects on the gastrointestinal tract in treating ulcers [48] and promoting healing by increasing mucus production in the stomach [48,49].

In the antioxidant defense system, T-AOC serves as an important measure of the antioxidant capacity of bioactive substances [50]. The enzymes such as SOD and GSH-Px work together to convert reactive oxygen species (ROS) into harmless substances, thereby protecting cells from oxidative damage [51]. Additionally, MDA content is an important parameter reflecting the antioxidant capacity of the body, which can reflect the rate and intensity of lipid peroxidation in the body and also indirectly reflect the degree of tissue peroxidation damage [52]. The T-AOC in the ED was significantly lower than that in the control group and APR treatment group (TD). All antioxidant indexes of TD mice showed that APR treatment could significantly enhance the ability of mice to resist oxidative stress and clear free radicals; this is consistent with previous results using APR to treat rheumatoid arthritis [18]. Furthermore, in the given study, it was observed that the intestinal bacterial load in the APR treatment group decreased by about 100 times compared with that in the ED. This is consistent with the results of intestinal biopsies, and APR can alleviate intestinal damage caused by *E. coli* [48].

From the alterations in intestinal microecology, Bacteroidota and Firmicutes_D are the common phyla with high abundance among the three groups. Nevertheless, the abundance of Firmicutes_A in the TD was significantly elevated. Firmicutes, mostly Gram-positive bacteria, play a crucial role in host nutrition and metabolism through the synthesis of short-chain fatty acids (SCFAs) [53]. At the genus level, it was found that eight genera had abundance differences. The abundance of *Ligilactobacillus*, *UBA644*, *Paludicola*, *Cupidesulfovibrio*, and *Blautia_A_1417806* had all increased in the APR treatment group. Among these, *Ligilactobacillus* is a well-established probiotic that has been shown to have antibacterial and anti-inflammatory properties [54,55] and has been used to treat or relieve diarrhea, colitis, mastitis, and other conditions [56,57]. *Paludicola* has rarely been reported in recent years, but it has been confirmed that it can produce propionate [58]. As a short-chain fatty acid, propionate has anti-inflammatory and immune homeostasis regulation functions [59] and can maintain the stability and health of intestinal microecology [60]. Similarly, *Blautia_A_1417806* is thought to have potential probiotic properties [61], having previously been found to help humans adapt to hypoxic environments [62], and it has also been found to produce butyrate, which is beneficial for gut health [63].

Similar results have been observed with berberin, which modulated gut microbiome by reducing the harmful bacteria and enhancing probiotic bacteria, thus increasing intestinal barrier function and metabolic health [64,65]. Berberine acts by directly reducing bacterial growth and modulating inflammatory pathways. Its effects on gut microbiota are well-documented, showing a significant increase in beneficial bacteria like *Lactobacillus* and *Bifidobacterium* [64,65]. These findings suggest that APR treatment enhances the abundance of beneficial gut microbiota, particularly SCFA-producing bacteria, which likely play a key role in alleviating inflammation and repairing intestinal damage caused by bacterial diarrhea, comparable to established results [64,65].

In terms of functional pathways, several significant differences were observed in the TD, including pathways related to pentose and glucuronate interconversions. Ascorbate and aldarate metabolism, tropane, piperidine, and pyridine alkaloid biosynthesis beta-lactam resistance, etc. Ascorbate metabolism, in particular, was linked to increased activity in pentose and glucuronate interconversions and D-galacturonate degradation, pathways that are known to enhance ascorbate synthesis [66]. Ascorbate plays a vital role in promoting iron absorption, boosting immunity, preventing scurvy, and acting as a potent antioxidant by binding ROS and converting to monodehydroascorbic acid to mitigate oxidative stress [67]. Additionally, pathways related to vitamin K synthesis, such as menaquinol biosynthesis, showed increased expression. The reduced form of vitamin K (vitamin K-hydroquinone, VKH2) has been identified as an antioxidant that can effectively inhibit ferroptosis [68]. These findings suggest that APR treatment enhances the expression of antioxidant functional pathways, thereby alleviating oxidative stress in mice. These findings suggest that APR treatment can alleviate oxidative stress in mice by increasing the expression of functional pathways in antioxidants [69].

While APR exhibited acute therapeutic efficacy against *E. coli*-induced intestinal injury, its clinical translation needs to address its stability, bioavailability, and regulatory measures. Coumarins and polysaccharides as bioactive compounds from APR require technical formulation strategies, e.g., nanoencapsulation and enteric coatings, to increase their stability and absorption [17,28]. Regulatory pathways need rigorous standardization of extracts, lengthy safety assessments, and adherence to good management practices (GMPs) protocols, as exemplified by successful TCM-derived therapeutics like artemisinin [9,10]. Future research should prioritize pharmacokinetic studies and collaborative regulatory frameworks to bridge traditional use with modern therapeutic validation.

5. Conclusions

This study showed that APR effectively neutralizes the acute intestinal damage caused by *E. coli* infection in mice. APR treatment significantly reduced bacterial load and alleviated oxidative burden by T-AOC, SOD, GSH-Px, and downregulated pro-inflammatory cytokines, e.g., IL-6, TNF-α, while upregulated the anti-inflammatory IL-10. Histopathological results confirmed APR's role in preserving intestinal mucosal integrity and villus structure. Furthermore, APR enriched the beneficial genera such as *Ligilactobacillus*, *Paludicola*, and *Blautia_A_1417806*, which are linked to SCFA production and antioxidant pathway activation. These findings explored the APR's therapeutic potential for acute bacterial diarrhea, underscoring its dual role in combating oxidative stress and inflammation while modulating gut microbiota beneficially. Future investigations should explore long-term treatment effects to assess sustained benefits and broader clinical applicability.

Author Contributions: Conceptualization: K.D., X.L. and K.L.; methodology: K.D., C.X. and Q.H.; software: K.L.; validation: K.D., C.X. and Q.H.; formal analysis: K.D., C.X. and Q.H.; investigation: K.L.; resources: K.D., C.X. and Q.H.; data curation: K.D. and K.L.; writing—original draft preparation: K.L., M.S., M.N. and X.L.; writing—review and editing: K.L., M.S., M.N. and X.L.; visualization: K.L.; supervision: K.L.; project administration: K.L.; funding acquisition: X.L. and K.L. All authors have read and agreed to the published version of the manuscript.

Funding: This study was supported by the Yichang Science and Technology Plan Projects (A23-3-063) and the Yichang Natural Science Research Project (A24-3-032).

Institutional Review Board Statement: The study was conducted by the Institutional Review Board of the ethics committee of Nanjing Agricultural University (NJAU.No20240226021).

Informed Consent Statement: Not applicable.

Data Availability Statement: All raw sequence data from animals were deposited in the NCBI Sequence Read Archive database under accession number PRJNA1196629 (https://dataview.ncbi.nlm.nih.gov/object/PRJNA1196629; accessed on 10 December 2024).

Conflicts of Interest: The authors declare no conflicts of interest.

References

1. Wang, L.-P.; Zhou, S.-X.; Wang, X.; Lu, Q.-B.; Shi, L.-S.; Ren, X.; Zhang, H.-Y.; Wang, Y.-F.; Lin, S.-H.; Zhang, C.-H.; et al. Etiological, Epidemiological, and Clinical Features of Acute Diarrhea in China. *Nat. Commun.* **2021**, *12*, 2464. [CrossRef] [PubMed]
2. López-Vélez, R.; Lebens, M.; Bundy, L.; Barriga, J.; Steffen, R. Bacterial Travellers' Diarrhoea: A Narrative Review of Literature Published over the Past 10 Years. *Travel. Med. Infect. Dis.* **2022**, *47*, 102293. [CrossRef] [PubMed]
3. He, Q.; Lu, S.; Wang, J.; Xu, C.; Qu, W.; Nawaz, S.; Ataya, F.S.; Wu, Y.; Li, K. *Lactobacillus salivarius* and Berberine Alleviated Yak Calves' Diarrhea via Accommodating Oxidation Resistance, Inflammatory Factors, and Intestinal Microbiota. *Animals* **2024**, *14*, 2419. [CrossRef]
4. Liu, M.; Ma, J.; Xu, J.; Huangfu, W.; Zhang, Y.; Ali, Q.; Liu, B.; Li, D.; Cui, Y.; Wang, Z.; et al. Fecal Microbiota Transplantation Alleviates Intestinal Inflammatory Diarrhea Caused by Oxidative Stress and Pyroptosis via Reducing Gut Microbiota-Derived Lipopolysaccharides. *Int. J. Biol. Macromol.* **2024**, *261*, 129696. [CrossRef]
5. Goodoory, V.C.; Ford, A.C. Antibiotics and Probiotics for Irritable Bowel Syndrome. *Drugs* **2023**, *83*, 687–699. [CrossRef]
6. Aslam, B.; Khurshid, M.; Arshad, M.I.; Muzammil, S.; Rasool, M.; Yasmeen, N.; Shah, T.; Chaudhry, T.H.; Rasool, M.H.; Shahid, A.; et al. Antibiotic Resistance: One Health One World Outlook. *Front. Cell Infect. Microbiol.* **2021**, *11*, 771510. [CrossRef]
7. Mekonnen, S.A.; Merenstein, D.; Fraser, C.M.; Marco, M.L. Molecular Mechanisms of Probiotic Prevention of Antibiotic-Associated Diarrhea. *Curr. Opin. Biotechnol.* **2020**, *61*, 226–234. [CrossRef]
8. Ramirez, J.; Guarner, F.; Bustos Fernandez, L.; Maruy, A.; Sdepanian, V.L.; Cohen, H. Antibiotics as Major Disruptors of Gut Microbiota. *Front. Cell Infect. Microbiol.* **2020**, *10*, 572912. [CrossRef]
9. Sun, Y.; Mao, W.; Cao, J.; Hao, P.G.; Su, J.; Yin, K.; Gu, K.; Zhao, H. Chinese Medicine Monomers Inhibit Biofilm Formation in Multidrug-Resistant *Pasteurella multocida* Isolated from Cattle Respiratory Infections. *Pak. Vet. J.* **2024**, *44*, 1095–1104. [CrossRef]
10. Zhu, Y.; Lu, S.; Cidan, Y.; Ali, M.; Zhang, X.; Pubu, P.; Kiani, F.A.; Saleem, M.U.; Basang, W.; Li, K. Protective Effects of Traditional Chinese Herbal Medicine Formulas (TCHMFs) ViaInfluencing Anti-Oxidative Capacity, Inflammatory Mediators, and Gut Microbiota inWeaned Yaks. *Pak. Vet. J.* 2025, *in press*. [CrossRef]
11. Meng, A.; Zhang, X.; Pubu, P.; Ali, M.; Wang, J.; Xu, C.; Almutairi, M.H.; Li, K. Protective Effect of Lentinan against LPS-Induced Injury in Mice Via Influencing Antioxidant Enzyme Activity, Inflammatory Pathways and Gut Microbiota. *Pak. Vet. J.* **2024**, *44*, 647–656. [CrossRef]
12. Hu, J.; Sun, J.; Zhong, Q.; Chen, S.; Yin, W.; Wei, X.; Li, L.; Li, K.; Ali, M.; Sun, W.; et al. *Edgeworthia gardneri* (Wall.) Meisn Mitigates CCL4-Induced Liver Injury in Mice by Modulating Gut Microbiota, Boosting Antioxidant Defense, and Reducing Inflammation. *Ecotoxicol. Environ. Saf.* **2025**, *293*, 118042. [CrossRef] [PubMed]
13. Wang, Y.; Zhang, Q.; Chen, Y.; Liang, C.-L.; Liu, H.; Qiu, F.; Dai, Z. Antitumor Effects of Immunity-Enhancing Traditional Chinese Medicine. *Biomed. Pharmacother.* **2020**, *121*, 109570. [CrossRef] [PubMed]
14. Wang, M.; Fu, R.; Xu, D.; Chen, Y.; Yue, S.; Zhang, S.; Tang, Y. Traditional Chinese Medicine: A Promising Strategy to Regulate the Imbalance of Bacterial Flora, Impaired Intestinal Barrier and Immune Function Attributed to Ulcerative Colitis through Intestinal Microecology. *J. Ethnopharmacol.* **2024**, *318*, 116879. [CrossRef]
15. Xu, C.; Ali, M.; Sun, J.; Li, X.; Fouad, D.; Iqbal, M.; Kulyar, M.F.-E.-A.; Wu, Y.; Li, K. Protective Effects of Abrus Cantoniensis Hance against Liver Injury through Modulation of Intestinal Microbiota and Liver Metabolites. *Ecotoxicol. Environ. Saf.* **2024**, *279*, 116495. [CrossRef]
16. Li, L.; Xiao, S.; Dai, X.; Tang, Z.; Wang, Y.; Ali, M.; Ataya, F.S.; Sahar, I.; Iqbal, M.; Wu, Y.; et al. Multi-Omics Analysis and the Remedial Effects of Swertiamarin on Hepatic Injuries Caused by CCl4. *Ecotoxicol. Environ. Saf.* **2024**, *282*, 116734. [CrossRef]
17. Lu, Y.; Wu, H.; Yu, X.; Zhang, X.; Luo, H.; Tang, L.; Wang, Z. Traditional Chinese Medicine of Angelicae Pubescentis Radix: A Review of Phytochemistry, Pharmacology and Pharmacokinetics. *Front. Pharmacol.* **2020**, *11*, 335. [CrossRef]
18. Chen, S.; Wang, Y.; Zhang, L.; Han, Y.; Liang, C.; Wang, S.; Qi, L.; Pang, X.; Li, J.; Chang, Y. Therapeutic Effects of Columbianadin from Angelicae Pubescentis Radix on the Progression of Collagen-Induced Rheumatoid Arthritis by Regulating Inflammation and Oxidative Stress. *J. Ethnopharmacol.* **2023**, *316*, 116727. [CrossRef]

19. Jiang, X.; Lu, Z.; Zhang, Q.; Yu, J.; Han, D.; Liu, J.; Li, P.; Li, F. Osthole: A Potential AMPK Agonist That Inhibits NLRP3 Inflammasome Activation by Regulating Mitochondrial Homeostasis for Combating Rheumatoid Arthritis. *Phytomedicine* **2023**, *110*, 154640. [CrossRef]
20. Ma, Z.; Peng, L.; Sheng, Y.; Chu, W.; Fu, Y. Anti-Inflammatory Effect of Columbianadin against D-Galactose-Induced Liver Injury In Vivo via the JAK2/STAT3 and JAK2/P38/NF-κB Pathways. *Pharmaceuticals* **2024**, *17*, 378. [CrossRef]
21. Wang, D.; Liu, Y.; Zhao, W. The Adjuvant Effects on Vaccine and the Immunomodulatory Mechanisms of Polysaccharides From Traditional Chinese Medicine. *Front. Mol. Biosci.* **2021**, *8*, 655570. [CrossRef]
22. Ren, C.; Luo, Y.; Li, X.; Ma, L.; Wang, C.; Zhi, X.; Zhao, X.; Li, Y. Pharmacological Action of *Angelica sinensis* Polysaccharides: A Review. *Front. Pharmacol.* **2025**, *15*, 1510976. [CrossRef] [PubMed]
23. Dilixiati, Y.; Aipire, A.; Song, M.; Nijat, D.; Wubuli, A.; Cao, Q.; Li, J. The Potential Role of Plant Polysaccharides in Treatment of Ulcerative Colitis. *Pharmaceutics* **2024**, *16*, 1073. [CrossRef] [PubMed]
24. Kuziel, G.A.; Rakoff-Nahoum, S. The Gut Microbiome. *Curr. Biol.* **2022**, *32*, R257–R264. [CrossRef]
25. Brody, H. The Gut Microbiome. *Nature* **2020**, *577*, S5. [CrossRef]
26. Morais, L.H.; Schreiber, H.L., IV; Mazmanian, S.K. The Gut Microbiota-Brain Axis in Behaviour and Brain Disorders. *Nat. Rev. Microbiol.* **2021**, *19*, 241–255. [CrossRef]
27. Lynch, S.V.; Pedersen, O. The Human Intestinal Microbiome in Health and Disease. *N. Engl. J. Med.* **2016**, *375*, 2369–2379. [CrossRef]
28. Healy, D.B.; Ryan, C.A.; Ross, R.P.; Stanton, C.; Dempsey, E.M. Clinical Implications of Preterm Infant Gut Microbiome Development. *Nat. Microbiol.* **2022**, *7*, 22–33. [CrossRef]
29. Xie, Q.; Zhang, Y.; Zhang, Z.; Gong, S.; Mo, Q.; Li, J. Characteristics and Dynamic Changes of Gut Microbiota in Cats with Colitis. *Pak. Vet. J.* **2024**, *44*, 414–422. [CrossRef]
30. Cidan, Y.; Lu, S.; Wang, H.; Wang, J.; Ali, M.; Fouad, D.; Ataya, F.S.; Zhu, Y.; Basang, W.; Li, K. Comparative Analysis of Microbiota in Jiani Yaks with Different Rib Structures. *Life* **2024**, *14*, 1458. [CrossRef]
31. Tanes, C.; Bittinger, K.; Gao, Y.; Friedman, E.S.; Nessel, L.; Paladhi, U.R.; Chau, L.; Panfen, E.; Fischbach, M.A.; Braun, J.; et al. Role of Dietary Fiber in the Recovery of the Human Gut Microbiome and Its Metabolome. *Cell Host Microbe* **2021**, *29*, 394–407.e5. [CrossRef]
32. Verdu, E.F.; Galipeau, H.J.; Jabri, B. Novel Players in Coeliac Disease Pathogenesis: Role of the Gut Microbiota. *Nat. Rev. Gastroenterol. Hepatol.* **2015**, *12*, 497–506. [CrossRef] [PubMed]
33. Aly, S.H.; Eldahshan, O.A.; Al-Rashood, S.T.; Binjubair, F.A.; El Hassab, M.A.; Eldehna, W.M.; Dall'Acqua, S.; Zengin, G. Chemical Constituents, Antioxidant, and Enzyme Inhibitory Activities Supported by In-Silico Study of n-Hexane Extract and Essential Oil of Guava Leaves. *Molecules* **2022**, *27*, 8979. [CrossRef] [PubMed]
34. Zhu, Y.; Cidan, Y.; Ali, M.; Lu, S.; Javed, U.; Cisang, Z.; Gusang, D.; Danzeng, Q.; Li, K.; Basang, W. Evaluating the Effect of Dietary Protein–Energy Ratios on Yak Intestinal Microbiota Using High-Throughput 16S rRNA Gene Sequencing. *Vet. Sci.* **2025**, *12*, 208. [CrossRef]
35. Stanzick, K.J.; Simon, J.; Zimmermann, M.E.; Schachtner, M.; Peterhoff, D.; Niller, H.-H.; Überla, K.; Wagner, R.; Heid, I.M.; Stark, K.J. DNA Extraction from Clotted Blood in Genotyping Quality. *Biotechniques* **2023**, *74*, 23–29. [CrossRef]
36. Zhou, X.; Liu, X.; Liu, M.; Liu, W.; Xu, J.; Li, Y. Comparative Evaluation of 16S rRNA Primer Pairs in Identifying Nitrifying Guilds in Soils under Long-Term Organic Fertilization and Water Management. *Front. Microbiol.* **2024**, *15*, 1424795. [CrossRef]
37. Plummer, E.; Bulach, D.; Carter, G.; Albert, M.J. Gut Microbiome of Native Arab Kuwaitis. *Gut Pathog.* **2020**, *12*, 10. [CrossRef]
38. Ali, Z.; Shahzadi, I.; Majeed, A.; Malik, H.M.T.; Waseem, S.; Ahmed, I.; Anis, R.A.; Saeed, S.; Anees, M. Comparative Analysis of the Serum Microbiome of HIV Infected Individuals. *Genomics* **2021**, *113*, 4015–4021. [CrossRef]
39. Ye, H.; Wen, Y.; Chen, Z.; Zhang, T.; Li, S.; Guan, M.; Zhang, Y.; Su, S. Relationship of Soil Microbiota to Seed Kernel Metabolism in *Camellia oleifera* Under Mulched. *Front. Plant Sci.* **2022**, *13*, 920604. [CrossRef]
40. Jin, Y.; Lu, L.; Tu, W.; Luo, T.; Fu, Z. Impacts of Polystyrene Microplastic on the Gut Barrier, Microbiota and Metabolism of Mice. *Sci. Total Environ.* **2019**, *649*, 308–317. [CrossRef]
41. Yu, X.; Jiang, W.; Kosik, R.O.; Song, Y.; Luo, Q.; Qiao, T.; Tong, J.; Liu, S.; Deng, C.; Qin, S.; et al. Gut Microbiota Changes and Its Potential Relations with Thyroid Carcinoma. *J. Adv. Res.* **2022**, *35*, 61–70. [CrossRef]
42. Douglas, G.M.; Maffei, V.J.; Zaneveld, J.R.; Yurgel, S.N.; Brown, J.R.; Taylor, C.M.; Huttenhower, C.; Langille, M.G.I. PICRUSt2 for Prediction of Metagenome Functions. *Nat. Biotechnol.* **2020**, *38*, 685–688. [CrossRef] [PubMed]
43. Luan, M.; Niu, M.; Yang, P.; Han, D.; Zhang, Y.; Li, W.; He, Q.; Zhao, Y.; Mao, B.; Chen, J.; et al. Metagenomic Sequencing Reveals Altered Gut Microbial Compositions and Gene Functions in Patients with Non-Segmental Vitiligo. *BMC Microbiol.* **2023**, *23*, 265. [CrossRef] [PubMed]

44. Lin, X.; Li, H. Obesity: Epidemiology, Pathophysiology, and Therapeutics. *Front. Endocrinol.* **2021**, *12*, 706978. [CrossRef] [PubMed]
45. York, A.G.; Skadow, M.H.; Oh, J.; Qu, R.; Zhou, Q.D.; Hsieh, W.-Y.; Mowel, W.K.; Brewer, J.R.; Kaffe, E.; Williams, K.J.; et al. IL-10 Constrains Sphingolipid Metabolism to Limit Inflammation. *Nature* **2024**, *627*, 628–635. [CrossRef]
46. Wang, Y.; Che, M.; Xin, J.; Zheng, Z.; Li, J.; Zhang, S. The Role of IL-1β and TNF-α in Intervertebral Disc Degeneration. *Biomed. Pharmacother.* **2020**, *131*, 110660. [CrossRef]
47. Han, Y.; Zhang, Y.-Y.; Pan, Y.-Q.; Zheng, X.-J.; Liao, K.; Mo, H.-Y.; Sheng, H.; Wu, Q.-N.; Liu, Z.-X.; Zeng, Z.-L.; et al. IL-1β-Associated NNT Acetylation Orchestrates Iron-Sulfur Cluster Maintenance and Cancer Immunotherapy Resistance. *Mol. Cell* **2023**, *83*, 1887–1902.e8. [CrossRef]
48. Yin, S.; You, T.; Tang, J.; Wang, L.; Jia, G.; Liu, G.; Tian, G.; Chen, X.; Cai, J.; Kang, B.; et al. Dietary Licorice Flavonoids Powder Improves Serum Antioxidant Capacity and Immune Organ Inflammatory Responses in Weaned Piglets. *Front. Vet. Sci.* **2022**, *9*, 942253. [CrossRef]
49. Zhang, Q.; Huang, H.; Qiu, M.; Wu, Z.; Xin, Z.; Cai, X.; Shang, Q.; Lin, J.; Zhang, D.; Han, L. Traditional Uses, Pharmacological Effects, and Molecular Mechanisms of Licorice in Potential Therapy of COVID-19. *Front. Pharmacol.* **2021**, *12*, 719758. [CrossRef]
50. Hu, W.; He, Z.; Du, L.; Zhang, L.; Li, J.; Ma, Y.; Bi, S. Biomarkers of Oxidative Stress in Broiler Chickens Attacked by Lipopolysaccharide: A Systematic Review and Meta-Analysis. *Ecotoxicol. Environ. Saf.* **2023**, *266*, 115606. [CrossRef]
51. Zhou, Y.; Jia, Z.; Wang, J.; Huang, S.; Yang, S.; Xiao, S.; Xia, D.; Zhou, Y. Curcumin Reverses Erastin-Induced Chondrocyte Ferroptosis by Upregulating Nrf2. *Heliyon* **2023**, *9*, e20163. [CrossRef]
52. Attia, M.M.; Soliman, S.M.; Mahmoud, M.A.; Salem, M.A. Oxidative Stress Markers, Immune-Regulating Cytokines, and the Pathological Evaluation of Sheep Co-Infected with *Oestrus ovis* and *Coenuruscerebralis*. *Microb. Pathog.* **2022**, *169*, 105613. [CrossRef] [PubMed]
53. Ma, L.; Ni, Y.; Wang, Z.; Tu, W.; Ni, L.; Zhuge, F.; Zheng, A.; Hu, L.; Zhao, Y.; Zheng, L.; et al. Spermidine Improves Gut Barrier Integrity and Gut Microbiota Function in Diet-Induced Obese Mice. *Gut Microbes* **2020**, *12*, 1–19. [CrossRef] [PubMed]
54. Chuandong, Z.; Hu, J.; Li, J.; Wu, Y.; Wu, C.; Lai, G.; Shen, H.; Wu, F.; Tao, C.; Liu, S.; et al. Distribution and Roles of *Ligilactobacillus murinus* in Hosts. *Microbiol. Res.* **2024**, *282*, 127648. [CrossRef] [PubMed]
55. He, T.; Hu, X.; Mi, J.; Hu, H.; Wang, H.; Qi, X.; Gao, L.; Zhang, Y.; Liu, C.; Wang, S.; et al. *Ligilactobacillus salivarius* XP132 with Antibacterial and Immunomodulatory Activities Inhibits Horizontal and Vertical Transmission of *Salmonella pullorum* in Chickens. *Poult. Sci.* **2024**, *103*, 104086. [CrossRef]
56. Lukasik, J.; Dierikx, T.; Besseling-van der Vaart, I.; de Meij, T.; Szajewska, H. Multispecies Probiotic for the Prevention of Antibiotic-Associated Diarrhea in Children: A Randomized Clinical Trial. *JAMA Pediatr.* **2022**, *176*, 860–866. [CrossRef]
57. Yan, S.; Yu, L.; Tian, F.; Zhao, J.; Chen, W.; Chen, H.; Zhai, Q. *Ligilactobacillus salivarius* CCFM 1266 Modulates Gut Microbiota and GPR109a-Mediated Immune Suppression to Attenuate Immune Checkpoint Blockade-Induced Colitis. *Food Funct.* **2023**, *14*, 10549–10563. [CrossRef]
58. Akasaka, H.; Ueki, A.; Hanada, S.; Kamagata, Y.; Ueki, K. *Propionicimonas paludicola* Gen. Nov., Sp. Nov., a Novel Facultatively Anaerobic, Gram-Positive, Propionate-Producing Bacterium Isolated from Plant Residue in Irrigated Rice-Field Soil. *Int. J. Syst. Evol. Microbiol.* **2003**, *53*, 1991–1998. [CrossRef]
59. Duan, H.; Wang, L.; Huangfu, M.; Li, H. The Impact of Microbiota-Derived Short-Chain Fatty Acids on Macrophage Activities in Disease: Mechanisms and Therapeutic Potentials. *Biomed. Pharmacother.* **2023**, *165*, 115276. [CrossRef]
60. Yang, S.; Shang, J.; Liu, L.; Tang, Z.; Meng, X. Strains Producing Different Short-Chain Fatty Acids Alleviate DSS-Induced Ulcerative Colitis by Regulating Intestinal Microecology. *Food Funct.* **2022**, *13*, 12156–12169. [CrossRef]
61. Liu, X.; Mao, B.; Gu, J.; Wu, J.; Cui, S.; Wang, G.; Zhao, J.; Zhang, H.; Chen, W. Blautia-a New Functional Genus with Potential Probiotic Properties? *Gut Microbes* **2021**, *13*, 1–21. [CrossRef]
62. Su, Q.; Zhuang, D.-H.; Li, Y.-C.; Chen, Y.; Wang, X.-Y.; Ge, M.-X.; Xue, T.-Y.; Zhang, Q.-Y.; Liu, X.-Y.; Yin, F.-Q.; et al. Gut Microbiota Contributes to High-Altitude Hypoxia Acclimatization of Human Populations. *Genome Biol.* **2024**, *25*, 232. [CrossRef] [PubMed]
63. Bernard, R.; Shilts, M.H.; Strickland, B.A.; Boone, H.H.; Payne, D.C.; Brown, R.F.; Edwards, K.; Das, S.R.; Nicholson, M.R. The Relationship between the Intestinal Microbiome and Body Mass Index in Children with Cystic Fibrosis. *J. Cyst. Fibros.* **2024**, *23*, 242–251. [CrossRef] [PubMed]
64. Luo, T.; Che, Q.; Guo, Z.; Song, T.; Zhao, J.; Xu, D. Modulatory Effects of Traditional Chinese Medicines on Gut Microbiota and the Microbiota-Gut-x Axis. *Front. Pharmacol.* **2024**, *15*, 1442854. [CrossRef] [PubMed]
65. Zhang, R.; Gao, X.; Bai, H.; Ning, K. Traditional Chinese Medicine and Gut Microbiome: Their Respective and Concert Effects on Healthcare. *Front. Pharmacol.* **2020**, *11*, 538. [CrossRef]
66. Fujii, J.; Osaki, T.; Bo, T. Ascorbate Is a Primary Antioxidant in Mammals. *Molecules* **2022**, *27*, 6187. [CrossRef]

67. Kaźmierczak-Barańska, J.; Boguszewska, K.; Adamus-Grabicka, A.; Karwowski, B.T. Two Faces of Vitamin C-Antioxidative and Pro-Oxidative Agent. *Nutrients* **2020**, *12*, 1501. [CrossRef]
68. Mishima, E.; Ito, J.; Wu, Z.; Nakamura, T.; Wahida, A.; Doll, S.; Tonnus, W.; Nepachalovich, P.; Eggenhofer, E.; Aldrovandi, M.; et al. A Non-Canonical Vitamin K Cycle Is a Potent Ferroptosis Suppressor. *Nature* **2022**, *608*, 778–783. [CrossRef]
69. Wei, X.; Wen, Y.; Wei, Y.; Liang, X.; Ma, X.; Zhang, B.; Tang, X. External Therapy of Traditional Chinese Medicine for Treating Irritable Bowel Syndrome with Diarrhea: A Systematic Review and Meta-Analysis. *Front. Med.* **2022**, *9*, 940328. [CrossRef]

veterinary sciences

MDPI

Article

Dietary *Bacillus toyonensis* BCT-7112^T Supplementation Influences Performance, Egg Quality, Ammonia Emission, and Cecal Microbiome in Laying Ducks

Tossaporn Incharoen [1,2], Rangsun Charoensook [1,2,*], Wandee Tartrakoon [1], Sonthaya Numthuam [1], Yutthana Sunanta [3], Guillermo Jimenez [4] and Juan J. Loor [5]

1 Division of Animal Science and Feed Technology, Department of Agricultural Sciences, Faculty of Agriculture, Natural Resources and Environment, Naresuan University, Phitsanulok 65000, Thailand; tossaporni@nu.ac.th (T.I.); wandeeta@nu.ac.th (W.T.); sonthayan@nu.ac.th (S.N.)

2 Center of Excellence in Nonlinear Analysis and Optimization, Faculty of Science, Naresuan University, Phitsanulok 65000, Thailand

3 Faculty of Animal Science and Technology, Maejo University, Chiang-Mai 50290, Thailand; yutthana_s@mju.ac.th

4 Rubinum S.A., 08191 Rubí, Catalonia, Spain; g.jimenez@groupandersen.com

5 Division of Nutritional Sciences, Department of Animal Sciences, University of Illinois, Urbana, IL 61801, USA; jloor@illinois.edu

* Correspondence: rangsunc@nu.ac.th; Tel.: +66-63-856-3556

Simple Summary: Duck egg production is crucial to Thailand's poultry industry, and supplying eggs is crucial to Thai and Chinese cuisine. While traditional duck farming occurs outdoors on rice fields, modern intensive farming prioritizes productivity and disease control, often relying on antibiotics. However, growing health and environmental concerns necessitate alternatives like probiotics. This study explored the effects of *Bacillus toyonensis* BCT-7112^T, a non-toxigenic, spore-forming bacterium, on Khaki Campbell laying ducks. The results show improved egg weight and shell thickness, reduced ammonia emissions, and increased cecal microbiota diversity. Our findings suggest *B. toyonensis* BCT-7112^T enhances egg production, gut health, and microbial diversity, offering a sustainable feed additive for intensive duck farming.

Abstract: This study evaluated the effects of *Bacillus toyonensis* BCT-7112^T on laying duck performance, egg quality, ammonia emission, and cecal microbiota. Two hundred twenty 32-week-old Khaki Campbell ducks were assigned to four dietary treatments (0, 200, 500, or 1000 ppm Toyocerin® 10^9 premixture, containing 1×10^9 *B. toyonensis* BCT-7112^T CFU/g) with five replicate pens of eleven birds each. Data on productivity, egg quality, and ammonia emissions were collected weekly, and cecal microbiota were analyzed at 44 weeks of age. Supplementation with 1000 ppm *B. toyonensis* BCT-7112^T significantly increased average egg weight ($p < 0.001$) and eggshell thickness ($p = 0.007$). Egg mass also improved at higher supplementation levels ($p < 0.05$), though feed intake, feed conversion, daily egg production, and most egg quality parameters were unaffected ($p > 0.05$). Ammonia emissions in litter decreased significantly ($p < 0.05$) with probiotic inclusion. Cecal microbiota analysis revealed higher diversity in ducks fed 1000 ppm, with a notable shift in predominant phyla from Bacteroidetes (35.12%) and Firmicutes (34.93%) in the controls to Bacteroidetes (40.52%), Firmicutes (34.08%), and Deferribacteres (9.54%) in the treated ducks. The findings suggest that 1000 ppm *B. toyonensis* BCT-7112^T enhances egg production, eggshell quality, and microbial diversity while reducing ammonia emissions in laying duck systems.

Keywords: ammonia emission; *Bacillus toyonesis*; egg performance; gut microbiome; laying ducks

1. Introduction

Duck egg production is one of Thailand's most important commercial poultry product sectors. Eggs from ducks are a popular ingredient in many recipes for both Thai and Chinese cuisines. In Thailand, ducks are traditionally raised outdoors on harvested rice fields [1]. However, in order to enhance productivity and prevent avian influenza infection from wild birds, poultry producers are increasingly turning to indoor industrial farming [2,3]. The way of raising laying ducks is changing from free-range on harvest paddy fields to intensive or semi-intensive systems. Because they are raised in a setting with high density and high production pressure, poultry may be exposed to various stressors and pathogen infections, which could affect their health and productivity. As a result, feeding antibiotic growth promoters (AGPs) is a frequent practice in the poultry production industry. However, the long-term usage of AGPs may result in bacterial resistance and antibiotic residues in poultry products and the environment [3,4]. Furthermore, with high-density rearing in intensive systems, ammonia damage is expected to be a significant issue in poultry production [5].

Since the ban on AGPs in the livestock industry, using probiotics as feed supplements in animal agriculture has grown dramatically over the last decade. Because of their ability to form spores, certain non-pathogenic Gram-positive lactic acid bacteria (LAB), such as Bacillus sp., are appealing for use as a probiotic additive in livestock feed [6,7]. *Bacillus toyonensis* BCT-7112^{T} (previously known as *Bacillus cereus* var. *toyoi*) is a non-toxigenic, Gram-positive, facultatively anaerobic, spore-forming bacterium in the Bacillus cereus group. It has been utilized since 1975 as a feed additive for livestock and aquaculture in various nations worldwide [3,8]. Several studies reported that *Bacillus toyonensis* improves performance, immune stimulation, intestinal health, and nutrient absorption as well as reducing enteric infections, decreasing ammonia emission, and increasing the variety of microbial communities in the gastrointestinal tract of several target species, including rabbits [9], swine [10,11], ruminants [12,13], chickens [14], quails [15], turkeys [16,17], meat-type duck [3], and aquatic animals [18,19]. The effect of *Bacillus toyonensis* on laying ducks, on the other hand, has not been investigated.

The productive performance, egg quality, gut health, and ammonia emissions of ducks are all linked to their intestinal microflora [5]. Furthermore, the gut microbiota promotes intestinal barrier integrity, stimulates more commensals, and excludes pathogens to protect the host's health, nutritional metabolism, and absorption, development of the immune system, and infection defense [3,20,21]. The conventional methods of microbial identification are frequently insufficient for identifying a large number of microorganisms present in habitats with microbial diversity. As a result of the rapid advancement of biotechnology, molecular identification based on genomic sequencing has garnered broad interest. In poultry, 16S rRNA gene sequencing is commonly applied in investigations of microbial diversity and community composition in the gut [3,5]. Thus, the objectives of this study were to investigate the effect of probiotic supplementation on performance, egg quality, ammonia emission, and cecal microbiota in laying ducks.

2. Materials and Methods

2.1. Ethics Statement

This investigation was conducted at the Poultry Nutrition and Feed Testing Unit affiliated with the Faculty of Agriculture, Natural Resources, and Environment at Naresuan University, Phitsanulok, Thailand. All animal procedures employed in the current research were approved and regulated by the Naresuan University Agricultural Animal Care and Use Committee (NUAACUC; reference number 61 01 009), following the Ethical Principles and Guidelines for the Use of Animals, National Research Council of Thailand.

2.2. Research Design and Animal Information

Three hundred 18-week-old Khaki Campbell laying ducks were purchased from a commercial layer duck farm (Charoenrat Farm, Phitsanulok province, Thailand). All birds were maintained in a temperature-controlled house using an evaporative system. They were reared in a litter and a slatted floor system equipped with a water bath and nest. The total number of eggs laid from each pen was recorded daily to determine the egg production percentage. Feed and drinking water were provided ad libitum during the pre-experimental period. At 32 weeks old, 220 ducks with laying and body weight uniformity were selected and assigned to four dietary treatments with five replicates per treatment and 11 birds per replicate. The experimental diets were a basal control (Table 1) supplemented with 0, 200, 500, or 1000 ppm (mg/kg) Toyocerin® 10^9 premixture (containing 1×10^9 *Bacillus toyonensis* BCT-7112^T CFU/g, Rubinum SA, Barcelona, Spain). The dosage levels were based on our previous study and manufacturer recommendations [3]. The basal diet with 18% crude protein level (Table 1) was formulated to satisfy the nutritional requirements for Khaki Campbell laying ducks, according to Likittrakulwong et al. [1]. Each dietary group provided feed and drinking water for ad libitum intake. The photoperiod was set at 17L: 7D through the experimental period (32 to 44 weeks old).

Table 1. Ingredients and calculated chemical composition of a basal diet (g/kg, as-fed basis unless stated otherwise).

Ingredients	g/kg, As-Fed Basis
Wheat grain	280.0
Casava chip	130.0
Wheat bran	125.5
Broken rice	42.0
Molasses	30.0
Rice bran oil	19.8
Rapeseed meal	40.0
Palm kernel meal	30.0
Soybean meal	97.0
Extruded soybean	40.0
Pork meal, 530 g/kg CP	44.0
Fish meal, 620 g/kg CP	31.0
Calcium carbonate	81.2
Salt	2.0
L-Lysine	1.0
DL-Methionine	2.0
Mycotoxin adsorbents	1.8
Pigment	0.2
Premix [1]	2.5
Total	1000.0
Calculated Chemical Composition Analysis [2]	

Table 1. *Cont.*

Ingredients	g/kg, As-Fed Basis
Metabolizable energy, kcal/kg	2850
Crude protein	180.0
Crude fiber	40.2
Calcium	35.1
Available phosphorus	4.3
Total lysine	9.6
Total methionine	4.0
Total methionine and cystine	7.2

[1] Vitamin–mineral premix provided per kilogram of diet: vitamin A (trans-retinyl acetate), 12,000 IU; vitamin D3 (cholecalciferol), 3000 IU; vitamin E (all-rac-tocopherol-acetate), 12 mg; vitamin K3 (bisulfate menadione complex), 3.6 mg; vitamin B1, 1.4 mg; vitamin B2, 5.4 mg; vitamin B6, 4.2 mg; vitamin B12 (cyanocobalamin), 0.02 mg; nicotinic acid, 9 mg; pantothenic acid (D-calcium pantothenate), 9 mg; folic acid, 0.6 mg; biotin, 45 mg; choline chloride, 210 mg; selenium, 0.18 mg; cobalt, 0.3 mg; iodine, 1.08 mg; zinc, 60 mg; iron, 54 mg; manganese, 96 mg; copper, 12 mg; [2] The nutrient values were calculated based on the analyzed nutrient values according to NRC [22].

2.3. *Data Collection for Laying Performance and Quality*

Egg production and weight were recorded daily from each pen, including intact, dirty, and broken eggs. The average daily egg production was calculated based on weekly results as the following formula: [total number of eggs per pen/(number of hens per pen × days)] × 100. The total egg weight of each pen was divided by the total number of eggs per pen and expressed as the average egg weight. Egg mass multiplied by average egg weight by average daily egg production percentage. Feed intake was obtained by deducting the total feed weight offered during the week by the remaining feed weight. Average daily feed intake was modified for mortality and was calculated using the following equation: [total feed intake per pen/(number of hens per pen × days)]. The feed conversion ratio was calculated by dividing the average daily feed intake by egg mass.

An egg from each pen was collected weekly to analyze the egg quality during the 12-week feeding period as follows: yolk ratio, albumen ratio, eggshell ratio, albumen height, eggshell thickness, yolk color, and Haugh units. Dirty and cracked eggs were sorted out and excluded from analysis. After the breakout, eggshells, albumen, and yolk were separated and weighed using a digital precision scale. The ratio of eggshell, albumen, and yolk was calculated based on the weight of the whole egg. Yolk color, albumen height, and the Haugh unit were measured by an egg multitester (EMT-7300, Robotmation Co., Ltd., Tokyo, Japan). The shell thickness was measured at the sharp, blunt ends and equator after removing the shell membranes using a digital micrometer (MW200-01DBL, Moore & Wright Co., Ltd., Sheffield, UK).

2.4. *Measurement of Ammonia Emission in Litter*

The ammonia emission was measured at the litter level during the whole feeding trial (12-week experimental period), according to the modified method of Yamauchi et al. [23]. Twelve samples of fresh litter (rice hull) were collected weekly from 12 surface locations within each pen using a sampling spoon, and each one was transferred to a vinyl bag. Thus, a total of 60 litter samples were obtained per treatment every week (12 samples/pen × 5 pens/treatment, $n = 60$). From each litter sample, a total of 100 g was transferred to a tight container with a cap and a gap sealed with paraffin wax. Each container was placed in an incubator at 37 °C for one hour before being allowed to cool to room temperature for 30 min. A portable gas detector was used to measure the ammonia emission in each litter sample (Crowcon Portable Gas Detection; Crowcon Detection Instruments, Oxfordshire, UK). The gas collector tube was inserted into the air sampling hole, and the maximum ammonia emission values were recorded in ppm units. The ammonia emission was expressed as the average value of the 60 litter samples collected per treatment.

2.5. DNA Extraction and PCR Amplification

At the end of the feeding trial, five ducks from the control and treatment groups that showed the best laying performances were randomly chosen (one duck per replicate). The ducks were individually euthanized for cecal content collection and immediately transferred to a sterile 1.5 mL microcentrifuge tube and maintained at −80 °C until DNA extraction [3]. DNA was isolated from 200 mg of cecal contents from each duck using a QIAamp DNA Stool Mini Kit (Qiagen, Valencia, CA, USA) with modifications to improve DNA concentration. Isolated DNA concentrations and agarose gel electrophoresis were measured by spectrophotometry (NanoDrop 2000, Thermo Science, Wilmington, DE, USA) at OD A260/A280 ratio >1.8. The extracted DNA was normalized to a 50 ng/mL concentration and stored at −20 °C. The diluted DNA was then used as a template for the PCR amplification of the bacterial 16S rRNA genes, and the V3–V4 variable regions were amplified with specific primers for bacterial diversity analysis [3]. Equal amounts of purified amplicons were individualized for next-generation sequencing (NGS).

2.6. Determination of Cecal Microbiome by Next-Generation Sequencing (NGS)

According to the manufacturer's instructions, the sequencing library and barcode were constructed using the TruSeq® DNA PCR-sample-free preparation kit (Illumina, San Diego, CA, USA). The sequencing was conducted on the Illumina MiSeq platform, which produced 250 bp paired-end readings. Paired-end reads were assigned to samples using their unique barcodes and then truncated by removing the barcode and primer sequences. Paired-end reads were merged using FLASH (V1.2.7), a fast and accurate analysis tool designed to merge paired-end reads when at least some of them overlap with reads generated from the opposite end of the same DNA fragment, and the splicing sequences were referred to as raw tags. Specific filtering conditions were applied to the raw tags using the Qiime (V1.9.1) (http://qiime.org/scripts/split_libraries_fastq.html; accessed on 8 March 2022) quality-controlled process to obtain high-quality clean tags. The tags were compared against the reference database (Gold database) using the UCHIME algorithm (https://drive5.com/usearch/manual/chimeras.html; accessed on 8 March 2022) to detect and remove chimera sequences. The remaining sequences were considered effective clean reads.

All clean reads cluster analyses were performed by the Uparse software (Uparse v7.0.1001, http://drive5.com/uparse/; accessed on 8 March 2022) using all the effective tags. Sequences with ≥97% similarity were assigned to the same Operation Taxonomic Unit (OTUs) and the highest frequency sequence was selected as the representative for each OTU sequence. The Mothur software was used for each representative sequence to annotate species at each taxonomic rank (kingdom, phylum, class, order, family, genus, species) using the SILVA Database's SSUrRNA database. The phylogenetic relationship of all OTU representative sequences was obtained using MUSCLE version 3.8.31 (http://www.drive5.com/muscle/; accessed on 8 March 2022), which can compare multiple sequences rapidly. The amount of OTU information was normalized using a standard sequence number corresponding to the sample with the fewest sequences. Lastly, alpha and beta diversity analyses were performed using the output normalized data.

Alpha diversity was applied to analyze the complexity of biodiversity for a sample through six indices, including observed species, Chao1, Shannon, Simpson, ACE, and good coverage. All these indices in our samples were calculated with QIIME version 1.7.0 and displayed with R software version 2.15.3. The unweighted UniFrac distance matrix was calculated for beta diversity metrics and the unweighted pair-group method with arithmetic means (UPGMA) cluster trees was applied using QIIME version 1.9.1. Principal component analysis, principal coordinate analysis (PCoA), and nonmetric multidimensional scaling figures were generated using R scripts [3].

2.7. Statistical Analysis

Egg performance, egg quality, and ammonia emissions from litter were analyzed using a one-way analysis of variance (ANOVA) with the Statistical Package for the Social Sciences (SPSS), version 17.0 (SPSS Inc., Chicago, IL, USA). The results are presented as group means along with the pooled standard error of the mean (SEM). Treatment differences were determined using Duncan's multiple range test, with statistical significance set at $p < 0.05$.

3. Results

3.1. Laying Productivity and Egg Quality

The effects of different levels of dietary *Bacillus toyonensis* BCT-7112^T on the laying productivity and egg quality of laying ducks are reported in Table 2. Increasing dietary *Bacillus toyonensis* BCT-7112^T levels led to a tendency for egg mass to be higher compared with the control group, and responses were highest ($p < 0.05$) with 1000 ppm *Bacillus toyonensis* BCT-7112^T. Compared to the control-fed group, all laying ducks *fed Bacillus toyonensis* BCT-7112^T had a significant improvement ($p < 0.001$) in average egg weight. However, in terms of the average daily egg production, feed conversion ratio, and feed intake, there were no significant differences ($p > 0.05$) among dietary treatments. Feeding *Bacillus toyonensis* BCT-7112^T increased eggshell thickness significantly ($p = 0.007$) compared to the control group. In contrast, no significant changes were detected for the yolk-to-albumen ratio, eggshell-to-albumen ratio, albumen height, Haugh units, or yolk color in response to *Bacillus toyonensis* BCT-7112^T compared to the control diet.

Table 2. Laying productivity and egg quality of laying ducks fed dietary *Bacillus toyonensis* BCT-7112^T from 32 to 44 weeks of age.

Item	Dietary *Bacillus toyonensis* BCT-7112^T, ppm				SEM [1]	*p*-Value
	Control $n = 55$	200 $n = 55$	500 $n = 55$	1000 $n = 55$		
Egg performance						
Average daily egg production, %	80.3	81.0	80.4	82.9	1.21	0.876
Average egg weight, g/egg	69.3 ^{a}	70.3 ^{b}	70.8 ^{b}	71.1 ^{b}	0.16	<0.001
Egg mass, g/h/d	55.7 ^{a}	57.0 ^{ab}	57.0 ^{ab}	59.0 ^{b}	0.40	0.031
Average daily feed intake, g/h/d	142.2	143.5	139.7	141.1	1.68	0.894
Feed conversion ratio	2.55	2.51	2.45	2.39	0.04	0.494
Egg quality						
Yolk ratio, %	33.0	33.3	32.5	32.5	0.10	0.400
Albumen ratio, %	56.1	55.8	56.6	56.6	0.11	0.150
Eggshell ratio, %	10.9	10.9	10.9	10.9	0.04	0.998
Eggshell thickness, µm	344.5 ^{b}	353.4 ^{a}	351.5 ^{a}	352.1 ^{a}	1.00	0.007
Albumen height, mm	5.1	5.1	5.0	5.1	0.05	0.937
Haugh units	62.0	62.2	61.4	62.7	0.43	0.779
Yolk color	14.7	14.8	14.8	14.7	0.02	0.243

^{a,b} Different superscripts letters in the same row indicate significant differences ($p < 0.05$); [1] SEM, standard error of the mean.

3.2. Quantity of Ammonia Emission in Litter

The ammonia emission (NH_3) in litter responses is depicted in Figure 1. Compared to the control group, all groups supplemented with *Bacillus toyonensis* BCT-7112^T had significantly reduced ammonia emission in the litter ($p < 0.05$).

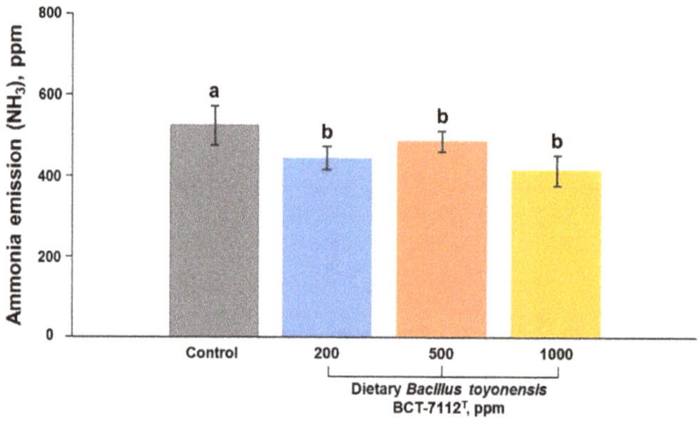

Figure 1. Average values of ammonia (NH_3) emission in the litter (ppm) of laying ducks fed dietary *Bacillus toyonensis* BCT-7112^T (200, 500, 1000 ppm) from 32 to 44 weeks of age compared to the control group ($n = 60$ per group). Error bars represent the standard error of the mean (SEM = 10.57). Different superscripts letters indicate significant differences ($p < 0.05$).

3.3. Analysis of Microbial Communities in Laying Duck Cecum

The gut microbiome analysis in this study was performed in the control group and the 1000 ppm *Bacillus toyonensis* BCT-7112^T group, which had the best laying performance among the supplemented groups. Basic information from various samples, such as effective tag data, low-frequency tag data, and tags annotation data, was collected to construct OTUs (Figure 2). To examine the species diversity in each sample, all effective tags were grouped into OTUs based on 97% DNA sequence similarity. The summary of classified sequence and sequence number for each sample are reported in Figure 3. Six different alpha diversity indices, including observed species, Shannon, Simpson Chao1, ACE, and Good's coverage, were used to access the sequence richness of the sample and analyze diversity (Table 3).

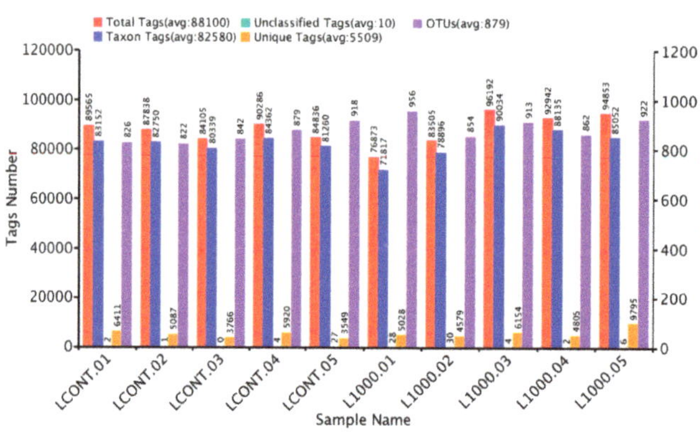

Figure 2. Summarization of the tags and OTU number of each sample in control (LCONT) and 1000 ppm (L1000) groups.

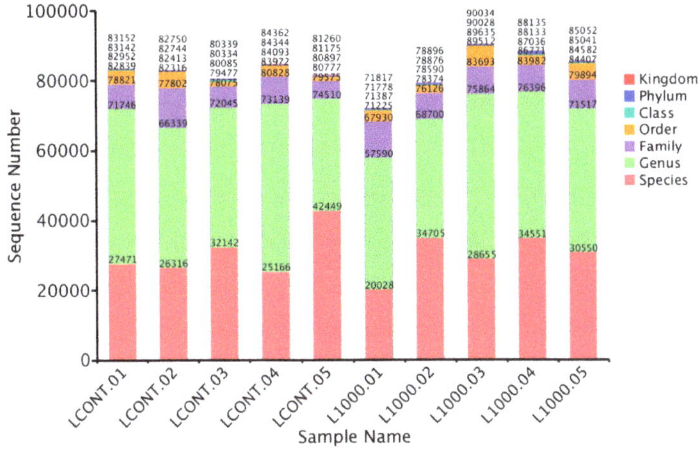

Figure 3. Summarization of the classified sequence and sequence number of each sample in the control (LCONT) and 1000 ppm (L1000) groups.

Table 3. Alpha diversity index statistics by sample.

Treatments	Observed Species	Shannon	Simpson	Chao1	ACE
Control	803.00	6.22	0.94	857.35	857.91
1000 ppm	1760.80	7.46	0.97	1894.17	1878.61
SEM	28.11	0.12	0.17	20.57	53.31
p-value	<0.01	0.02	0.13	<0.01	<0.01

ACE = abundance-based coverage estimator. Values are mean ± pooled SEM (n = 5).

The microbial population in the control group was dominated by the following phyla: Bacteroidetes (35.12%), Firmicutes (34.93%), Fusobacteria (7.38%), Proteobacteria (6.72%), and Deferribacteres (7.05%). In contrast, Bacteroidetes (40.52%), Firmicutes (34.08%), Proteobacteria (9.90%), Deferribacteres (9.54%), and Fusobacteria (2.24%) were the predominant phyla of the cecal microbiota in the 1000 ppm group (Figure 4).

The top genera in the control group were *Bacteroides* (17.17%), *Fusobacterium* (12.42%), *Prevotella* 7 (12.42%), *Mucispirillum* (7.05%), *Faecalibacterium* (5.25%), and *Anaerobiospirillum* (2.27%). In contrast, *Bacteroides* (22.23%), *Mucispirllum* (9.54%), *Prevotella 7* (5.55%), *Anaerobiospirillum* (4.43%), *Faecalibacterium* (3.83%), and *Fusobacterium* (2.24%) were predominant in the microbiota from the 1000 ppm group (Figure 5). The explicit comparison of microbial communities according to their composition is known as beta diversity. In order to analyze the microbial communities between each set of community samples, the beta diversity evaluates their differences. The microbial communities in the cecum were visually depicted through a principal component analysis (PCA) (Figure 5). These data suggested that the gut microbiome may have changed due to dietary and environmental stressors, with the group fed 1000 ppm *Bacillus toyonensis* BCT-7112^{T} having a greater microbial diversity compared to the control.

The R software was used to cluster and plot the heatmap results comparing the 35 most abundant genera between the control and 1000 ppm groups. When comparing the two groups, genera with a comparatively greater abundance in the corresponding samples are displayed in red, while those with relatively lower abundance are shown in blue (Figure 6). The heatmap results comparing the 1000 ppm and control groups revealed a slightly higher relative abundance at the phylum level of *Proteobacteria*, *Actinobacteria*, *Synergistetes*, *Thermomicrobia*, and *Spirochaetes* in the 1000 ppm group. In contrast, *Verrucomicrobia*, *Euryarchaeota*, *Saccaribacteria*, *Chloroflexi*, *Atribacteria*, and *Fusobacteria* were more abundant in the control group (Figure 6).

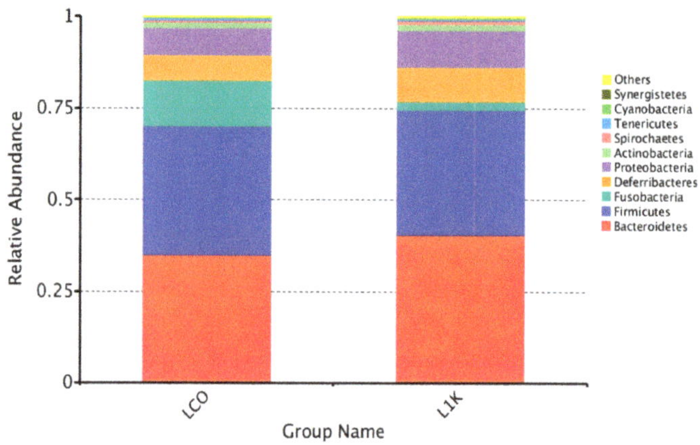

(**A**)

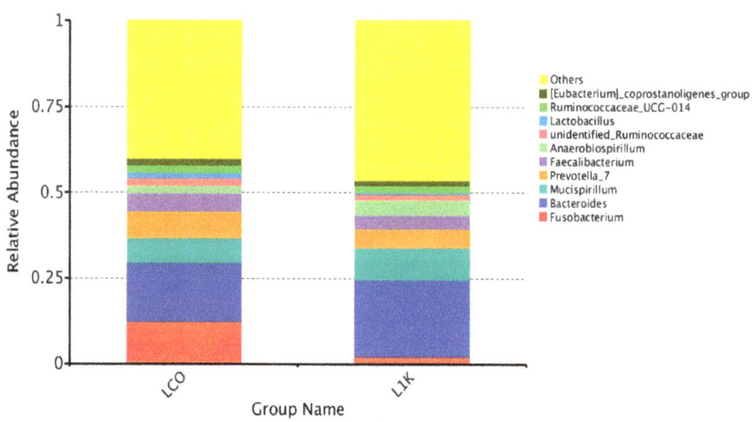

(**B**)

Figure 4. Relative abundance of cecal bacteria in laying ducks at the phylum (**A**) and genus (**B**) levels in the control (LCO) and 1000 ppm (L1K) groups.

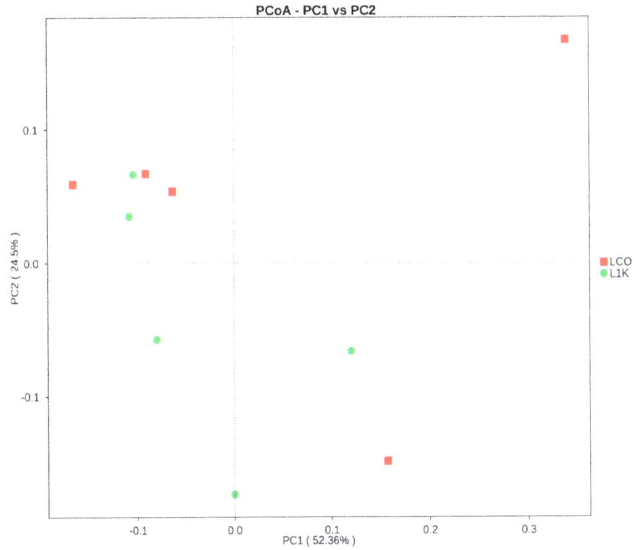

Figure 5. Principal component analysis (PCA) of the microbial community of the laying ducks in the control (LCO) and 1000 ppm (L1K) groups.

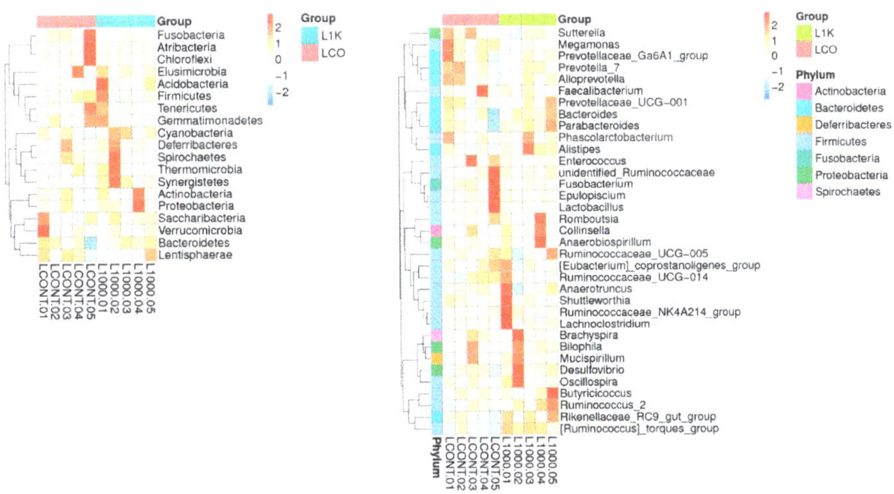

Figure 6. Heatmap analysis of 35 most abundant phyla and genera of each bird was carried out in the control (LCON.01–LCON.05) group with basal diet and 1000 ppm (L1000.01–1000.05 ppm) group with the B. toyonensis BCT-7112^{T} supplement. Notes: Plots by sample name on the X-axis and the Y-axis represent the phylum. The absolute value of 'z' represents the distance between the raw score and the mean of the standard deviation. 'z' is negative when the raw score is below the mean, and vice-versa (rank from −2 to 2).

4. Discussion

Numerous alternatives for antibiotics are being investigated in experiments to determine their efficacy in both humans and animals. One alternative that is often utilized as a feed supplement in the poultry farming industry is probiotics, i.e., a non-antibiotic method with a number of positive impacts on the modulation of productive performance in poultry, including growth performance, egg performance, immunological response, nutrient digestion, and gut health [3,15,16,24,25].

As the results reveal, the inclusion of dietary *B. toyonensis* BCT-7112^T supplementation resulted in a remarkable enhancement in some measures of egg production in laying ducks. Suswoyo et al. [26] reported that Tegal laying ducks fed a probiotic mixture (*Lactobacillus* sp. and *Saccharomyces* sp.) showed the highest egg mass ($p < 0.05$). Based on a study in White layers conducted by Abdelqader et al. [23], the inclusion of *B. subtilis* PB6 at a dosage of 0.5–1.0 g/kg in the basal diet of hens resulted in significant improvements in both egg mass and egg weight. Egg mass also was significantly increased in hens fed a diet containing *B. subtilis* yb-114246 [27].

It is possible that the increased egg production could be a consequence of a boost in nutrient utilization [28], which is related to the positive effect of probiotics on the beneficial gut microbes [29] and gut architecture [30]. Furthermore, the potential mechanisms through which probiotics enhance these parameters can be ascribed to promoting dietary nutrient digestion and utilization through the upregulation of host digestive enzyme activities [31,32]. Consistent with our recent research, we observed that feeding *B. toyonensis* BCT-7112^T improved the gastrointestinal health of commercial meat ducks by modulating the beneficial bacteria associated with nutritional digestibility [3]. According to our findings, the beneficial impacts on egg production and average egg weight were caused by feeding *B. toyonensis* BCT-7112^T. These occurrences could be attributed to improve nutrient absorption and metabolism, as probiotics have been shown to enhance gut microbiota composition and digestive enzyme activity, ultimately facilitating better nutrient utilization.

The lack of an effect on average daily feed intake, feed conversion ratio, and average daily egg production was similar to those of previous studies, e.g., Fathi et al. [33] in which *B. subtilis* did not affect egg production, feed intake, or FCR in laying hens. The dietary supplementation of *B. subtilis* TLRI 211-1 also did not alter the feed intake and egg production of Leghorn layers [34]. In a long-term experiment carried out by Guo et al. [35], it was shown that the inclusion of 10^5 to 10^8 CFU/kg *B. subtilis* CGMCC 1.921 in the diet of Hy-Line brown laying hens had no effect on feed intake or egg production rate. Adding *B. subtilis* ATCC PTA-6737 to the laying diet also had no statistically significant influence on FCR, daily feed intake, average egg weight, or laying rate [36]. Likewise, a feeding trial conducted on Pekin ducks also demonstrated that a dry direct-fed microbial diet containing *B. subtilis* (2.2×10^8 CFU/g) had no significant impact on weight gain, feed intake, and FCR during the 1–21 day period [37]. Our study suggests that the absence of changes in the overall egg productivity could be due to the fact that, while probiotics enhance nutrient absorption, their influence on reproductive hormone regulation and follicle recruitment may be limited. Additionally, the production stage, breed characteristics, and environmental conditions may modulate the extent to which probiotics impact the laying performance.

In other studies, a combination of *B. subtilis* and *Clostridium butyricum* as dietary supplements resulted in substantial enhancements in daily feed intake, egg production rate, and body weight of laying ducks throughout the later stages of production [38]. The supplementation of Brown-Nick layers' feed with a commercial probiotic (BioPlus 2B) increased egg production and FCR substantially ($p < 0.05$) [38]. Furthermore, the inclusion of *Clostridium butyricum* (5×10^4 or 1×10^5 CFU/g) in the diet improved egg performance of laying hens in the late-productive phase [39]. Clearly, data from the current study may differ from previous work due to factors such as the use of different bacterial strains, supplemental dose, environment, diet structure, animal models, breed, and age [33,40–42].

The lack of negative influence on all egg quality parameters, including yolk ratio, albumen ratio, eggshell ratio, albumen height, Haugh units, and yolk color, suggests that *B. toyonensis* BCT-7112^T can be used as a feed supplement in laying duck production at levels ranging from 200 to 1000 ppm without a negative impact on the overall egg qualities. Furthermore, the increased eggshell thickness due to feeding *B. toyonensis* BCT-7112^T supports data from previous studies with the dietary probiotic *B. subtilis* in laying hens [33], aged laying hens [43], and in broiler breeder hens [28,44], thus suggesting a similar effect of both probiotics. The results support data from previous studies indicating that adding *B. subtilis* to laying diets led to an increase in eggshell thickness and using the probiotic *B. subtilis* C-3102 as a feed additive in the basal diet improved eggshell thickness in aged laying hens. Also, supplementation with *B. subtilis* PB6 led to an increase in eggshell thickness in broiler breeder chickens. It is common knowledge that the eggshells of avian species consist primarily of calcium, along with various trace minerals. Increased calcium carbonate deposition in the eggshell can be linked to a pH reduction in the digestive system, which increases the solubility and digestibility of phosphorus and calcium [45]. Chang and Yu [46] noted that *Bacillus species* have the ability to generate short-chain fatty acids (SCFAs) by breaking down complex polysaccharides to modulate the gut microbiota composition due to decreased intestinal pH. Thus, it is possible that the enhancement in eggshell thickness in the current study was related to the reduction in intestinal pH caused by SCFA secretion during microbial fermentation, leading to an amplified calcium deposition on the shell glands.

However, while *B. subtilis* and *B. toyonensis* share common probiotic properties, their effects and underlying mechanisms can differ based on the host species and specific health parameters targeted. In order to have a complete understanding of these differences, further comparative research is necessary.

Decreasing NH_3 emissions is a current challenge for poultry producers. Manure nitrogen content, temperature, pH, humidity, ventilation rate, and air velocity are the main factors determining NH_3 volatility in poultry litter [47]. The production of NH_3 in livestock housing systems is primarily attributed to the nitrogen (e.g., protein, urea, and uric acid) present in the undigested excreta from protein-rich feedstuffs [48]. Approximately 60–70% of the nitrogen present in poultry manure is excreted as uric acid and urea [49]. There have been worldwide attempts to mitigate NH_3 emissions, employing adsorbent and acidifying additives [50,51], and decreasing the CP content in the diet [52]. Recent studies on the effects of nutrition on ammonia production in poultry have focused mainly on broilers and laying hens, but the influence of probiotics in ducks has rarely been investigated. In this context, probiotics can encourage the digestive activity of proteases and peptidases, promoting the absorption and transport of small peptides and amino acids through epithelial cells [53], and decreasing uric acid–nitrogen excretion [48]. Based on our results, including *B. toyonensis* BCT-7112^T could reduce the amount of NH_3 emission released from the laying duck litter. The observed mechanism appears to be associated with enhanced protein digestibility and nitrogen utilization efficiency, consequently resulting in reduced excretion of nitrogenous waste products.

Recently, it has been reported that ammonia causes changes in the abundance of cecal microflora and a decline in the productive performance of laying ducks [5]. Vilela et al. [54] demonstrated that uric acid degradation by microbial uricase in excreta was the primary cause of NH_3 synthesis. Similarly, Santoso et al. [55] observed that suppressing urease-producing microbiota by feeding a dried culture of *B. subtilis* significantly decreased NH_3 concentration. Multi-strain probiotics (*B. subtilis, P. guilliermondii*, and *L. plantarum*) also reduced NH_3 production by approximately 46% during in vitro gas fermentation using the cecal contents of laying hens [56]. Our study did not investigate gut microbial urease activity, nitrogen metabolism pathways, feed digestibility, or protein utilization. Future studies should conduct further analyses to understand the observed effects better.

The cecum of poultry is the primary location of fermentation due to the presence of a diverse microbial community, and it plays an essential function in the health and intestinal development of this species [57]. The cecal microbiome analysis was similar to previous studies, which indicated that Firmicutes, Proteobacteria, and Bacteroidetes are the predominant taxa in the duck cecum and ileum, comprising over 80% of the gut microbiota [3,58–60]. As evidenced by the OTU number of the intestinal microflora in the probiotic group in the current experiment being significantly higher than that of the control group, feeding probiotic *B. toyonensis* was advantageous in modulating the colonization and diversity of microbial communities. In terms of the diversity of microbial communities in the host gut, a moderate level of variety is beneficial for a stable gut microbiota. Not all gut microorganisms are useful: some are harmful, while others are conditioned. Some bacteria have little influence on the host, constituting the diversity and richness of the gut microflora [57].

In the current study, Bacteroides was among the most prevalent bacterial taxa in the *B. toyonensis* probiotic group. They are necessary for breaking down complicated compounds into smaller components for host growth [61]. They have been known for breaking down long-chain polysaccharides, and their metabolism can result in the production of acetate, propionate, or succinate, which actively improves the intestinal environment for beneficial microbes. Our findings show that supplementing *B. toyonensis* BCT-7112^{T} in laying duck diets did not significantly increase the proportions of the genera *Lactobacillus* and *Bacillus* in the gut. However, Bacteroides species can acidify nutritional broths in vitro almost as efficiently as *Bacillus and Lactobacillus* [62,63]. Both are known for producing organic acids, such as lactic acid, which play a crucial role in maintaining gut microbiome balance. The increase in these acid-producing bacteria can create an acidic environment in the gut, inhibiting the growth of pathogens and promoting digestive health [3,61,63]. Our result is comparable to that of Sun et al. [57], who observed that adding compound probiotics to duck diets increased the number of OTUs and the abundance of Bacteroidetes. We also observed that adding 1000 ppm of *B. toyonensis* BCT-7112^{T} to the diets decreased the amount of harmful bacteria, such as Fusobacterium. Kollarcikova et al. [64] reported that the cecal microbiota of poorly performing adult hens was characterized by a high abundance of Fusobacterium (around 20% of cecal microbiota). Moreover, other studies reported that Fusobacterium is a proinflammatory bacterium [65].

Our findings indicate that supplementing with *B. toyonensis* BCT-7112^{T} enhanced the relative abundance of beneficial bacteria while reducing the relative abundance of harmful bacteria. However, to properly understand the impact of probiotic *B. toyonensis* BCT-7112^{T} on the relationship of the gut microbiome, transcriptome, and metabolome in ducks, we recommend performing a shotgun metagenomic study and RNA-seq analysis using next-generation sequencing.

5. Conclusions

In conclusion, the supplementation of *Bacillus toyonensis* BCT-7112^{T} in laying duck diets at the different inclusion levels tested significantly increased the average egg weight, egg mass, and eggshell thickness and decreased ammonia emission in the litter. At 1000 ppm (1000 mg/kg), *Bacillus toyonensis,* BCT-7112^{T}, significantly increased the egg mass, and the beneficial bacteria involved in nutrient digestibility were more predominant. These positive effects took place while pathogenic bacteria numbers decreased. Overall, *Bacillus toyonensis* BCT-7112^T has the potential to serve as an effective probiotic and may function as an alternative to antibiotic growth promoters (AGPs) in the production of laying ducks. In order to confirm the effects observed in this study, further studies should be performed on laying ducks using different scenarios (e.g., different duck breeds, breeding environments, and management modes).

Author Contributions: Conceptualization, T.I., R.C., and G.J.; methodology, T.I., R.C., and G.J.; software, T.I. and R.C.; formal analysis, T.I. and R.C.; data curation, T.I. and R.C.; writing—original draft preparation, T.I. and R.C.; writing—review and editing, T.I., R.C., W.T., S.N., Y.S., G.J., and J.J.L.; supervision, T.I., R.C., and G.J.; project administration, R.C.; funding acquisition, T.I. and R.C. All authors have read and agreed to the published version of the manuscript.

Funding: Financial support for this study was provided by Rubinum S.A., Spain, and Center of Excellence for Agricultural and Livestock Innovation, Faculty of Agriculture Natural Resources and Environment, Naresuan University, Phitsanulok, Thailand (grant number: R2561A064).

Institutional Review Board Statement: This study was conducted at the Poultry Nutrition and Feed Testing Unit, Faculty of Agriculture, Natural Resources, and Environment, Naresuan University, Phitsanulok Province, Thailand. All animal procedures were approved by the Naresuan University Agricultural Animal Care and Use Committee (NUAACUC; Approval ID 61 01 009) and followed the ethical guidelines for animal experimentation set by the National Research Council of Thailand.

Informed Consent Statement: Not applicable.

Data Availability Statement: Data produced in this study are available from the corresponding. authors upon reasonable request.

Acknowledgments: The authors express their gratitude to the Researcher from the Center of Excellence for Agricultural and Livestock Innovation, Faculty of Agriculture, Natural Resources, and Environment, Naresuan University, Phitsanulok, Thailand, for support. We also acknowledge Rubinum S.A., Spain, for providing facilities, products, and knowledge exchange. Additionally, we sincerely appreciate the valuable contributions of graduate and undergraduate students from the Division of Animal Science and Feed Technology at Naresuan University.

Conflicts of Interest: G.J. was employed by Rubinum S.A., Spain. The remaining authors declare that the research was conducted without any relationships that could be construed as a potential conflict of interest.

References

1. Likittrakulwong, W.; Moonsatan, S.; Incharoen, T. Enhancement of tibia bone and eggshell hardness through the supplementation of bio-calcium derived from fish bone mixed with chelated trace minerals and vitamin D3 in laying duck diet. *Vet. Anim. Sci.* **2021**, *8*, 100204. [CrossRef] [PubMed]
2. Charoensook, R.; Knorr, C.; Brenig, B.; Gatphayak, K. Thai pigs and cattle production, genetic diversity of livestock and strategies for preserving animal genetic resources. *Maejo Int. J. Sci. Technol.* **2013**, *7*, 113–132.
3. Pechrkong, T.; Incharoen, T.; Hwanhlem, N.; Kaewkong, W.; Subsoontorn, P.; Tartrakoon, W.; Numthuam, S.; Jiménez, G.; Charoensook, R. Effect of *Bacillus toyonensis* BCT-7112^T supplementation on growth performance, intestinal morphology, immune-related gene expression, and gut microbiome in Barbary ducks. *Poult. Sci.* **2023**, *102*, 102991. [CrossRef] [PubMed]
4. Incharoen, T.; Charoensook, R.; Onoda, S.; Tartrakoon, W.; Numthuam, S.; Pechkong, T. The effects of heat-killed *Lactobacillus plantarum* L-137 supplementation on growth performance, intestinal morphology, and immune-related gene expression in broiler chickens. *Anim. Feed Sci. Technol.* **2019**, *257*, 114272. [CrossRef]
5. Tao, Z.; Xu, W.; Zhu, C.; Zhang, S.; Shi, Z.; Song, W.; Liu, H.; Li, H. Effects of ammonia on intestinal microflora and productive performance of laying ducks. *Poult. Sci.* **2019**, *98*, 1947–1959. [CrossRef]
6. Hong, H.A.; Duc, L.H.; Cutting, S.M. The use of bacterial spore formers as probiotics. *FEMS Microbiol. Rev.* **2005**, *29*, 813–835. [CrossRef]

7. Hwanhlem, N.; Salaipeth, L.; Charoensook, R.; Kanjan, P.; Maneerat, S. Lactic Acid Bacteria from Gamecock and Goat Originating from Phitsanulok, Thailand: Isolation, Identification, Technological Properties and Probiotic Potential. *J. Microbiol. Biotechnol.* **2022**, *32*, 355–364. [CrossRef]
8. Jiménez, G.; Urdiain, M.; Cifuentes, A.; López-López, A.; Blanch, A.R.; Tamames, J.; Kämpfer, P.; Kolstø, A.B.; Ramón, D.; Martínez, J.F.; et al. Description of *Bacillus toyonensis* sp. nov., a novel species of the *Bacillus cereus* group, and pairwise genome comparisons of the species of the group by means of ANI calculations. *Syst. Appl. Microbiol.* **2013**, *36*, 383–391. [CrossRef]
9. Pinheiro, V.; Mourão, J.; Jiménez, G. Influence of Toyocerin® (*Bacillus cereus* var. *Toyoi*) on breeding performance of primiparous rabbit does. *World Rabbit Sci.* **2007**, *15*, 178–188.
10. Lodemann, U.; Lorenz, B.; Weyrauch, K.; Martens, H. Effects of *Bacillus cereus* var. *Toyoi* as probiotic feed supplement on intestinal transport and barrier function in piglets. *Arch. Anim. Nutr.* **2008**, *62*, 87–106.
11. Kantas, D.; Papatsiros, V.; Tassis, P.; Giavasis, I.; Bouki, P.; Tzika, E. A feed additive containing *Bacillus toyonensis* (Toyocerin) protects against enteric pathogens in postweaning piglets. *J. Appl. Microbiol.* **2014**, *118*, 727–738. [CrossRef]
12. Männer, K. Efficacy of *Bacillus cereus* var. *toyoi* on performance, faecal microflora and immune response in rearing calves. In *5th European Spores Conference Proceedings (Abstracts for Oral Presentations)*; Royal Holloway University of London: London, UK, 2012; p. 26.
13. Roos, T.B.; de Moraes, C.M.; Sturbelle, R.T.; Dummer, L.A.; Fischer, G.; Leite, F.P.L. Probiotics *Bacillus toyonensis* and *Saccharomyces boulardii* improve the vaccine immune response to Bovine herpesvirus type 5 in sheep. *Res. Vet. Sci.* **2018**, *117*, 260–265. [CrossRef]
14. Vilà, B.; Fontgibell, A.; Badiola, I.; Esteve-Garcia, E.; Jiménez, G.; Castillo, M.; Brufau, J. Reduction of *Salmonella enterica* var. *Enteritidis* colonization and invasion by *Bacillus cereus* var. *toyoi* inclusion in poultry feeds. *Poult. Sci.* **2009**, *88*, 975–979. [PubMed]
15. Abou-Kassem, D.E.; Badr, M.; Abdel-Moneim, A.-M.; Mahgoub, S.; Elaraby, G.M.; Taha, A.E.; Elshafie, M.M.; Alkhawtani, D.; El-Hack, M.E.; Ashour, E.A. Growth, carcass characteristics, meat quality and microbial aspects of growing quail fed diets enriched with two different types of probiotics (*Bacillus toyonensis* and *Bifidobacterium bifidum*). *Poult. Sci.* **2020**, *100*, 84–93. [CrossRef] [PubMed]
16. Batkowska, J.; Brodacki, A.; Tomczyk, G. The Influence of Probiotic Bacteria (*Bacillus Toyoi*) on Livability and Performance of Young Meat-Type Turkeys. *Rev. Bras. Ciência Avícola* **2015**, *17*, 433–438. [CrossRef]
17. Chamani, M. Efficacy of Bactocell® and Toyocerin® as probiotics on growth performance, blood parameters and intestinal morphometry of Turkey poults. *Iran. J. Appl. Anim. Sci.* **2016**, *6*, 211–218.
18. de Souza, D.M.; Suita, S.M.; Leite, F.P.L.; Romano, L.A.; Wasielesky, W.; Ballester, E.L.C. The use of probiotics during the nursery rearing of the pink shrimp *Farfantepenaeus brasiliensis* (Latreille, 1817) in a zero-exchange system. *Aquac. Res.* **2011**, *43*, 1828–1837. [CrossRef]
19. Gisbert, E.; Castillo, M.; Skalli, A.; Andree, K.B.; Badiola, I. *Bacillus cereus* var. *toyoi* promotes growth, affects the histological organization and microbiota of the intestinal mucosa in rainbow trout fingerlings. *J. Anim. Sci.* **2013**, *91*, 2766–2774.
20. Vasaï, F.; Ricaud, K.B.; Cauquil, L.; Daniel, P.; Peillod, C.; Gontier, K.; Tizaoui, A.; Bouchez, O.; Combes, S.; Davail, S. *Lactobacillus sakei* modulates mule duck microbiota in ileum and ceca during overfeeding. *Poult. Sci.* **2014**, *93*, 916–925. [CrossRef]
21. Zhao, Y.; Li, X.; Sun, S.; Chen, L.; Jin, J.; Liu, S.; Song, X.; Wu, C.; Lu, L. Protective role of dryland rearing on netting floors against mortality through gut microbiota-associated immune performance in Shaoxing ducks. *Poult. Sci.* **2019**, *98*, 4530–4538. [CrossRef]
22. NRC. *Nutrient Requirements of Poultry*; The National Academy Press: Washington, DC, USA, 1994.
23. Yamauchi, K.; Ruttavut, J.; Takenoyama, S. Effect of dietary bamboo charcoal powder including vinegar liquid on chicken performance and histological alterations of intestine. *J. Anim. Feed Sci.* **2010**, *19*, 257–268. [CrossRef]
24. Abdelqader, A.; Irshaid, R.; Al-Fataftah, A.R. Effects of dietary probiotic inclusion on performance, eggshell quality, cecal microflora composition, and tibia traits of laying hens in the late phase of production. *Trop. Anim. Health Prod.* **2013**, *45*, 1017–1024. [CrossRef] [PubMed]
25. Li, C.L.; Wang, J.; Zhang, H.J.; Wu, S.G.; Hui, Q.R.; Yang, C.B.; Fang, R.J.; Qi, G.H. Intestinal morphologic and microbiota responses to dietary *Bacillus* spp. in a broiler chicken model. *Front. Physiol.* **2018**, *9*, 1968. [CrossRef]
26. Suswoyo, I.; Ismoyowati, I.; Tugiyanti, E. Study on probiotic as antibiotic replacement to improve egg production in commercial duck farms. *Tech. BioChemMed* **2021**, *2*, 65–72. [CrossRef]
27. Yang, J.; Zhan, K.; Zhang, M. Effects of the use of a combination of two *Bacillus* species on performance, egg quality, small intestinal mucosal morphology, and cecal microbiota profile in aging laying hens. *Probiotics Antimicrob. Proteins* **2020**, *12*, 204–213. [CrossRef] [PubMed]
28. Souza, O.; Adams, C.; Rodrigues, B.; Krause, A.; Bonamigo, R.; Zavarize, K.; Stefanello, C. The impact of *Bacillus subtilis* PB6 and chromium propionate on the performance, egg quality and nutrient metabolizability of layer breeders. *Animals* **2021**, *11*, 3084. [CrossRef] [PubMed]

29. Xu, C.; Wei, F.; Yang, X.; Feng, Y.; Liu, D.; Hu, Y. Lactobacillus salivarius CML352 isolated from Chinese local breed chicken modulates the gut microbiota and improves intestinal health and egg quality in late-phase laying hens. *Microorganisms* **2022**, *10*, 726. [CrossRef]
30. Song, D.; Wang, Y.W.; Lu, Z.X.; Wang, W.W.; Miao, H.J.; Zhou, H.; Wang, L.; Li, A.K. Effects of dietary supplementation of microencapsulated *Enterococcus fecalis* and the extract of Camellia oleifera seed on laying performance, egg quality, serum biochemical parameters, and cecal microflora diversity in laying hens. *Poult. Sci.* **2019**, *98*, 2880–2887. [CrossRef]
31. Elmer, G.W. Probiotics: "living drugs". *Am. J. Health Syst. Pharm.* **2001**, *58*, 1101–1109. [CrossRef]
32. Lutful Kabir, S.M. The role of probiotics in the poultry industry. *Int. J. Mol. Sci.* **2009**, *10*, 3531–3546. [CrossRef]
33. Fathi, M.; Al-Homidan, I.; Al-Dokhail, A.; Ebeid, T.; Abou-Emera, O.; Alsagan, A. Effects of dietary probiotic (*Bacillus subtilis*) supplementation on productive performance, immune response and egg quality characteristics in laying hens under high ambient temperature. *Ital. J. Anim. Sci.* **2018**, *17*, 804–814. [CrossRef]
34. Tsai, M.Y.; Shih, B.L.; Liaw, R.B.; Chen, W.T.; Lee, T.Y.; Hung, H.W.; Hung, K.H.; Lin, Y.F. Effect of dietary supplementation of *Bacillus subtilis* TLRI 211-1 on laying performance, egg quality and blood characteristics of Leghorn layers. *Anim. Biosci.* **2023**, *36*, 609–618. [CrossRef] [PubMed]
35. Guo, J.R.; Dong, X.F.; Liu, S.; Tong, J.M. Effects of long-term *Bacillus subtilis* CGMCC 1.921 supplementation on performance, egg quality, and fecal and cecal microbiota of laying hens. *Poult. Sci.* **2017**, *96*, 1280–1289. [CrossRef]
36. Sobczak, A.; Kozłowski, K. The effect of a probiotic preparation containing *Bacillus subtilis* ATCC PTA-6737 on egg production and physiological parameters of laying hens. *Ann. Anim. Sci.* **2015**, *15*, 711–723. [CrossRef]
37. Mitin, H.; Zulkifli, I.; Che Jamri, M.H.; Zamzuri, N.A.; Samian, N.A.; Hussein, A.N.; Goh, Y.M.; Sazili, A.Q. Alleviation of catching and crating stress by dietary supplementation of *Bacillus subtilis* in Pekin ducks. *Animals* **2022**, *12*, 3479. [CrossRef]
38. Cao, Y.; Xun, M.; Ren, S.; Wang, J. Effects of dietary organic acids and probiotics on laying performance, egg quality, serum antioxidants and expressions of reproductive genes of laying ducks in the late phase of production. *Poult. Sci.* **2022**, *101*, 102189. [CrossRef]
39. Kurtoglu, V.; Kurtoglu, F.; Seker, E.; Coskun, B.; Balevi, T.; Polat, E.S. Effect of probiotic supplementation on laying hen diets on yield performance and serum and egg yolk cholesterol. *Food Addit. Contam.* **2004**, *21*, 817–823. [CrossRef] [PubMed]
40. Zhan, H.Q.; Dong, X.Y.; Li, L.L.; Zheng, Y.X.; Gong, Y.J.; Zou, X.T. Effects of dietary supplementation with *Clostridium butyricum* on laying performance, egg quality, serum parameters, and cecal microflora of laying hens in the late phase of production. *Poult. Sci.* **2019**, *98*, 896–903. [CrossRef]
41. Yıldırım, A.; Kraimi, N.; Constantin, P.; Mercerand, F.; Leterrier, C. Effects of tryptophan and probiotic supplementation on growth and behavior in quail. *Poult. Sci.* **2020**, *99*, 5206–5213. [CrossRef]
42. Chen, P.; Xu, T.; Zhang, C.; Tong, X.; Shaukat, A.; He, Y.; Liu, K.; Huang, S. Effects of probiotics and gut microbiota on bone metabolism in chickens: A review. *Metabolites* **2022**, *12*, 1000. [CrossRef]
43. Nishiyama, T.; Nakagawa, K.; Imabayashi, T.; Iwatani, S.; Yamamoto, N.; Tsushima, N. Probiotic *Bacillus subtilis* C-3102 improves eggshell quality after forced molting in aged laying hens. *J. Poult. Sci.* **2021**, *58*, 230–237. [CrossRef] [PubMed]
44. Darsi, E.; Zhaghari, M. Effects of *Bacillus subtilis* PB6 supplementation on productive performance, egg quality and hatchability in broiler breeder hens under commercial farm condition. *J. Appl. Anim. Res.* **2021**, *49*, 109–117. [CrossRef]
45. Swiatkiewicz, S.; Arczewska-Wlosek, A. Prebiotic fructans and organic acids as feed additives improving mineral availability. *World Poult. Sci. J.* **2012**, *68*, 269–279. [CrossRef]
46. Chang, W.Y.; Yu, Y.H. Effect of Bacillus species-fermented products and essential oils on growth performance, gut morphology, cecal short-chain fatty acid levels, and microbiota community in broilers. *Poult. Sci.* **2022**, *101*, 101970. [CrossRef]
47. Swelum, A.A.; El-Saadony, M.T.; El-Hack, M.E.A.; Ghanima, M.M.A.; Shukry, M.; Alhotan, R.A.; Hussein, E.O.S.; Suliman, G.M.; Ba-Awadh, H.; Ammari, A.A.; et al. Ammonia emissions in poultry houses and microbial nitrification as a promising reduction strategy. *Sci. Total Environ.* **2021**, *781*, 146978. [CrossRef]
48. Such, N.; Mezőlaki, Á.; Rawash, M.A.; Tewelde, K.G.; Pál, L.; Wágner, L.; Schermann, K.; Poór, J.; Dublecz, K. Diet composition and using probiotics or symbiotics can modify the urinary and faecal nitrogen ratio of broiler chicken's excreta and also the dynamics of in vitro ammonia emission. *Animals* **2023**, *13*, 332. [CrossRef] [PubMed]
49. Nahm, K.H. Evaluation of the nitrogen content in poultry manure. *World Poult. Sci. J.* **2003**, *59*, 77–88. [CrossRef]
50. De Laune, P.B.; Moore, P.A., Jr.; Daniel, T.C.; Lemunyon, J.L. Effect of chemical and microbial amendments on ammonia volatilization from composting poultry litter. *J. Environ. Qual.* **2004**, *33*, 728–734. [CrossRef]
51. Wlazło, Ł.; Nowakowicz-Dębek, B.; Kapica, J.; Kwiecień, M.; Pawlak, H. Removal of ammonia from poultry manure by aluminosilicates. *J. Environ. Manag.* **2016**, *183*, 722–725. [CrossRef]
52. van Emous, R.A.; Winkel, A.; Aarnink, A.J.A. Effects of dietary crude protein levels on ammonia emission, litter and manure composition, N losses, and water intake in broiler breeders. *Poult. Sci.* **2019**, *98*, 6618–6625. [CrossRef]

53. Wang, J.; Ji, H. Influence of probiotics on dietary protein digestion and utilization in the gastrointestinal tract. *Curr. Protein Pept. Sci.* **2019**, *20*, 125–131. [CrossRef] [PubMed]
54. Vilela, M.d.O.; Gates, R.S.; Souza, C.F.; Teles Junior, C.G.S.; Sousa, F.C. Nitrogen transformation stages into ammonia in broiler production: Sources, deposition, transformation, and emission into the environment. *DYNA* **2020**, *87*, 221–228. [CrossRef]
55. Santoso, U.; Ohtani, S.; Tanaka, K.; Sakaida, M.; Ohtani, S.; Tanaka, K.; Sakaida, M. Dried *Bacillus subtilis* culture reduced ammonia gas release in poultry house. *Asian-Australas. J. Anim. Sci.* **1999**, *12*, 806–809. [CrossRef]
56. Mi, J.; Chen, X.; Liao, X. Screening of single or combined administration of 9 probiotics to reduce ammonia emissions from laying hens. *Poult. Sci.* **2019**, *98*, 3977–3988. [CrossRef]
57. Sun, H.; Du, X.; Zeng, T.; Ruan, S.; Li, G.; Tao, Z.; Xu, W.; Lu, L. Effects of compound probiotics on cecal microbiome and metabolome of Shaoxing duck. *Front. Microbiol.* **2022**, *12*, 813598. [CrossRef]
58. Best, A.A.; Porter, A.L.; Fraley, S.M.; Fraley, G.S. Characterization of gut microbiome dynamics in developing Pekin ducks and impact of management system. *Front. Microbiol.* **2017**, *7*, 2125. [CrossRef]
59. Wang, S.; Chen, L.; He, M.; Shen, J.; Li, G.; Tao, Z.; Wu, R.; Lu, L. Different rearing conditions alter gut microbiota composition and host physiology in Shaoxing ducks. *Sci. Rep.* **2018**, *8*, 7387. [CrossRef] [PubMed]
60. Naumova, N.B.; Alikina, T.Y.; Zolotova, N.S.; Konev, A.V.; Pleshakova, V.I.; Lescheva, N.A.; Kabilov, M.R. Bacillus-based probiotic treatment modified bacteriobiome diversity in duck feces. *Agriculture* **2021**, *11*, 406. [CrossRef]
61. Lan, P.T.N.; Sakamoto, M.; Sakata, S.; Benno, Y. *Bacteroides barnesiae* sp. nov., *Bacteroides salanitronis* sp. nov. and *Bacteroides gallinarum* sp. nov., isolated from chicken caecum. *Int. J. Syst. Evol. Microbiol.* **2006**, *56*, 2853–2859. [CrossRef]
62. Line, J.E.; Hiett, K.L.; Guard-Bouldin, J.; Seal, B.S. Differential carbon source utilization by *Campylobacter jejuni* 11,168 in response to growth temperature variation. *J. Microbiol. Methods* **2010**, *80*, 198–202. [CrossRef]
63. Polansky, O.; Sekelova, Z.; Faldynova, M.; Sebkova, A.; Sisak, F.; Rychlik, I. Important metabolic pathways and biological processes expressed by chicken cecal microbiota. *Appl. Environ. Microbiol.* **2015**, *82*, 1569–1576. [CrossRef] [PubMed]
64. Kollarcikova, M.; Kubasova, T.; Karasova, D.; Crhanova, M.; Cejkova, D.; Sisak, F.; Rychlik, I. Use of 16S rRNA gene sequencing for prediction of new opportunistic pathogens in chicken ileal and cecal microbiota. *Poult. Sci.* **2019**, *98*, 2347–2353. [CrossRef] [PubMed]
65. Bai, Y.; Wang, R.; Yang, Y.; Li, R.R.; Wu, X.T. Folic Acid Absorption Characteristics and Effect on Cecal Microbiota of Laying Hens. *Front. Vet. Sci.* **2021**, *8*, 720851. [CrossRef] [PubMed]

veterinary sciences

MDPI

Article

The CRISPR/Cas9-Mediated Knockout of *VgrG2* in Wild Pathogenic *E. coli* to Alleviate the Effects on Cell Damage and Autophagy

Tian-Ling Pan [1,†], Jin-Long Cha [1,†], Hao Wang [2,†], Jing-Song Zhang [1], Jin-Long Xiao [1], Jue Shen [1], Meng Zhou [2], Yue Li [1], Jin-Zhi Ma [1], Kai-Yuan Zhao [1], Yong-Kang Zhang [1], Peng Xiao [1,*] and Hong Gao [1,*]

1 College of Veterinary Medicine, Yunnan Agricultural University, Kunming 650201, China; 13759420973@163.com (T.-L.P.); 18287258920@163.com (J.-L.C.); zhangjs228612@163.com (J.-S.Z.); xiaojl_h@163.com (J.-L.X.); nicetomeet_jue@163.com (J.S.); 18184868232@163.com (Y.L.); 15108713682@163.com (J.-Z.M.); z17716069408@163.com (K.-Y.Z.); zhangyongkang0313@163.com (Y.-K.Z.)

2 College of Food Science and Technology, Yunnan Agricultural University, Kunming 650201, China; wanghaoxu0117@163.com (H.W.); zhoumeng@mail.kiz.ac.cn (M.Z.)

* Correspondence: 2021005@ynau.edu.cn (P.X.); 1985001@ynau.edu.cn (H.G.)

† These authors contributed equally to this work.

Simple Summary: This study focuses on the efficacy of CRISPR/Cas9 technology in wild-type *E. coli*, a major pathogen responsible for foodborne illnesses and intestinal inflammation. The primary objective was to investigate the role of *VgrG2*, a key virulence factor, in enhancing *E. coli* pathogenicity. We screened *E. coli* strains harboring the *VgrG2* gene, which were also susceptible to both kanamycin and spectinomycin, and successfully knocked out the *VgrG2* gene using a dual-plasmid CRISPR/Cas9 system. The results showed that *VgrG2* activated the mTOR signaling pathway, inhibited *mTOR* and *p62* mRNA levels, upregulated autophagy-related genes, and increased the expression of the LC3-II protein. These findings enhance our understanding of the molecular mechanisms by which *VgrG2* contributes to cellular damage and provide valuable insights into genome editing strategies for pathogenic bacteria, offering potential approaches for controlling bacterial infections and their associated health impacts.

Abstract: CRISPR/Cas9, as a well-established gene editing technology, has been applied in numerous model organisms, but its application in wild-type *E. coli* remains limited. Pathogenic wild-type *E. coli*, a major cause of foodborne illnesses and intestinal inflammation in humans and animals, poses a significant global public health threat. The valine-glycine repeat protein G (VgrG) is a key virulence factor that enhances *E. coli* pathogenicity. In this study, PCR was used to identify 50 strains carrying the virulence gene *VgrG2* out of 83 wild pathogenic *E. coli* strains, with only one strain sensitive to kanamycin and spectinomycin. A homologous repair template for *VgrG2* was constructed using overlap PCR. A dual-plasmid CRISPR/Cas9 system, combining pTarget (spectinomycin resistance) and pCas (kanamycin resistance) with Red homologous recombination, was then used to induce genomic cleavage and knock out *VgrG2*. PCR and sequencing confirmed the deletion of a 1708 bp fragment of the *VgrG2* gene in wild-type *E. coli*. IPEC-J2 cells were infected with *E. coli*-WT and *E. coli* Δ*VgrG2*, and treated with the mTOR inhibitor rapamycin to study the effects of *VgrG2* on the mTOR signaling pathway. The qPCR results showed that *VgrG2* activated the mTOR pathway, suppressed *mTOR* and *p62* mRNA levels, and upregulated the autophagy-related genes and LC3-II protein expression. In conclusion, we utilized CRISPR/Cas9 technology to achieve large-fragment deletions in wild-type *E. coli*, revealing that *VgrG2* activates the mTOR signaling pathway and upregulates autophagy markers.

These findings offer new insights into *E. coli* genome editing and clarifies the pathogenic mechanisms through which *VgrG2* induces cellular damage.

Keywords: wild-type *E. coli*; CRISPR/Cas9; *VgrG2* gene; large fragment deletion; mTOR signaling pathway

1. Introduction

The CRISPR/Cas9 system serves as an immune defense mechanism that originated in prokaryotic organisms [1,2], composed of CRISPR sequences and Cas9 proteins [3,4]. It functions by recognizing foreign DNA sequences and incorporating them into the CRISPR spacer. The Cas9 nuclease identifies protospacer adjacent motifs (PAM) on foreign DNA and cleaves the exogenous DNA, thus protecting the host cell [5,6]. This system offers a greater precision, simplicity, and efficiency compared to traditional gene editing systems like TALEN and ZFN [7,8]. When applied for gene editing, the CRISPR/Cas9 system must be expressed in the target cells to produce the Cas9 protein and guide RNA (gRNA). The gRNA then directs Cas9 to recognize the target DNA sequence and the PAM (Protospacer Adjacent Motif) sequence, enabling the cleavage of the target DNA and generating a double-stranded break (DSB). When a double-strand break (DSB) occurs, cells initiate self-repair mechanisms to prevent death, with homologous recombination repair (HDR) being one of the key pathways for DNA repair. In the presence of homologous sequences, cells utilize the HDR mechanism for DSB repair. However, the HDR system is primarily found in eukaryotes and is largely absent in prokaryotes, which limits the application of this technology in wild-type *E. coli* [7]. Therefore, when using this system for gene editing, an HDR template must be added to the vector to facilitate the repair of broken DNA. Jiang et al. constructed a two-plasmid system that incorporated a homologous recombination mechanism. They co-transfected the donor sequence along with the construction plasmid into *E. coli* MG1655, successfully achieving gene knockout in this strain using the CRISPR/Cas9 system [9]. With this protocol, the CRISPR/Cas9 system enables the precise and efficient rejection of specific target genes within target strains. The co-transformation strategy of the donor sequence and plasmid not only optimizes the protocol, but also significantly reduces the complexity of the technical operation. Similarly, Wang et al. successfully induced site-directed mutations of the proB gene by using pCas9 and pTargetF+ vectors in Escherichia coli BL21. This method significantly reduces the likelihood of the repeat cleavage of gRNA/Cas9 and preserves the integrity of the PAM sequence, preventing the introduction of unnecessary silencing mutations in the genome [10]. At present, gene editing research in genetically engineered bacteria is relatively well established, but the research in wild-type *E. coli* has rarely been reported. In particular, research focusing on the deletion of large fragments of the virulence genes in wild-type *E. coli* is even more limited.

The Type VI Secretion System (T6SS) was first identified in Pseudomonas aeruginosa, and the presence of this secretory system has subsequently been observed in some gram-negative bacteria. A key component of T6SS is VgrG, which forms a phage-like tail needle structure together with the Hcp protein. This structure facilitates the injection of effector molecules into adjacent target cells through a conduit formed by PAAR (Proline-Alanine-Alanine-Arginine) proteins [11–13]. The VgrG protein is located at the tip of the T6SS pipeline [14]. Its structure comprises a variable active region, a gp5 region, and a gp27 region, which are similar to the tail spike of a T4 bacteriophage. This structure is responsible for penetrating the cell membrane [15], allowing the effector protein to reach specific cellular regions where it interacts with target molecules, thereby affecting host cell functions [16].

Bacterial toxins play a vital role in enhancing pathogenicity and triggering host immune responses. Lipopolysaccharide (LPS) and bacterial lipoprotein are key components of gram-negative bacteria. LPS is a major component of the outer membrane of gram-negative bacteria and has been shown to activate inflammatory pathways and induce host immune responses, and bacterial lipoproteins also promote pathogenesis by interacting with host receptors and modulating immune responses [17,18]. The *VgrG2* is a key factor in bacterial virulence and plays an important role in the interaction between bacteria and host cells. The mTOR signaling pathway is a critical intracellular signaling pathway that plays a role in a variety of physiological and pathological processes and has a significant regulatory role in the initiation and progression of autophagy [19]. Surviladze et al. discovered that human papillomavirus type 16 inhibits autophagy by activating the mTOR pathway, which facilitates its entry into host cells [20]. The mTOR pathway is also involved in autophagy induced by coxsackievirus B3 (CVB3) in HeLa cells [21]. Additionally, Yuying et al. demonstrated that Vibrio parahaemolyticus *VgrG2* can induce autophagy in macrophages [22]. However, the role of *E. coli VgrG2* in the mTOR signaling pathway and autophagy is still unclear.

In this study, we used CRISPR/Cas9 combined with Red homologous recombination to achieve large fragment deletions of the *VgrG2* virulence gene in wild-type *E. coli* and explored the effect of the *VgrG2* gene on the mTOR signaling pathway. Our analysis of key genes in the mTOR pathway revealed that *E. coli VgrG2* can activate the mTOR pathway and upregulate autophagy marker signals. These findings provide insights into the cellular damage mechanisms associated with *E. coli VgrG2* and further elucidate the pathogenic mechanisms underlying *E. coli*-induced host damage.

2. Materials and Methods

2.1. Cells, Strains and Plasmids

IPEC-J2 cells, derived from porcine immortalized small intestinal epithelial cells (Guangzhou Jennio Biotech, Guangzhou, China). The cells were cultured in DMEM supplemented with a 10% fetal bovine serum (FBS), penicillin, and streptomycin at 37 °C in a 5% CO_2 incubator.

The eighty-three wild-type *E. coli* strains used in this experiment were isolated from the feces of diarrheal piglets at the Shaba Pig Farm in Lufeng County, Chuxiong Autonomous Prefecture, Yunnan Province, China. All strains were cultured in an LB medium at 37 °C.

Plasmids pTargetF (MC_0000012) with spectinomycin resistance and pCas (MC_0000011) with kanamycin resistance were obtained from Molecular Cloud.

2.2. PCR Detection of the Virulence Gene VgrG2

Based on the *E. coli VgrG2* gene sequence from GenBank (Accession No. CP006632.1), amplification primers were designed using the NCBI online tool: F: 5′-CGGATTACGTTTCACGCTGG-3′ and R: 5′-TTCATTAAACCCGCTGCCCT-3′. Genomic DNA was extracted from *E. coli* using a bacterial genomic DNA extraction kit (Tiangen Biochemical Technology, Beijing, China), and strains containing the *VgrG2* gene were identified by PCR with the designed *VgrG2*-F/R primers.

2.3. Resistance Screening

Strains carrying the *VgrG2* gene were inoculated in the LB medium until an optical density (OD_{600}) of 0.6 was reached. Subsequently, 2 μL of this culture was transferred to

10 mL of the LB medium containing 50 mg/L kanamycin and mixed thoroughly. The OD_{600} was measured initially and again after 12 h of incubation at 37 °C with shaking at 160 rpm. Using a similar method, *E. coli* strains sensitive to 50 mg/L of both spectinomycin and kanamycin were further screened.

2.4. Construction of Knockout Vectors

The sgRNA primer sequence for *VgrG2* was designed using an online tool (http://spot.colorado.edu/~slin/cas9.html, accessed on 17 February 2025) as follows: sgRNA-*VgrG2*-SpeI-F: 5′-CTAGT*CCTGCCACCGGACGCGTTT*GGTTTTAGAGCTAGAAATAGC-3′ and sgRNA-*VgrG2*-R: 5′-GATGATCTTGCTTCATCTAGAGAATTCAAAAAAAGCACC-3′. The underlined bases represent the cleavage sites for SpeI, and the italicized bases indicate the overlapping sequences used to link the donor sequence via overlapping PCR. The designed sgRNA sequences were synthesized by a commercial company. The pTargetF plasmid was used as the template to amplify sgRNA-*VgrG2*.

To facilitate bacterial gene repair following cleavage, a homologous recombination template is required for targeted repair. Wild-type *E. coli* DNA as a template, and primers Arm-up-F/Arm-up-R and Arm-down-F/Arm-down-SalI-R, were used to amplify the upstream and downstream homology arms of the *VgrG2* knockout fragment, constructing the repair templates labeled L-Donor and R-Donor. The L-Donor and R-Donor were then ligated via overlap PCR. Subsequently, using this ligated donor and sgRNA-*VgrG2* as templates, and sgRNA-*VgrG2*-SpeI-F and R-Donor-SalI-R2 as primers, the final ligation was performed using overlap PCR, resulting in the product sgRNA-*VgrG2*-donor. Primer Sequences: L1: 5′-TGAAGCAAGATCATCCGGGC-3′; L2: 5′-GTGAGTCGACATCTGGCGTT-3′; R1: 5′-CAGATGTCGACTCACTAAAGTATGGCGGTCCATTGTC-3′; and R2: 5′-TCGACTGTTGGTCGCCAGGTAAAGA-3′. The underlined bases indicate the SalI restriction sites.

The ligated sgRNA-*VgrG2*-donor and pTargetF plasmids were treated with SalI and SpeI endonucleases to generate compatible ends. The pTargetF plasmid was then ligated with T4 ligase and sgRNA-*VgrG2*-donor. The ligation product was electroporated into *E. coli* DH5α, plated on LB agar containing spectinomycin, and incubated overnight at 37 °C. The following day, positive clones were screened, and plasmid DNA was extracted from monoclonal colonies. PCR identification was performed using primers pTargetF-JD-F and R. The products that matched the expected size of the target fragment were sent for sequencing verification. The validated vector was named pTargetF-sgRNA-*VgrG2*-donor. The identification primer sequences were as follows: F: 5′-ATTACCGCCTTTGAGTGAGC-3′ and R: 5′-GGATAACAGGGTAATAGATC-3′.

2.5. CRISPR/Cas9 Gene Editing Strategy for E. coli

To facilitate transfection, wild-type *E. coli* was first prepared as electrocompetent cells. The pCas plasmid was then electroporated into these cells and plated on LB agar containing kanamycin to select the positive clones. PCR identification was performed on the selected clones. The strains containing the pCas plasmid were made competent again, during which arabinose was added to induce the expression of the λ-Red protein from the pCas plasmid. The constructed vector, PTargetF-sgRNA-*VgrG2*-donor, was electroporated into *E. coli* containing the pCas plasmid and plated on LB agar containing both kanamycin and spectinomycin for selection. Monoclonal colonies were picked and identified by PCR using Δ*VgrG2*-JD-F/R primers, followed by sequencing. The primer sequences for Δ*VgrG2*-JD-F/R were as follows: F: 5′-TGCACATTATTGGAGGGGCA-3′ and R: 5′-TCCGTCACGCCGGTGATTT-3′.

2.6. Determination of the Growth Curve and Relative Cell Growth Rate

After incubating the target strain overnight, 200 μL of the culture medium is inoculated into 250 mL of the LB medium. The OD_{600} of the cultured strain is measured every 2 h for 24 h using a spectrophotometer. The growth curve of *E. coli* is plotted with time on the *x*-axis and mean OD_{600} on the *y*-axis. IPEC-J2 cells are infected with *E. coli*-WT and *E. coli* Δ*VgrG2*, and the relative growth rate of IPEC-J2 cells over the same 24 h period is assessed, with the blank group exhibiting a growth rate of 100%.

2.7. Cell Infection Experiments

To investigate the cell damage mechanism of *E. coli VgrG2*, 2×10^6 IPEC-J2 cells were seeded into 6-well plates. Each well was supplemented with 2 mL of a cell culture medium containing 1 mL of the *E. coli* solution (OD_{600} = 0.6). The experimental setup included four groups: a blank group (control), *E. coli*-WT infection group, *E. coli*Δ*VgrG2* infection group, and a rapamycin treatment group. Following infection, cells and cell supernatants were collected at 0.5 h, 3 h, 6 h, 9 h, 12 h, and 24 h for further analysis.

2.8. RT-qPCR Detection

The total RNA was extracted from cultured cells using the RNAiso Plus reagent (Takara, Dalian, China). Genomic DNA (gDNA) contamination was eliminated, and complementary DNA (cDNA) was synthesized using the PrimeScript RT Master Mix Kit with gDNA Eraser Kit (Takara, Dalian, China). Real-time PCR assays were conducted on a Bio-Rad CFX PCR instrument (Hercules, CA, USA) using TB Green Premix Ex Taq™ II. The mRNA expression levels of the *mTOR*, *ULK1*, *Beclin-1*, *Atg12*, *Atg5*, *Atg3*, *P62*, and *LC3* genes, which are associated with the mTOR signaling pathway, were normalized using the ΔΔCt method with β-actin as the reference gene. The primer sequences are provided in Table S1.

2.9. Cellular Immunofluorescence Detection

Pre-treated IPEC-J2 cells were fixed onto glass coverslips with 4% paraformaldehyde for 15 min, followed by permeabilization with 0.1% Triton X-100 for 10 min. After permeabilization, the cells were blocked with 5% bovine serum albumin (BSA) for 1 h. Specific primary antibodies (LC3-II Antibody, Proteintech Group, Inc., Wuhan, China, Cat. No. 14600-1-AP) were incubated with the cells overnight at 4 °C. Following incubation with the primary antibodies, the cells were washed three times with PBS and then incubated with the corresponding fluorescently labeled secondary antibody (Goat anti-rabbit IgG) for 1 h at room temperature in the dark. The cells were then stained with DAPI for 5 min. Fluorescent images of the stained cells were visualized using a fluorescence microscope (Olympus Corporation, Tokyo, Japan), and a quantitative analysis of the images was performed using ImageJ 1.8 software.

2.10. Statistical Analysis

Statistical analysis was based on a minimum of three biological replicates. Data were expressed as the mean ± SD and analyzed using either Student's *t*-test or two-way ANOVA. Significance was defined as *p*-values less than 0.05.

3. Results

3.1. Identification and Resistance Screening of the Virulence Gene VgrG2 in Pathogenic E. coli

A total of 83 wild-type *E. coli* strains were isolated, and PCR amplification revealed that 50 strains contained the *VgrG2* gene (Figure 1a–d). These 50 *E. coli* strains were subsequently inoculated into the LB medium containing kanamycin (50 mg/L). Of these, three strains were found to be susceptible to kanamycin, representing 6% (3/50) of the total. The three kanamycin-susceptible strains were then inoculated into an LB medium containing spectinomycin (50 mg/L), and only one strain was sensitive to both kanamycin and spectinomycin, resulting in a dual-susceptibility rate of 33.33% (1/3). This dual-sensitive strain was selected as the target strain for *VgrG2* gene knockout.

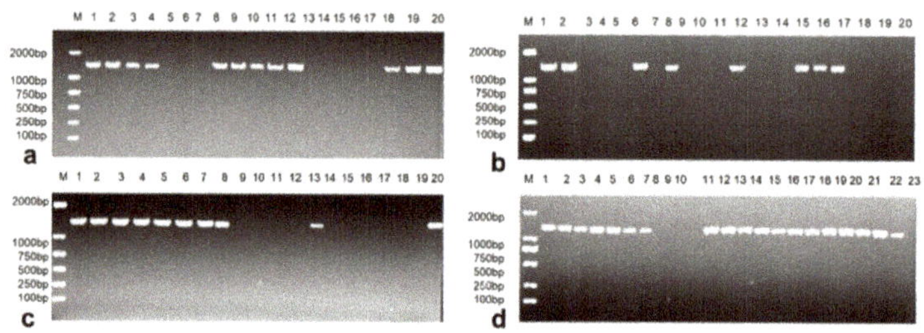

Figure 1. Detection of the *E. coli VgrG2* gene. (**a**–**d**). A total of 50 *E. coli* strains containing the *VgrG2* gene were screened via PCR. A total of 83 wild-type *E. coli* strains were isolated. Lanes 1–20 in (**a**–**c**) represent strains 1–20, 21–41, and 42–61, respectively, and lanes 1–23 in (**d**) represent strains 62–83. (M: 2000 bp DNA marker and lanes 1–23 represent the experimental strains. The target fragment size is 1435 bp).

3.2. Construction of the E. coli-VgrG2 Targeting Vector pTargetF-sgRNA-VgrG2-Donor

The pTargetF plasmid was linearized by double digestion with SalI and SpeI enzymes, resulting in a product size of 2037 bp (Figure 2a). The sgRNA-*VgrG2*-donor was then ligated with the digested plasmid. The ligation product was electroporated into *E. coli* DH5α. The transformed bacteria were plated on LB plates containing spectinomycin and incubated overnight at 37 °C (Figure 2b). Single colonies were subsequently picked for PCR verification. A successful construction was indicated by a PCR product size of 1330 bp, while an unsuccessful attempt produced a product size of 337 bp (Figure 2c). Sequencing results were consistent with the expected sequence of the inserted sgRNA-*VgrG2*-donor (Figure 2d), confirming the successful construction of the pTargetF-sgRNA-*VgrG2*-donor plasmid.

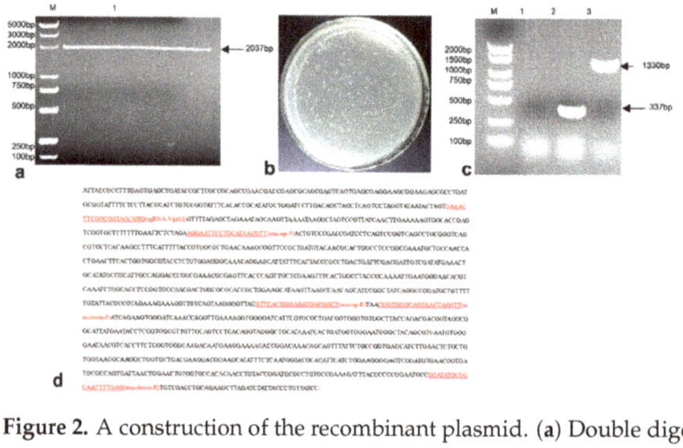

Figure 2. A construction of the recombinant plasmid. (**a**) Double digestion of the pTargetF plasmid with SalI and SpeI (M: 2000 bp DNA marker, lane 1: pTargetF plasmid double digested with SalI and SpeI); (**b**) colonies on the agar plate following transformation with the ligation product; (**c**) PCR verification of the pTargetF-*sgRNA-VgrG2*-donor plasmid construction (M: 2000 bp DNA marker, lane 1: negative control, lane 2: pTargetF plasmid, lane 3: pTargetF-*sgRNA-VgrG2*-donor plasmid); and (**d**) sequencing results of the PTargetF-*sgRNA-VgrG2*-donor plasmid.

3.3. Construction of the E. coli VgrG2 Virulence Gene Deletion Strain

The PCas plasmid was electroporated into competent cells prepared from wild-type *E. coli*. Positive transformants were selected and subjected to PCR amplification, yielding a product size of 483 bp (Figure 3a). The pTargetF-sgRNA-*VgrG2*-donor plasmid was then electroporated into wild-type *E. coli* containing the PCas plasmid. Single colonies were picked for PCR analysis. A successful knockout was indicated by a PCR product size of 1294 bp, whereas an unsuccessful attempt produced a product size of 3002 bp (Figure 3b). Sequencing results confirmed the successful construction of the *VgrG2* gene deletion strain, which exhibited a total deletion of 1708 bp (Figure 3c).

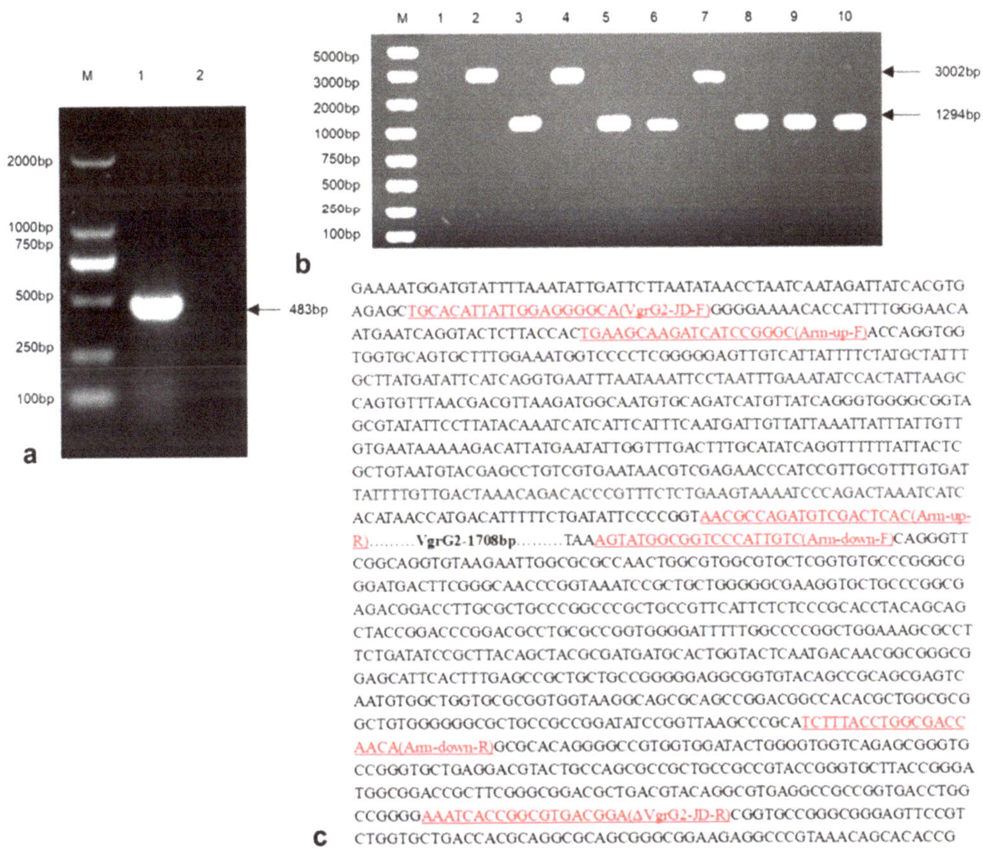

Figure 3. A construction of *E. coli ΔVgrG2* strains. (**a**) PCR verification of the PCas plasmid transformation results (M: 2000 bp DNA marker, lane 1: PCas plasmid transformed bacterial colonies, lane 2: negative control). (**b**) PCR verification of the *E. coli VgrG2* gene knockout strain construction (M: 5000 bp DNA marker, lane 1: negative control, lane 2: parent strains, lanes 3–10: experimental strains). (**c**) Sequencing results of the *E. coli VgrG2* gene knockout strain construction.

3.4. The Effect of the VgrG2 Virulence Gene Deletion on the E. coli Growth Rate

E. coli WT and *E. coli ΔVgrG2* strains were inoculated into a liquid culture medium and incubated with shaking, with the optical density recorded at different time points. Growth curves were plotted with culture time on the *x*-axis and optical density on the *y*-axis. The results showed that the growth trends of the strains were similar before and after the *VgrG2* gene knockout (Figure 4a), indicating that the deletion of the *VgrG2* virulence gene did not affect the growth activity of *E. coli*.

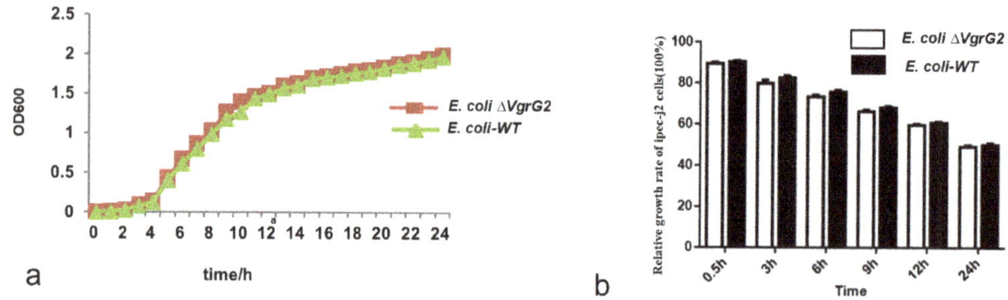

Figure 4. The effect of *VgrG2* gene knockout on strain and cell growth. (**a**) Growth curves of *E. coli-WT* and *E. coli* Δ*VgrG2*. (**b**) Relative growth rates of cells in the *E. coli-WT* and *E. coli* Δ*VgrG2* groups.

3.5. The Effect of *E. coli-VgrG2* on IPEC-J2 Cell Growth

E. coli-WT and *E. coli* Δ*VgrG2* were used to infect IPEC-J2 cells. The relative growth rates of IPEC-J2 cells were not significantly affected by the *VgrG2* deletion at any of the measurement time points (Figure 4b).

3.6. *E. coli VgrG2* Activates the mTOR Signaling Pathway in IPEC-J2 Cells

Programmed cell death is a significant mode of cell death, with autophagy representing one mechanism of this process. It has been established that mTOR promotes anabolism while inhibiting the induction of autophagy [23]. To further investigate the impact of *VgrG2* on the mTOR signaling pathway, we assessed the mRNA levels of key genes involved in this pathway at various time points following *E. coli* infection. The results demonstrated that the expression of autophagy-related genes *ULK1*, *Beclin-1*, *Atg5*, *Atg12*, *Atg3*, and *LC3* was up-regulated in IPEC-J2 cells (Figure 5b–g), whereas the expression of *mTOR* and *P62* was down-regulated (Figure 5a,h), an effect that was diminished in *E. coli* Δ*VgrG2*-infected cells. Collectively, these findings suggest that *E. coli* infection activates the mTOR signaling pathway in IPEC-J2 cells, with *VgrG2* enhancing this activation.

3.7. The Effect of *E. coli VgrG2* on the Expression of the Autophagy Marker Protein LC3-II

LC3-II is a well-established marker of autophagy, where cytoplasmic proteins and organelles are sequestered by autophagosomes, with LC3-II being recruited to the autophagosome membrane during this process [24]. To assess the effect of *E. coli VgrG2* on LC3-II expression, we performed an immunofluorescence analysis in IPEC-J2 cells infected with either *E. coli*-WT or *E. coli* Δ*VgrG2*. Fluorescence microscopy revealed that LC3-II was recruited to the autophagosomal membrane, forming distinct fluorescent spots. A quantitative analysis of LC3-II fluorescence indicated an increase in the fluorescence intensity over time in infected IPEC-J2 cells, with *E. coli VgrG2* significantly enhancing LC3-II fluorescence (Figure 6a,b). In summary, our findings demonstrate that *E. coli* infection upregulates autophagy marker signals in IPEC-J2 cells and that *E. coli VgrG2* further promotes the production of these signals.

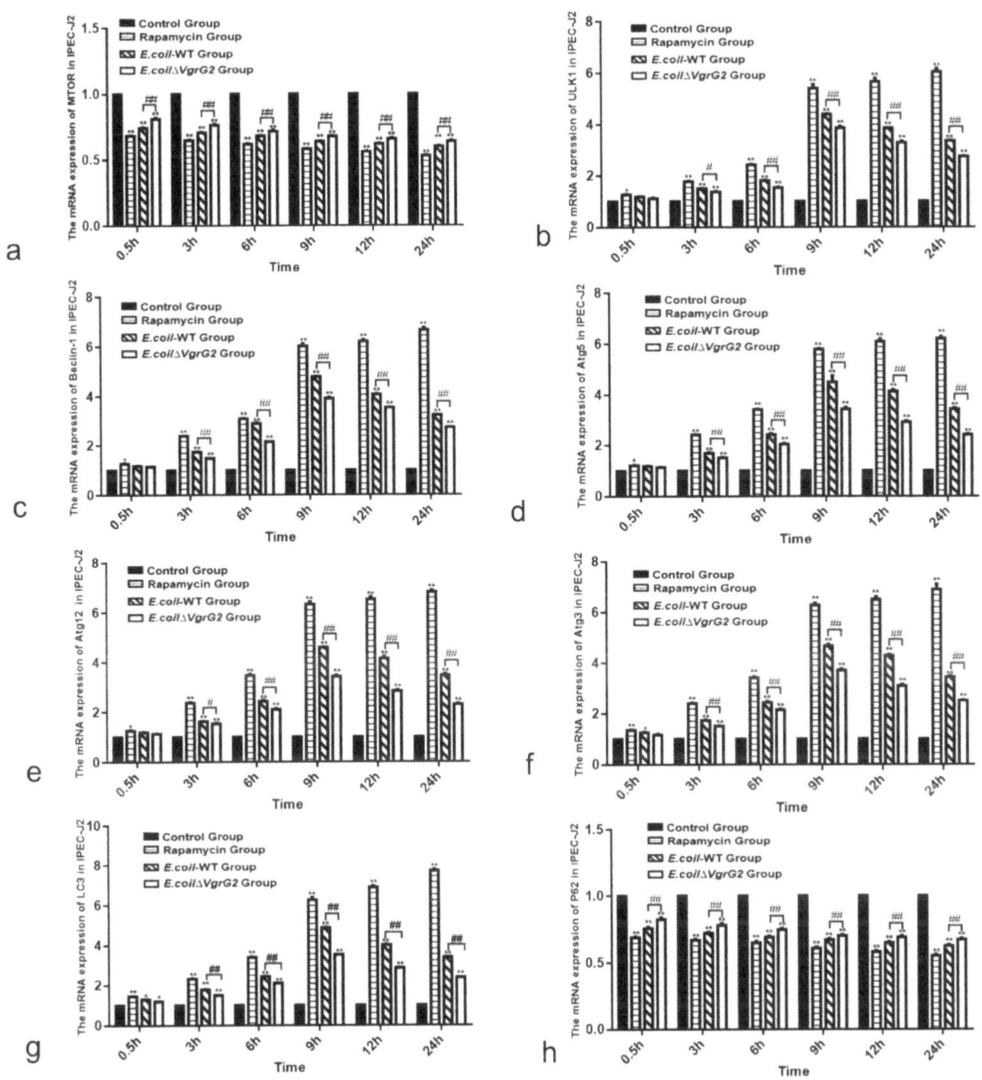

Figure 5. Detection of the expression levels of key genes in the mTOR signaling pathway at various time points using qPCR. (**a**) Relative expression levels of the *mTOR* gene in IPEC-J2 cells for each group at different time points. (**b**) Relative expression levels of the *ULK1* gene in IPEC-J2 cells for each group at different time points. (**c**) Relative expression levels of the *Beclin-1* gene in IPEC-J2 cells for each group at different time points. (**d**) Relative expression levels of the Atg5 gene in IPEC-J2 cells for each group at different time points. (**e**) Relative expression levels of the *Atg12* gene in IPEC-J2 cells for each group at different time points. (**f**) Relative expression levels of the *Atg3* gene in IPEC-J2 cells for each group at different time points. (**g**) Relative expression levels of the *LC3* gene in IPEC-J2 cells for each group at different time points. (**h**) Relative expression levels of the *P62* gene in IPEC-J2 cells for each group at different time points. All data are shown as the mean ± SD. Statistical analysis was performed using at least three biological replicates, analyzed using either Student's *t*-test or two-way ANOVA. Significant differences compared to the control group: ** $p < 0.01$, * $p < 0.05$. Significant differences between the *E. coli-WT* group and the *E. coli ∆VgrG2* group: ## $p < 0.01$, # $p < 0.05$.

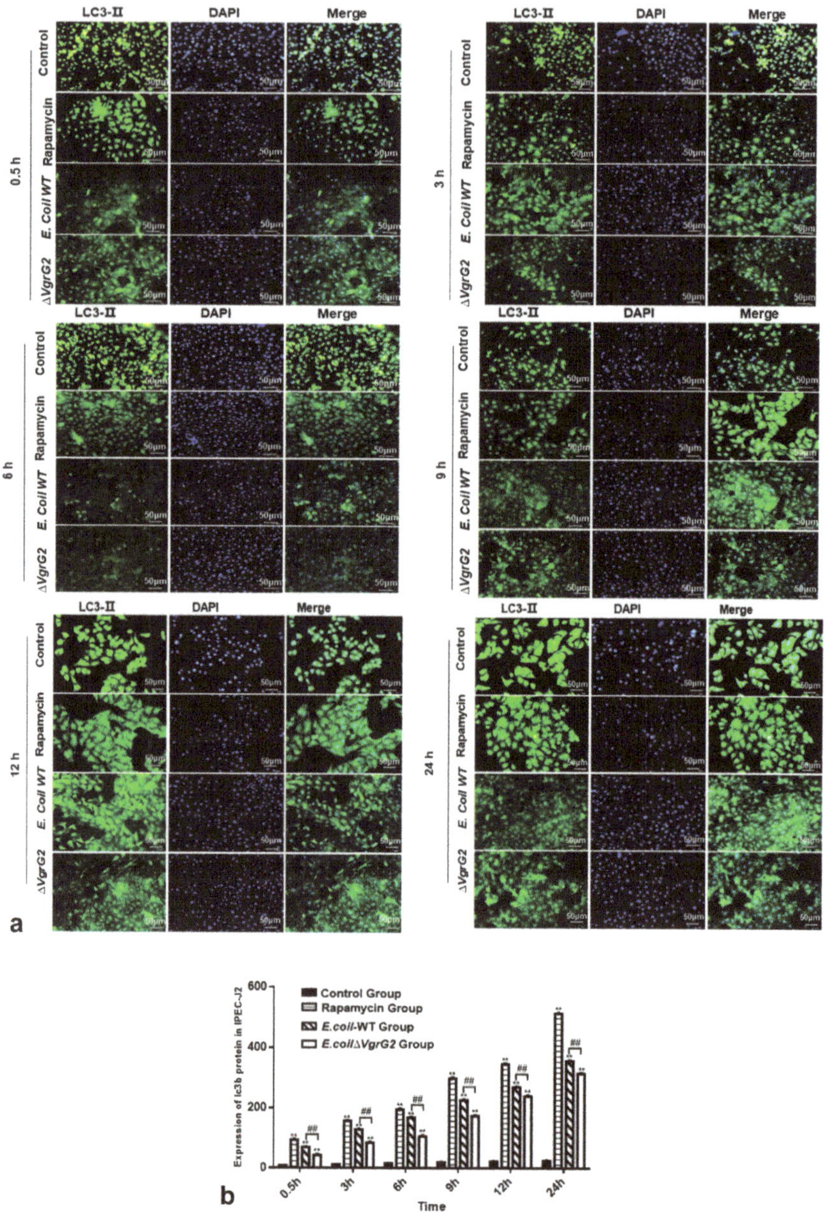

Figure 6. (**a**) Immunofluorescence staining to determine the expression of the LC3-II protein in IPEC-J2 cells for each group at various time points (400×) (where DAPI is the nuclear staining image, LC3-II is shown in green fluorescence, and Merge is the overlay of the first two images). (**b**) Total optical density sum values of the LC3-II protein expression in IPEC-J2 cells for each group at different time points. All data are shown as the mean ± SD. Statistical analysis was performed using at least three biological replicates, analyzed using either Student's *t*-test or two-way ANOVA. Significant differences compared to the control group: ** $p < 0.01$. Significant differences between the *E. coli-WT* group and the *E. coli ΔVgrG2* group: ## $p < 0.01$.

4. Discussion

In recent years, the rapid expansion of the pig farming industry has been accompanied by a rise in the incidence and drug resistance of *E. coli* infections, posing a significant threat to food safety and public health. Understanding the pathogenic mechanisms of *E. coli* is crucial for advancing the pig farming industry and safeguarding human health. Bacterial pathogens often use various mechanisms and pathways to infect hosts, cause tissue damage, and disrupt immune responses. Many of these pathogens rely on secretion systems to facilitate these processes. Our study focused on the *E. coli VgrG2* protein and found that among 83 pathogenic *E. coli* strains, 50 contained the *VgrG2* gene, resulting in a positive rate of 60.24%. This correlation suggests that *VgrG2* may play a role in the pathogenicity of *E. coli*. Resistance screening further revealed that only one strain out of the 83 was sensitive to both kanamycin and spectinomycin. This highlights the prevalent issue of drug resistance in *E. coli* within veterinary clinical settings and underscores the ease with which highly resistant and pathogenic strains can develop. This poses a considerable challenge to veterinary practice and public health.

The CRISPR/Cas9 gene editing system operates through the combined functions of sgRNA guidance, vector resistance labeling, and Cas9 specificity. Vector resistance labeling allows for the selection of positive clones on antibiotic-resistant plates, facilitating the verification of successful transfection. Guided by the sgRNA, the Cas9 protein precisely targets and cleaves a specific DNA sequence, creating a double-strand break. Research has demonstrated that CRISPR/Cas9 technology, with modifications to the pCas/pTargetF system, can be used for gene editing in *E. coli* strains CGMCC3705, K-12 MG1655, and DH5α, greatly reducing the time required for bacterial gene editing [25]. Although CRISPR/Cas9 technology in *E. coli* is well established, most of the research has been conducted with genetically engineered strains, with relatively few studies focusing on gene editing in wild-type *E. coli*. This presents challenges for studying gene functions in wild-type *E. coli* due to difficulties such as plasmid transformation, low transformation efficiency, and the more complex environment of wild-type strains compared to standard strains, which increases the risk of plasmid loss after transformation. In our study, we employed a two-plasmid system and prepared wild-type *E. coli* as competent cells for electroporation, facilitating the introduction of exogenous plasmids and improving transformation efficiency. We used LB plates containing kanamycin for the screening and identification of positive clones, which allowed for the rapid and accurate confirmation of strains transformed with the pCas plasmid. The constructed vector, pTargetF-sgRNA-*VgrG2*-donor, was electroporated into *E. coli* containing the pCas plasmid, and selection was performed on plates containing both kanamycin and spectinomycin. This dual-resistance screening enhanced the specificity and reliability of the selection process. The current results indicate a significant improvement in plasmid transfection efficiency, enabling successful CRISPR/Cas9-mediated gene editing in wild-type *E. coli*. This advancement addresses the challenges of genome editing in wild-type *E. coli* and offers new insights for studying gene function in wild-type strains.

VgrG2 is considered a core component of the T6SS. In some strains, the T6SS gene cluster includes multiple *VgrG* genes with distinct roles. For example, Vibrio parahaemolyticus has been shown to induce autophagy in macrophages, with *VgrG2* deletion leading to reduced autophagy [22]. At physiological levels, autophagy helps cells manage various stress conditions, such as starvation, hypoxia, mitochondrial damage, and pathogen infection, thereby promoting cell survival. However, the negative aspects of autophagy are also becoming increasingly recognized; it can promote cell death by activating certain intracellular signaling pathways [26–29]. To explore the role of the *VgrG2* gene in wild-type *E. coli*, we used CRISPR/Cas9 technology to construct a Δ*VgrG2* strain and compared its growth with that of the wild-type (WT) strain. A growth curve analysis revealed that the growth trends of WT and Δ*VgrG2* strains were similar at all time points, indicating that *VgrG2* does not affect *E. coli* growth. We also investigated the impact of *VgrG2* on the mTOR signaling pathway by examining the transcription levels of key genes, including *mTOR*, *ULK1*, *Beclin-1*, *P62*, *LC3*, *Atg3*, *Atg5*, and *Atg12*, following infection, and assessed the expression of the autophagy marker protein LC3-II. Our results indicated that *E. coli* infection significantly modulated the expression of key genes in the mTOR signaling pathway, with *VgrG2* playing a crucial role in this process. LC3-II, a marker of autophagy, mediates membrane extension and fusion, leading to the formation of autophagosomes [30,31]. Rapamycin, an mTOR inhibitor, promotes autophagy [32]. After treating IPEC-J2 cells with rapamycin, we observed an increased LC3-II protein expression. Similarly, *E. coli* infection elevated LC3-II levels. However, in the *E. coli* Δ*VgrG2* infection group, we observed lower levels of LC3-II, suggesting that *VgrG2* likely promotes autophagy in IPEC-J2 cells, potentially by influencing the mTOR signaling pathway and upregulating LC3-II expression. While this study focused on the impact of *E. coli VgrG2* on key mTOR signaling pathway genes, the specific mechanisms by which *VgrG2* induces cellular autophagy warrant further investigation.

Supplementary Materials: The following supporting information can be downloaded at https://www.mdpi.com/article/10.3390/vetsci12030249/s1; Table S1: Primers used for PCR amplification.

Author Contributions: Conceptualization, T.-L.P. and J.-L.C.; methodology, H.W. and J.-S.Z.; experiments, J.-L.X., J.S., M.Z. and Y.L.; data analysis, J.-Z.M., K.-Y.Z. and Y.-K.Z.; writing—original draft, J.-L.C.; writing—review and editing, H.W.; funding acquisition, P.X. and H.G. All authors have read and agreed to the published version of the manuscript.

Funding: This work was supported by the National Natural Science Foundation of China (Grant No. 31660704; 31960692).

Institutional Review Board Statement: Not applicable.

Informed Consent Statement: Not applicable.

Data Availability Statement: All data can be obtained from the corresponding author.

Acknowledgments: We would like to express our heartfelt thanks to all the individuals who contributed to this research. Their support, insight, and encouragement throughout the research process were invaluable. We are also grateful for the resources provided by our lab, which have contributed to the success of this work.

Conflicts of Interest: The authors declare no conflicts of interest.

References

1. Horvath, P.; Barrangou, R. CRISPR/Cas, the immune system of bacteria and archaea. *Science* **2010**, *327*, 167–170. [CrossRef] [PubMed]
2. Shah, S.Z.; Rehman, A.; Nasir, H.; Asif, A.; Tufail, B.; Usama, M.; Jabbar, B. Advances In Research On Genome Editing Crispr-Cas9 Technology. *J. Ayub Med. Coll. Abbottabad JAMC* **2019**, *31*, 108–122. [PubMed]
3. Redman, M.; King, A.; Watson, C.; King, D. What is CRISPR/Cas9? *Arch. Dis. Child. Educ. Pract. Ed.* **2016**, *101*, 213–215. [CrossRef] [PubMed]
4. El-Mounadi, K.; Morales-Floriano, M.L.; Garcia-Ruiz, H. Principles, Applications, and Biosafety of Plant Genome Editing Using CRISPR-Cas9. *Front. Plant Sci.* **2020**, *11*, 56. [CrossRef] [PubMed]
5. Hsieh-Feng, V.; Yang, Y. Efficient expression of multiple guide RNAs for CRISPR/Cas genome editing. *aBIOTECH* **2020**, *1*, 123–134. [CrossRef]
6. Makarova, K.S.; Grishin, N.V.; Shabalina, S.A.; Wolf, Y.I.; Koonin, E.V. A putative RNA-interference-based immune system in prokaryotes: Computational analysis of the predicted enzymatic machinery, functional analogies with eukaryotic RNAi, and hypothetical mechanisms of action. *Biol. Direct* **2006**, *1*, 7. [CrossRef]

7. Zhang, B.; Wang, H.; Zhao, W.; Shan, C.; Liu, C.; Gao, L.; Zhao, R.; Ao, P.; Xiao, P.; Lv, L.; et al. New insights into the construction of wild-type Saba pig-derived *Escherichia coli* irp2 gene deletion strains. *3 Biotech* **2021**, *11*, 408. [CrossRef]
8. Janik, E.; Niemcewicz, M.; Ceremuga, M.; Krzowski, L.; Saluk-Bijak, J.; Bijak, M. Various Aspects of a Gene Editing System-CRISPR-Cas9. *Int. J. Mol. Sci.* **2020**, *21*, 9604. [CrossRef]
9. Jiang, Y.; Chen, B.; Duan, C.; Sun, B.; Yang, J.; Yang, S. Multigene editing in the *Escherichia coli* genome via the CRISPR-Cas9 system. *Appl. Environ. Microbiol.* **2015**, *81*, 2506–2514. [CrossRef]
10. Wang, X.; He, J.; Le, K. Making point mutations in *Escherichia coli* BL21 genome using the CRISPR-Cas9 system. *FEMS Microbiol. Lett.* **2018**, *365*, fny060. [CrossRef]
11. Mougous, J.D.; Cuff, M.E.; Raunser, S.; Shen, A.; Zhou, M.; Gifford, C.A.; Goodman, A.L.; Joachimiak, G.; Ordoñez, C.L.; Lory, S.; et al. A virulence locus of *Pseudomonas aeruginosa* encodes a protein secretion apparatus. *Science* **2006**, *312*, 1526–1530. [CrossRef] [PubMed]
12. Monjarás Feria, J.; Valvano, M.A. An Overview of Anti-Eukaryotic T6SS Effectors. *Front. Cell. Infect. Microbiol.* **2020**, *10*, 584751. [CrossRef] [PubMed]
13. Wettstadt, S.; Filloux, A. Manipulating the type VI secretion system spike to shuttle passenger proteins. *PLoS ONE* **2020**, *15*, e0228941. [CrossRef] [PubMed]
14. De Maayer, P.; Venter, S.N.; Kamber, T.; Duffy, B.; Coutinho, T.A.; Smits, T.H. Comparative genomics of the Type VI secretion systems of *Pantoea* and *Erwinia* species reveals the presence of putative effector islands that may be translocated by the VgrG and Hcp proteins. *BMC Genom.* **2011**, *12*, 576. [CrossRef] [PubMed]
15. Dutta, P.; Jijumon, A.S.; Mazumder, M.; Dileep, D.; Mukhopadhyay, A.K.; Gourinath, S.; Maiti, S. Presence of actin binding motif in VgrG-1 toxin of Vibrio cholerae reveals the molecular mechanism of actin cross-linking. *Int. J. Biol. Macromol.* **2019**, *133*, 775–785. [CrossRef]
16. Ma, J.; Sun, M.; Pan, Z.; Lu, C.; Yao, H. Diverse toxic effectors are harbored by vgrG islands for interbacterial antagonism in type VI secretion system. *Biochim. Biophys. Acta Gen. Subj.* **2018**, *1862*, 1635–1643. [CrossRef]
17. Xu, T.; Liu, R.; Lu, X.; Wu, X.; Heneberg, P.; Mao, Y.; Jiang, Q.; Loor, J.; Yang, Z. Lycium barbarum polysaccharides alleviate LPS-induced inflammatory responses through PPARγ/MAPK/NF-κB pathway in bovine mammary epithelial cells. *J. Anim. Sci.* **2022**, *100*, skab345. [CrossRef]
18. Braun, V.; Hantke, K. Lipoproteins: Structure, Function, Biosynthesis. *Subcell. Biochem.* **2019**, *92*, 39–77.
19. Xu, L.; Brink, M. mTOR, cardiomyocytes and inflammation in cardiac hypertrophy. *Biochim. Biophys. Acta* **2016**, *1863*, 1894–1903. [CrossRef]
20. Surviladze, Z.; Sterk, R.T.; DeHaro, S.A.; Ozbun, M.A. Cellular entry of human papillomavirus type 16 involves activation of the phosphatidylinositol 3-kinase/Akt/mTOR pathway and inhibition of autophagy. *J. Virol.* **2013**, *87*, 2508–2517. [CrossRef]
21. Chang, H.; Li, X.; Cai, Q.; Li, C.; Tian, L.; Chen, J.; Xing, X.; Gan, Y.; Ouyang, W.; Yang, Z. The PI3K/Akt/mTOR pathway is involved in CVB3-induced autophagy of HeLa cells. *Int. J. Mol. Med.* **2017**, *40*, 182–192. [CrossRef] [PubMed]
22. Yu, Y.; Fang, L.; Zhang, Y.; Sheng, H.; Fang, W. VgrG2 of type VI secretion system 2 of Vibrio parahaemolyticus induces autophagy in macrophages. *Front. Microbiol.* **2015**, *6*, 168. [CrossRef] [PubMed]
23. Kim, Y.C.; Guan, K.L. mTOR: A pharmacologic target for autophagy regulation. *J. Clin. Investig.* **2015**, *125*, 25–32. [CrossRef]
24. Tanida, I.; Ueno, T.; Kominami, E. LC3 and Autophagy. *Methods Mol. Biol.* **2008**, *445*, 77–88. [PubMed]
25. Li, Q.; Sun, B.; Chen, J.; Zhang, Y.; Jiang, Y.; Yang, S. A modified pCas/pTargetF system for CRISPR-Cas9-assisted genome editing in *Escherichia coli*. *Acta Biochim. Biophys. Sin.* **2021**, *53*, 620–627. [CrossRef]
26. Li, X.; He, S.; Ma, B. Autophagy and autophagy-related proteins in cancer. *Mol. Cancer* **2020**, *19*, 12. [CrossRef]
27. Denton, D.; Kumar, S. Autophagy-dependent cell death. *Cell Death Differ.* **2019**, *26*, 605–616. [CrossRef]
28. Green, D.R.; Levine, B. To be or not to be? How selective autophagy and cell death govern cell fate. *Cell* **2014**, *157*, 65–75. [CrossRef]
29. Liu, Y.; Levine, B. Autosis and autophagic cell death: The dark side of autophagy. *Cell Death Differ.* **2015**, *22*, 367–376. [CrossRef]
30. Renna, F.J.; Herrera Lopez, M.; Manifava, M.; Ktistakis, N.T.; Vaccaro, M.I. Evaluating Autophagy Levels in Two Different Pancreatic Cell Models Using LC3 Immunofluorescence. *J. Vis. Exp. JoVE* 2023. [CrossRef]
31. Kabeya, Y.; Mizushima, N.; Ueno, T.; Yamamoto, A.; Kirisako, T.; Noda, T.; Kominami, E.; Ohsumi, Y.; Yoshimori, T. LC3, a mammalian homologue of yeast Apg8p, is localized in autophagosome membranes after processing. *EMBO J.* **2000**, *19*, 5720–5728. [CrossRef] [PubMed]
32. Li, Y.; Cheng, Y.; Zhou, Y.; Du, H.; Zhang, C.; Zhao, Z.; Chen, Y.; Zhou, Z.; Mei, J.; Wu, W.; et al. High fat diet-induced obesity leads to depressive and anxiety-like behaviors in mice via AMPK/mTOR-mediated autophagy. *Exp. Neurol.* **2022**, *348*, 113949. [CrossRef] [PubMed]

veterinary sciences

MDPI

Article

N-(3-Oxododecanoyl)-Homoserine Lactone Induces Intestinal Barrier Damage in Piglets via the Lipid Raft-Mediated Apoptosis Pathway

Yang Yang [1,2,*,†], Xin Zhang [1,2,†], Jin Yang [3,†], Ziyan Wu [1,2], Junpeng Li [1,2], Ruilong Song [1,4], Chuang Meng [4] and Guoqiang Zhu [1,2,*]

1 College of Veterinary Medicine, Yangzhou University, Yangzhou 225009, China; zhangx41@sbtjt.com (X.Z.); 18255676558@163.com (Z.W.); wei77917weijie@163.com (J.L.); 006018@yzu.edu.cn (R.S.)

2 Jiangsu Co-Innovation Center for Important Animal Infectious Diseases and Zoonoses, and Joint Laboratory of International Cooperation on Prevention and Control Technology of Important Animal Diseases and Zoonoses of Jiangsu, Yangzhou 225009, China

3 Northern Jiangsu People's Hospital Affiliated to Yangzhou University, Yangzhou 225009, China; 18051062957@yzu.edu.cn

4 Jiangsu Key Lab of Zoonosis/Jiangsu Co-Innovation Center for Prevention and Control of Important Animal Infectious Diseases and Zoonoses, Yangzhou University, Yangzhou 225009, China; 006503@yzu.edu.cn

* Correspondence: yy@yzu.edu.cn (Y.Y.); yzgqzhu@yzu.edu.cn (G.Z.)

† These authors contributed equally to this work.

Simple Summary: Quorum sensing (QS) not only regulates bacterial virulence through signaling molecules, but is also capable of causing direct damage to host cells. Previous studies have found a high correlation between N-acyl homoserine lactone (AHL) of the QS in the piglet intestine and piglet health. In a mouse model, AHL signaling N-(3-oxododecanoyl)-homoserine lactone (OdDHL) induced apoptosis of intestinal cells and damage to intestinal barrier. In a porcine intestinal epithelial cell model, OdDHL also was confirmed to induce apoptosis. The direct role and mechanism of OdDHL in the digestive tract of porcine animal models have not been elucidated. For the first time in a piglet model, we have verified that AHL induces apoptosis and disrupts the intestinal barrier, and we have also verified the protective effect of QS inhibitors. The present study also investigates the central role of cellular lipid rafts in the critical phase in which OdDHL adheres to cells and exerts functions. In cells where lipid rafts were eliminated, the ability of OdDHL adherence to cell surface was significantly reduced, and apoptosis induction was similarly impaired. This result indicates an important role for lipid rafts in relevant processes. This study provides basic data for the antimicrobial application of QSI in pig farming.

Abstract: Quorum sensing (QS) is a process by which bacteria sense their population density and regulate behavior accordingly. QS not only regulates bacterial virulence but also directly influences host cells. Previous studies have shown that QS is strongly associated with piglet intestinal health, but the mechanism is not yet clear. For the first time, we have confirmed in a piglet animal model that OdDHL directly damages intestinal cells in weaned piglets, disrupting the intestinal barrier. We also provide a preliminary exploration of the underlying mechanism of these effects. TUNEL assays confirmed that damage to the piglet intestinal barrier coincided temporally and spatially with dysregulated apoptosis. Lipid rafts, key components of the cell membrane, are involved in many biological processes, including the activation of apoptosis-related proteins. Following the disruption of the lipid raft structure in IPEC-J2 cells, the apoptosis rate under OdDHL stimulation decreased by 50%. These data demonstrate that lipid rafts mediate the attachment of OdDHL to porcine intestinal cells; then, OdDHL induces apoptosis in porcine intestinal cells through the mitochondrial and death receptor pathways, thereby compromising the integrity of the porcine intestinal barrier. This study provides foundational insights into the role of QS in piglet intestinal diseases.

Keywords: quorum sensing; OdDHL; IPEC-J2; apoptosis; lipid raft

1. Introduction

The intestine is the site where digestion, nutrient absorption, and microorganism colonization occur in animals [1]. The intestinal microbiota is an indicator of animal health. The intestinal mucosal barrier is an important defense against bacterial invasion and prevents the entry of toxins into the body [2]. The permeability of the mucosal layer increases after intestinal barrier damage. This disrupts the ion balance in the body and opens the gateway for bacteria and toxins to enter the bloodstream, which in turn leads to diseases such as bacteremia and septicemia [3]. The intestinal villi form an essential part of the intestinal mucosal barrier [4]. They promote nutrient absorption, aid in intestinal motility, and resist external threats [5,6].

In pig farming, the incidence of piglet diarrhea is 20–50%. Diarrhea leads to weakened immunity in piglets, high secondary infection rates, and increased mortality, resulting in significant economic losses for the swine industry [7]. Some pathogens use metabolites produced during quorum sensing (QS) to evade immune detection of host intestinal epithelial cells. These pathogens exert their pathogenicity by regulating the expression of virulence factors, ultimately causing diarrhea in the host [8–18]. QS is a unique mechanism of bacterial cell–cell communication via secreted, hormone-like signaling molecules. Most bacteria produce special signaling molecules known as autoinducers (AIs) during growth. Bacteria detect AI signals in the environment. When the AI concentration reaches the threshold level, the expression of virulence genes is triggered, leading to physiological changes. Previous studies have suggested that bacterial QS molecules, especially N-(3-oxododecanoyl)-homoserine lactone (OdDHL) of the N-acyl homoserine lactone (AHL)-mediated QS system, are significant risk factors for intestinal health in low-birth-weight piglets [19].

OdDHL was found to be structurally and functionally similar to mammalian lipid hormones, whereby OdDHL has lipid-soluble and membrane-permeable properties that allow it to bind to mammals and exert its activity, allowing it to enter osteoblasts, epithelial cells, and fibroblasts in a short period of time [15]. OdDHL has a strong pro-apoptotic effect on a wide range of cells, but not all cell types are sensitive [18]. OdDHL disrupts the tight junction proteins and periplasm of intestinal cells, damaging the intestinal barrier [19]. It induces cells to release a variety of pro-inflammatory cytokines and immunomodulatory factors, such as IL-6 and IL-8 from human corneal epithelial cells, and IL-1 and IL-8 from endothelial and mesenchymal cells [20]. However, the function and mechanism of OdDHL in the porcine model have never been resolved.

Various AHL quorum-sensing molecules have been detected in pig feces, including C4-HSL, C6-HSL, C8-HSL, 3OC8-HSL, C10-HSL, 3OC10-HSL, C12-HSL, 3OC12-HSL, and 3OC14-HSL [19]. Low-birth-weight piglets often suffer from intestinal dysfunction and high mortality, resulting in significant economic losses for the swine industry. At 50 days of age, the concentrations of OdDHL and 3OC14-HSL in the feces of low-birth-weight piglets are 2.5- and 2.24-fold higher than those of piglets born with normal weight [19]. In a mouse CT-26 cell model, OdDHL induces apoptosis and disrupts the intestinal epithelial barrier. Apoptosis is a genetically regulated, programmed cell death process [20]. Distinct from necrosis, apoptosis is crucial for maintaining cellular homeostasis [21]. As a properly regulated process, apoptosis serves to clear the "redundant" cells in the body, for example by inducing rapid cell death in intestinal cells after bacterial adhesion to prevent the spread of infection to adjacent cells, which helps the body to eliminate pathogens [22]. However, when apoptosis is disordered, excessive apoptotic activity damages the intestinal epithelial barrier and leads to increased mucosal permeability [23]. As a result, endotoxins and bacteria can enter the systemic circulation and cause systemic infection. OdDHL disrupts the intestinal microbiota, damages the intestinal barrier, and induces systemic inflammation in mice. Fecal microbiota transplantation (FMT) experiments showed that the phenotype of specific pathogen-free (SPF) mice could almost be reproduced in germ-free mice, manifesting as weight loss, colon damage, and changes in serum biochemical markers [24]. Our research group previously found that OdDHL directly stimulates porcine intestinal cells (IPEC-J2), causing changes in cell morphology and promoting cytoskeleton rearrangement [25].

QS molecules regulate the expression of virulence factors in many mammalian intestinal pathogens [8,10,11,25]. In animal husbandry, farmers seek all possible means to mitigate the harm caused by drug-resistant pathogens [26–30]. Antibiotics can kill bacteria, but the selective pressure that antibiotics put on bacteria has led to the emergence of resistance. Nowadays, the global research and development of new antibiotics is very rare, mainly due to the huge cost and long time period. QS can regulate bacterial virulence and biofilm formation, which are directly related to bacterial pathogenicity. QS inhibitors (QSIs) anchor QS regulatory pathway, inhibit bacterial virulence, and reduce bacterial pathogenicity without affecting bacterial survival, which will not cause survival pressure and lead to drug resistance [28]. Antimicrobial agents targeting bacterial QS are gaining attention as alternatives to antibiotics. In aquaculture, antibiotics are limited by the aquaculture environment, and QSI was reported to greatly reduce the infection of *Aeromonas hydrophila* in zebrafish and carp aquaculture [30]. Organic acids and plant extracts functioning as QSIs regulate the expression of efflux pump genes in bacteria and inhibit bacterial invasiveness and infectivity, thus providing a new strategy for combating microbial drug resistance [31–37]. Extensive evidence has alluded to the involvement of bacterial QS molecules, such as OdDHL, in the damage caused to the host intestinal barrier [38,39]. However, QSIs protect the piglet intestinal barrier from OdDHL damage, although their exact mechanism remains unclear. In our previous studies, OdDHL was found to induce changes in apoptosis-related pathways and the extracellular matrix [25].

This study is the first to explore the mechanism by which OdDHL directly damages the piglet intestinal barrier, confirming that bacterial QS is a significant risk factor for piglet intestinal health in the farming industry. This study also confirms, for the first time in a piglet animal model, the role of QSI in protecting the piglet intestinal barrier. Our findings preliminarily verify the crucial role of lipid rafts in the action of OdDHL on porcine intestinal epithelial cells. We investigated the relationships among QS, pathogens, and hosts, and provide a theoretical basis for using bacterial QS characteristics to aid agricultural farming.

2. Results

2.1. OdDHL Damages the Intestinal Barrier in Model Piglets

H&E-stained intestine tissues were collected from piglets. In the control group, the villi of the duodenum, jejunum, and ileum were dense and elongated, with a long villus length. In the OdDHL group, the villi in the duodenum, jejunum, and ileum appeared shorter and more sparse. No significant difference was observed between the QSI and control groups (Figure 1A–C). The lengths of the small intestinal villi in the duodenum, jejunum, and ileum of the OdDHL group were reduced by 10.5%, 11.2%, and 25.3%, respectively, compared to those in the control group (Figure 1D). The crypt depth in the duodenum, jejunum, and ileum of the OdDHL group were increased by 16.7%, 24.7%, and 21.3%, respectively, compared to those in the control group (Figure 1E). There was no significant difference in the length of the small intestinal villi or crypt depth between the QSI and control groups, demonstrating the protective effect of QSI on the intestinal barrier.

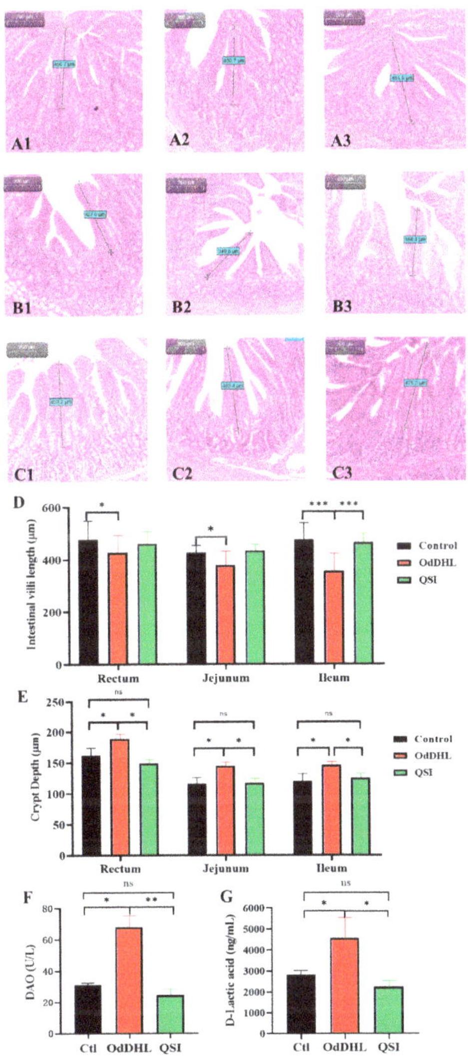

Figure 1. Effect of OdDHL on piglet intestinal barrier. (**A1–A3**): Hematoxylin–eosin staining results of duodenum, jejunum, and ileum samples from piglets in the Control group (piglets had free access to normal food and water). (**B1–B3**): H&E Staining results of duodenum, jejunum, and ileum samples from piglets in the OdDHL group (piglets were fed 0.5 L of 200 μmol/L OdDHL solution daily, while also having free access to water and regular pig feed). (**C1–C3**): H&E Staining results of duodenum, jejunum, and ileum samples from piglets in the QSI group (piglets were fed 0.5 L of 200 μmol/L OdDHL

solution daily, while also having free access to water and pig feed with 0.5% (*W*/*W*) AiiA-based QSI added). (**D**) The length of small intestine villi in Control, OdDHL, and QSI groups of piglet model. (**E**) The crypt depth in Control, OdDHL, and QSI groups of piglet model. (**F**) DAO concentration in pig blood in Control (Ctl), OdDHL, and QSI groups of piglet model. (**G**) D-lactic acid concentration in pig blood in Control, OdDHL, and QSI groups of piglet model. *: $p < 0.05$; **: $p < 0.01$; ***: $p < 0.001$; ns: no significance.

When the intestinal mucosal barrier is mechanically damaged, the levels of DAO and D-lactate increase. In the blood of piglets in the OdDHL group, the DAO and D-lactate concentrations were 2.19- and 1.52-fold higher, respectively, than those in the control group, while there were no significant differences between the DAO and D-lactate levels in the QSI and control groups (Figure 1F,G).

2.2. OdDHL Induces Apoptosis in Piglet Intestinal Epithelial Cells In Vivo and In Vitro

In the animal model, TUNEL was used to detect apoptosis in piglet intestinal tissues. As shown in Figure 2A, only lightly stained brown and blue areas were present in the duodenum, jejunum, and ileum of the control group, indicating normal apoptosis occurring as part of the physiological process. In contrast, in the OdDHL group, extensive and overlapping brown and blue areas were observed in the cell nuclei, indicating a significant increase in the apoptosis rate of intestinal cells (Figure 2B,D). In the QSI group, only normal apoptotic cells were observed, with no significant upregulation in apoptosis, thus confirming the protective effect of QSI on the small intestine of piglets (Figure 2C).

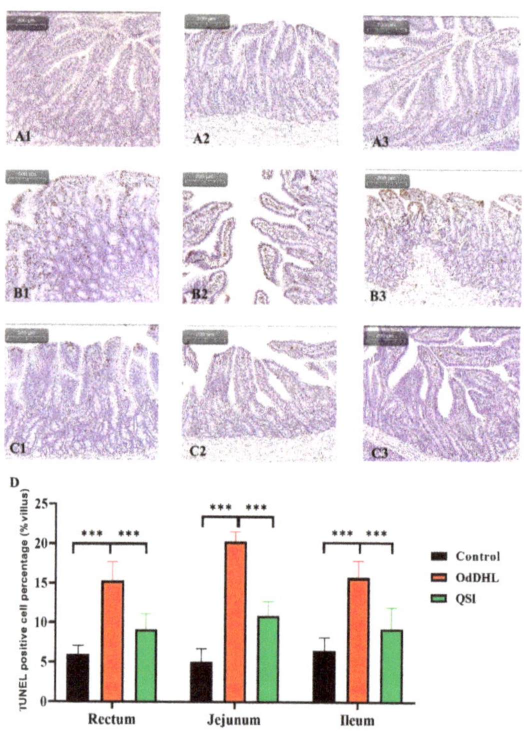

Figure 2. TUNEL test of the intestines of piglets. (**A1–A3**): TUNEL results of duodenum (**A1**), jejunum (**A2**), and ileum (**A3**) sample from piglets in Control group (piglets had free access to normal

food and water). (**B1–B3**): TUNEL results of duodenum (**B1**), jejunum (**B2**), and ileum (**B3**) sample from piglets in OdDHL group (piglet was fed 1 L of 200 μmol/L OdDHL solution daily, while also having free access to water and regular pig feed). (**C1–C3**): TUNEL results of duodenum (**C1**), jejunum (**C2**), and ileum (**C3**) sample from piglets in QSI group (piglet was fed 1 L of 200 μmol/L OdDHL solution daily, while also having free access to water and pig feed added with 0.5% (W/W) AiiA-based QSI). (**D**) TUNEL-positive cell percentage, obtained by calculating the proportion of TUNEL-positive cells out of all cells on 10 random villi. ***: $p < 0.001$.

Subsequently, in the porcine small intestinal epithelial cell model (IPEC-J2), different concentrations of OdDHL (25 μmol/L, 50 μmol/L, 100 μmol/L, 200 μmol/L) were applied, resulting in apoptosis rates of 7.41%, 9.52%, 11.5%, and 13.71%, respectively (Figure 3A). Compared to the negative control, the apoptosis rate increased dose-dependently, with a significant uptick in apoptosis following OdDHL induction. However, the correlation between the apoptosis rate and OdDHL (200 μmol/L) exposure time was weak and statistically insignificant (Figure 3B). Therefore, in subsequent experiments, the OdDHL concentration was set at 200 μmol/L, and the exposure time was 2 h.

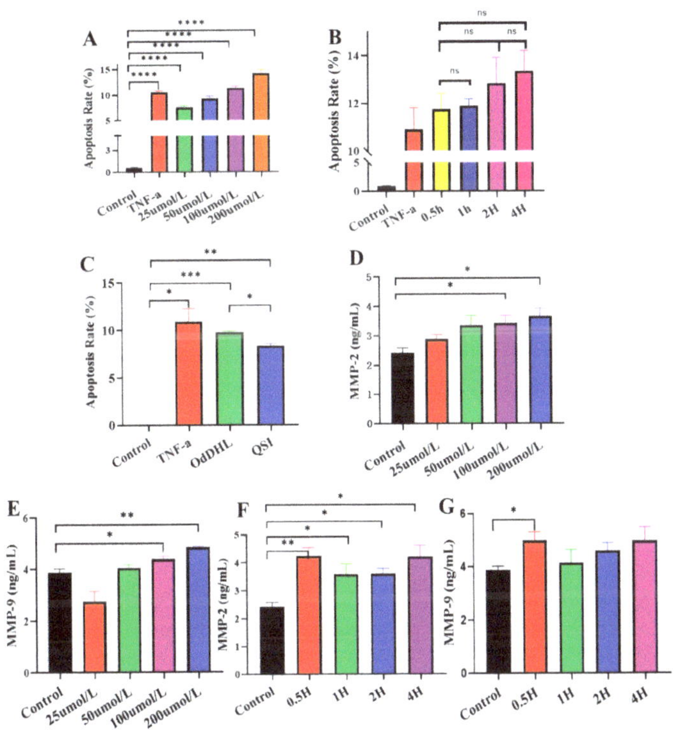

Figure 3. Cell apoptosis-related parameter detection. (**A**) Apoptosis rate under different concentrations of OdDHL on IPEC-J2 cell; apoptosis rate is the proportion of the number of apoptotic cells out of a given total number of cells in flow cytometry. (**B**) Apoptosis rate under different time of OdDHL on IPEC-J2 cell. (**C**) QSI quenching of the apoptosis rate from OdDHL; in Control group, PBS was added into cells; TNF-α was added as apoptosis inducer. (**D**) Effect of different concentrations of OdDHL on MMP-2 expression in IPEC-J2. (**E**) Effect of different concentrations of OdDHL on MMP-9 in IPEC-J2. (**F**) Effect of different duration of OdDHL on MMP-2 in IPEC-J2. (**G**) Effect of different duration of OdDHL on MMP-9 in IPEC-J2. *: $p < 0.05$; **: $p < 0.01$; ***: $p < 0.001$; ****: $p < 0.0001$; ns: no significance.

QSIs hydrolyze quorum-sensing signaling molecules to effectively lower the apoptosis rate of piglet intestinal epithelial cells. The results show that in the QSI group, the IPEC-J2 apoptosis rate decreased by 15.3%, indicating a significant reduction in OdDHL-induced apoptosis (Figure 3C).

MMP-2 and MMP-9 are closely associated with apoptosis [40]. OdDHL effectively and dose-dependently upregulated the expression of MMP-2 and MMP-9 (Figure 3D,E). The expression of MMP-2 and MMP-9 was weakly correlated with the duration of OdDHL exposure (Figure 3F,G).

2.3. *OdDHL Significantly Regulates the Expression of Apoptosis-Related Proteins*

The caspase family of proteins plays an essential role in apoptosis. Caspase-3 and -8 are involved in the death receptor-mediated apoptosis pathway, whereas caspase-3 and -9 are involved in the mitochondrial pathway of apoptosis [41]. The results showed that OdDHL significantly increased the activities of caspases 3, 8, and 9, whereas the stimulatory effect of OdDHL on these caspases was significantly inhibited after QSI treatment (Figure 4A–C).

Western blotting (Figure 4D,E) revealed that the expression of the apoptosis-related proteins CytC, Bid, BAX, Bcl-2, and FASCD95 was relatively low under normal conditions. After stimulation with the apoptosis inducer TNF-α, the expression of CytC, BAX, Bcl-2, and FASCD95 was significantly upregulated. When OdDHL was co-incubated with cells, the expression of BAX, Bcl-2, and FASCD95 was significantly increased, whereas no significant difference was observed in the expression of CytC and Bid. The results of the QSI group demonstrated that QSI effectively inhibited the stimulatory effect of OdDHL on the expression of apoptosis-related proteins.

2.4. *Lipid Rafts Are Closely Related to OdDHL-Induced Apoptosis in Porcine Intestinal Cells*

After M-β-CD treatment, the mean fluorescence intensity from CTB (Cholera Toxin Subunit B)-FITC (Absin, Shanghai, China) binding to GM1 decreased by 40%, indicating successful disruption of lipid raft on IPEC-J2. In the Chol group, fluorescence intensity was increased by 23% compared with M-β-CD group, which indicated that the lipid rafts had recovered to some extent. TNF-α, used as a positive control, induced an average apoptosis rate of 10.23%, but a rate of 14% in the OdDHL group. However, after lipid rafts in intestinal epithelial cells were removed with M-β-CD, the rate of OdDHL-induced apoptosis decreased by approximately 50%. In the Chol group, the addition of cholesterol partially restored the lipid rafts, and the apoptosis rate increased by approximately 12% compared to that in the M-β-CD group (Figure 5A).

In the M-β-CD group, the ability of OdDHL to induce caspase expression was similarly tested. After lipid raft removal, the activities of caspases 3, 8, and 9 all showed varying degrees of decline. In the Chol group, the apoptosis rate was increased (Figure 5B).

Western blotting experiments suggested that after lipid raft removal, IPEC-J2 cells showed lower CytC, Bid, BAX, and FASCD95 expression compared to those in the OdDHL-treated group, while Bcl-2 expression increased. These results indicate that lipid rafts participate in OdDHL-induced apoptosis in IPEC-J2 cells. In the Chol group, the expression of apoptosis-related proteins increased (Figure 5C,D).

FITC-OdDHL conjugates were co-incubated with IPEC-J2 cells. Fluorescence microscopy revealed that OdDHL effectively adhered to IPEC-J2 cells, with intense green fluorescence observed. However, after removing the lipid raft structure from the cells, the intensity of green fluorescence decreased significantly, proving that lipid rafts are directly involved in OdDHL adherence to cells. In the Chol group, a significant increase in green fluorescence was observed after partial lipid raft replenishment (Figure 5E).

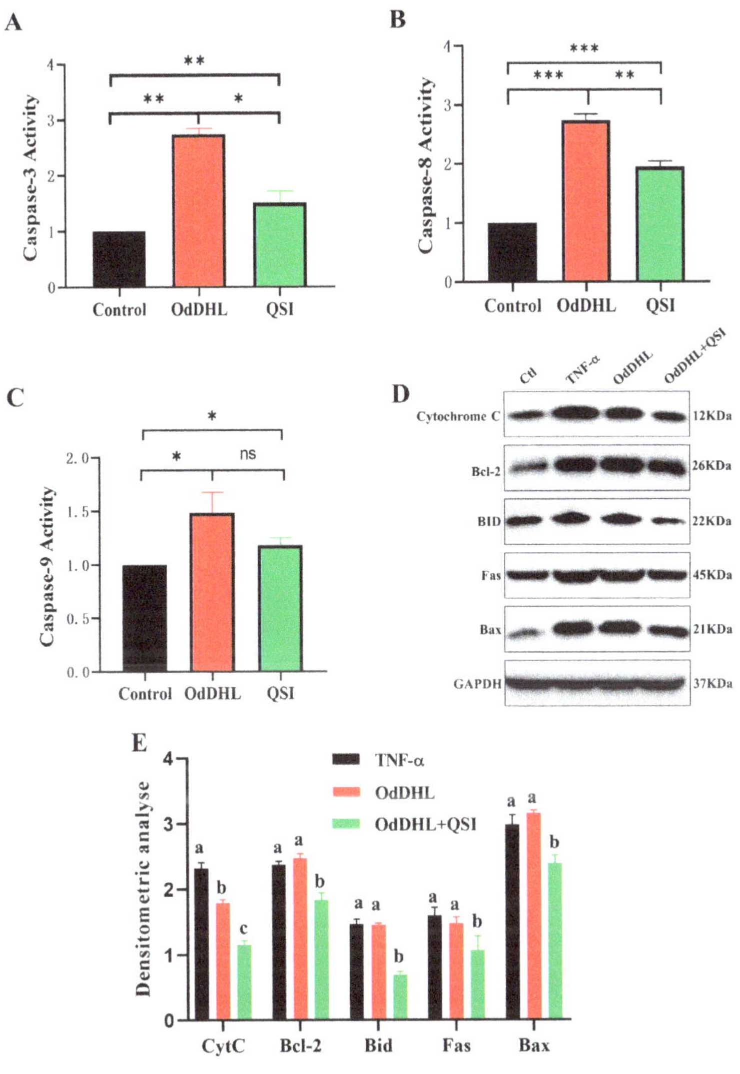

Figure 4. Caspase activity test and apoptosis-related protein expression level detection. (**A**) Caspase-3 activity in Control, OdDHL, and QSI groups of cell model. In Control group, PBS was added into cells. In OdDHL group, 200 µmol/L OdDHL were added into each cell well. In the QSI team, a mixture of OdDHL and QSI was added into cell wells. (**B**) Caspase-8 activity in Control, OdDHL, and QSI groups of cell model. (**C**) Caspase-9 activity in Control, OdDHL, and QSI groups of cell model; caspase activity was determined by calculating the multiplicity of OD_{450} values of samples/OD450 values of negative control. (**D**) Expression of apoptosis related proteins by Western blot. TNF-α was added as apoptosis inducer. (**E**) Densitometric analyses of apoptosis related proteins. Each value represents the mean ± SD (n = 3) considering the control sample value as 1. Different letters indicate significant differences between treatments. *: $p < 0.05$; **: $p < 0.01$; ***: $p < 0.001$; ns: no significance. The original images of the Western blot are published as supplementary material.

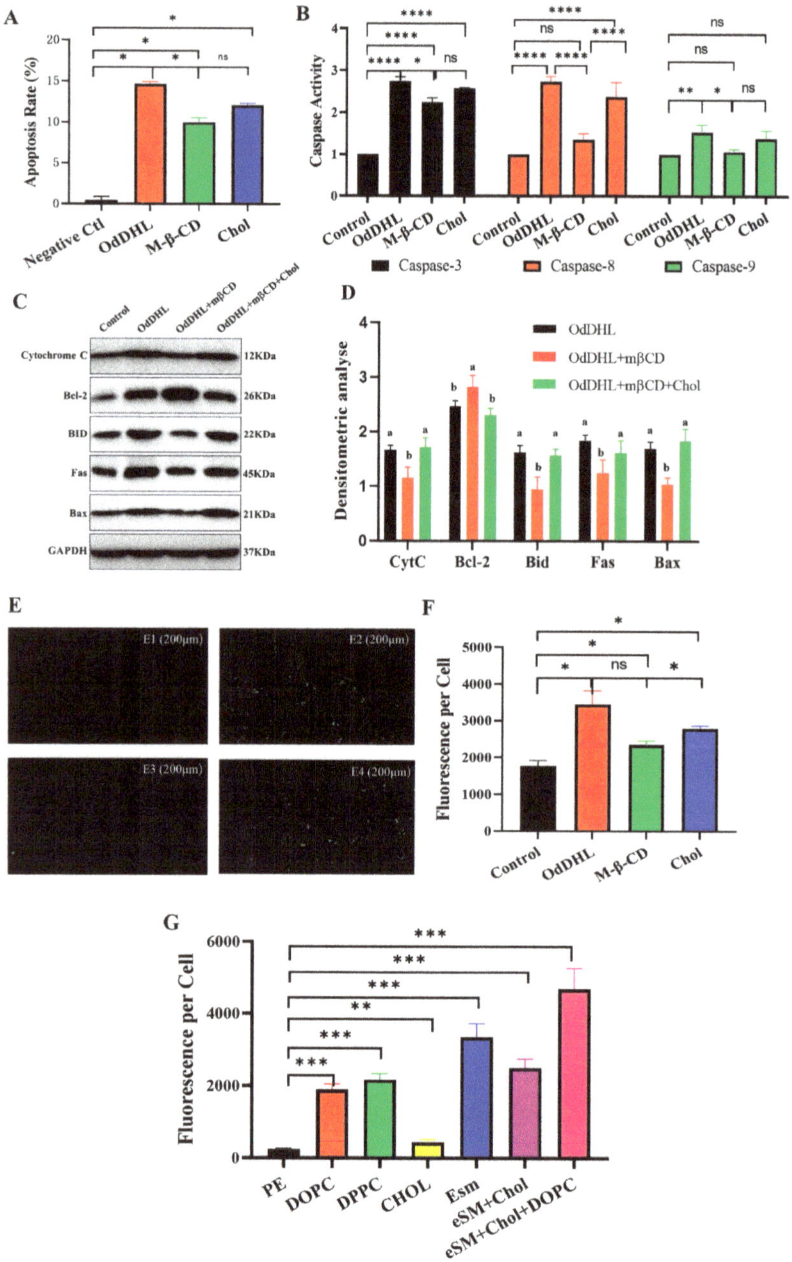

Figure 5. Detection of OdDHL-related parameters after removal of lipid rafts from IPEC-J2. (**A**) Detection of cell apoptosis rate. In negative Control group, PBS was added into cells. In positive control group, TNF-α was added as apoptosis inducer. In OdDHL group, 200 μmol/L OdDHL was added into each cell well. In M-β-CD group, to remove lipid rafts from IPEC-J2 surface, 2 mg/mL M-β-CD was added for 1 h incubation. In Chol group, cells were incubated with 400 μg/mL cholesterol for 1 h before M-β-CD treatment. (**B**) Detection of caspase activity. (**C**) Expression level of apoptosis-related proteins. (**D**) Densitometric analyses of apoptosis-related proteins. Each value represents the mean ± SD (n = 3) considering the control sample value as 1. Different letters indicate significant differences between

treatments. (**E**) Fluorescence microscope observation after co-incubation with FITC-OdDHL conjugates: E1—negative control; E2—OdDHL group; E3—M-β-CD group; E4—Chol group. (**F**) Fluorescence intensity of individual cells in flow cytometry. (**G**) Affinity of OdDHL to different lipids. *: $p < 0.05$; **: $p < 0.01$; ***: $p < 0.001$; ****: $p < 0.0001$; ns: no significant.

Flow cytometry was used to quantify the fluorescence intensity of individual cells in each group. In the M-β-CD group, the fluorescence intensity of single cells displaying green fluorescence (i.e., cells that adsorbed OdDHL) was significantly lower than that in the OdDHL group by approximately 32.71%. The Chol group showed significantly higher fluorescence intensity than the M-β-CD group (approximately 18.72% higher, Figure 5F).

FITC-OdDHL conjugates were used to measure the affinity and adsorption of OdDHL toward various lipids that comprise lipid rafts to determine the lipids that play a major role in OdDHL adherence and its action on porcine intestinal epithelial cells. Artificially synthesized lipids such as PE (phosphatidylethanolamine) were used as negative controls. The results show that OdDHL has a low affinity for simple lipids, particularly for PE and Chol (cholesterol). However, compared to the negative controls, the affinity of OdDHL toward the SM (sphingomyelin) and Chol/DOPC (1,2-dioleoyl-sn-glycero-3-phosphocholine) mixture [42] (commonly used to simulate lipid raft structures) was significantly higher (19.9 times higher, Figure 5G).

3. Discussion

Using weaned piglets fed purified QS signaling molecules (OdDHL) and QSI (AiiA) as an animal model, we explored the mechanism by which OdDHL damages the intestinal barrier and the protective effect of QSI on it. Under the stimulation of OdDHL, the small intestinal villi of piglets shrink, become sparser, and shorten in length. When the intestinal mucosal barrier is mechanically damaged, the levels of two key indicators, DAO and D-lactate, increase [43]. Our results revealed that, after administering OdDHL to piglets via drinking water, the peripheral blood levels of DAO and D-lactate significantly increased, confirming that the QS signaling molecule OdDHL can directly damage the intestinal mucosal barrier. Adding QSI to the feed effectively protected the intestinal barrier from OdDHL-induced damage.

A TUNEL assay was designed to detect apoptotic cells [44], and it revealed that following OdDHL treatment, large areas of apoptotic cells were clearly observed in the duodenum, jejunum, and ileum of piglets, whereas QSI effectively inhibited apoptosis. This finding demonstrates, for the first time, that QS molecules induce unregulated intestinal cell apoptosis in a piglet model and that OdDHL-induced apoptosis is the direct cause of intestinal barrier damage triggered by QS. Current research confirms that apoptosis can be categorized into exogenous and endogenous pathways. The exogenous pathway is regulated by stimulation of tumor necrosis factor receptor (TNFR) superfamily members. The endogenous pathway is characterized by mitochondrial outer membrane permeabilization and mitochondrial cytochrome c release, leading to the assembly of a caspase activation complex between caspase-9 and apoptotic bodies. Pathogen infection causes structural damage and loss of function in host cells and stimulates the host's recognition and immune response. Pathogen-induced apoptosis enhances or diminishes pathogenicity. Apoptosis is an active physiologic death process mediated by multiple genes. Apoptosis is a highly regulated energy-dependent cell death and is an important pathway for host to resist external infections. Apoptosis occurs as a result of a combination of pro- and anti-apoptotic genes [45–49].

Subsequently, we explored the mechanism underlying OdDHL-induced apoptosis in intestinal cells using an IPEC-J2 cell model. Flow cytometry using annexin-FITC assay showed that OdDHL directly induced apoptosis in IPEC-J2 cells. Under OdDHL induction, the activities of caspases 3, 8, and 9 were significantly increased, as determined by ELISA. However, QSI significantly inhibited the stimulatory effect of OdDHL on the activities of caspases 3, 8, and 9. During the early stages of apoptosis, Caspase-3 is activated to cleave cytoplasmic and nuclear substrates, leading to apoptosis. Caspase-3 and -8 are involved in the death receptor pathway of apoptosis, whereas caspase-3 and -9 are related to the mitochondrial apoptosis pathway [22]. These results suggest that OdDHL induces apoptosis through both the death receptor and mitochondrial pathways. The Western blot results showed that following OdDHL treatment, the expression of the apoptosis-related proteins CytC, BAX, and FASCD95 increased significantly, whereas QSI significantly inhibited the changes in the expression levels of apoptosis-related proteins. These results demonstrate that OdDHL, a bacterial QS molecule, induces apoptosis through multiple pathways and that its ability to induce apoptosis is lost after its degradation by QSI. MMPs degrade extracellular matrix and are closely related to apoptosis [40]. We demonstrated that the concentration of MMPs in IPEC-J2 cells changed depending on the dose of OdDHL applied. Therefore, OdDHL may indirectly influence apoptosis by affecting MMPs.

The cell membrane is a signaling platform for intercellular communication. Various proteins on the cell membrane can bind to extracellular signaling molecules and transmit signals to cells, inducing specific biological effects. Lipid rafts are an important component of cell membranes and are involved in many biological processes, including transmembrane signal transduction, apoptosis, Chol transport, and protein sorting during endocytosis and exocytosis [50,51]. Lipid rafts are cell membrane structures composed of Chol, glycosylphosphatidylinositol-anchored proteins, glycosphingolipids, SM, and unsaturated acyl chain phospholipids. Many studies have reported that lipid rafts are involved in cell invasion by non-enveloped viruses, such as rotavirus, rhinovirus, and enterovirus [52]. Additionally, lipid rafts play a role in the colonization of host cells by pathogens such as *Vibrio cholerae* and Enteropathogenic *Escherichia coli* [53]. The spatial structure of acyl homoserine lactones provides them with both hydrophilic and hydrophobic properties, but it is not clear whether they adhere to host intestinal cells via lipid rafts.

Because lipid rafts contain high Chol concentrations, they exhibit a liquid-ordered state that is between the liquid crystalline and gel phases. Removal of Chol with M-β-CD disrupts the phase of lipid rafts [54], thereby impairing their biological functions. In this study, FITC-OdDHL conjugates were applied to IPEC-J2 cells, and fluorescence microscopy and flow cytometry were used to qualitatively and quantitatively observe the green fluorescence intensity of the cells in each group. The results showed that after M-β-CD treatment, the number of OdDHL molecules adsorbed to the surface of IPEC-J2 cells was significantly reduced, whereas Chol supplementation restored the ability of the cells to adsorb FITC-OdDHL molecules. Flow cytometry was used to quantitatively analyze the fluorescence intensity of individual cells in each group. The results showed that after M-β-CD treatment, the ability of individual IPEC-J2 cells to adsorb OdDHL was significantly weakened. These results demonstrate that lipid rafts mediate the adsorption of OdDHL molecules to cells and that disrupting their structure significantly reduces the ability of cells to adsorb OdDHL molecules.

We further explored whether the adsorption of OdDHL to lipid rafts is due to its affinity for specific lipids or requires the mediation of a complete lipid raft structure. Various lipids found in lipid rafts were tested, with PE (a synthetic lipid, not present in lipid rafts) used as the control. FITC-OdDHL conjugates were applied to DPPC, PE, DOPC,

Chol, and SM. The results show that OdDHL exhibited a much lower affinity for individual lipids than simulated lipid raft structures (composed of SM, Chol, and DOPC) [42]. This confirms that mediation by specific lipid raft structures is required for OdDHL to exert its damaging effect on the intestinal barrier.

The QS system is closely related to bacterial virulence. In the context of the complete ban on antibiotic use in farming in China since 2020, antibacterial drugs targeting bacterial QS have gained increasing attention as alternatives to antibiotics [31,33,34,55]. Since the discovery of bacterial QS in the 1990s, research on its application in the population-level control of bacterial traits has deepened, and it has been closely linked to the synthesis of virulence factors in various pathogenic bacteria. QSIs are used to interfere with QS signals and suppress the expression of virulence genes. In this way, bacterial pathogenicity can be reduced without creating selective pressure, thereby preventing the rapid emergence of antibiotic resistance. Using a piglet model, we demonstrate for the first time that the bacterial QS molecule OdDHL is mediated by lipid rafts and induces intestinal cell apoptosis through both the death receptor and mitochondrial pathways, causing damage to the piglet intestinal barrier, while QSIs effectively protect the piglet intestinal barrier.

In conclusion, this study provides a preliminary explanation for specific roles of QS and QSI and explores the mechanism. To our knowledge, this is the first qualitative piglet model experiment to validate the effect of QS and QSI on the intestinal barrier. Although the research has a degree of novelty, the authors recognize the limitation of the experimental data. For the purpose of the qualitative experiment, only two piglets were arranged in each group in this manuscript. The authors believe that the data obtained from this number of experimental animals have been able to adequately qualitatively validate the function of QS, and that the results from piglets are in complete accordance with the results of cell model that have been repeated enough times. However, the authors also acknowledged that the insufficient number of experimental animals limited the continuation of deeper digging into the relevant conclusions. We believe that this study provides preliminary data on the function and mechanism of QS and QSI in the piglet model, and also believe that the current data can provide inspiration for subsequent replicated animal experiments on adequate piglet numbers, which could thoroughly elucidate the potential function of QS in pig farming and the prospect of QSI application, and could provide theoretical support for the prevention and control of infectious diseases in livestock and poultry under antibiotic restriction policies.

4. Materials and Methods

4.1. Ethics Statement

All animal experiments were conducted in compliance with the Guidelines for the Care and Use of Laboratory Animals issued by the Ministry of Science and Technology of the People's Republic of China. The animal experiments were approved by the Animal Welfare and Ethics Committee of Yangzhou University (Approval Number: 202103003) and were conducted in compliance with the ethical standards of animal welfare and the Jiangsu Provincial Animal Management Committee. The experiments were performed at the Animal Experiment Center of the College of Veterinary Medicine, Yangzhou University.

4.2. Experimental Animals

Six weaned piglets (Duroc × [Yorkshire × Landrace]) used in this study were purchased from Jiangsu Gangwan Agricultural Science & Technology Group, Yangzhou, China (average initial body weight: 7.0 ± 1.0 kg, age: 28 ± 1 days). The piglets were divided into the control, OdDHL, and QSI groups, with two piglets per group. The authors performed several repetitions of a pig epithelial cell model to determine a suitable working concentration (200 μM) of OdDHL [25]. In order to avoid the stress associated with gavage administration, the dosing drinkers were used to precisely control the amount of solution that piglets drank freely. The control group of piglets had free access to regular food and water. In the OdDHL and QSI groups, each piglet was fed 0.5 L of 200 μmol/L OdDHL (Changzhou Wanyao Biotechnology Co., Changzhou, China) solution daily through rationed waterers. After drinking 0.5 L of solution, piglets were able to drink regular water freely. The piglets in the OdDHL group had free access to regular pig feed. In the QSI group, 0.5% (W/W) AiiA-based QSI (1000 U/g, Beijing Challenge Agricultural Technology Co., Beijing, China) was added to the feed. The pigsty was cleaned daily and disinfected regularly to remove any external influences. After 5 days of feeding, blood was collected from the anterior vena cava of the piglets and stored at 4 °C. The piglets were euthanized with a sodium pentobarbital overdose of 40 mg/kg body weight, followed by exsanguination. The duodenum, jejunum, and ileum of piglets from each group were collected and fixed in 4% paraformaldehyde (Yuanye Bio-Technology, Shanghai, China).

4.3. Cell Lines

The IPEC-J2 (intestinal porcine enterocyte) cell line (DSMZ ACC 701) was obtained from the Faculty of Veterinary Medicine in the Department of Microbiology at Yangzhou University [56]. The IPEC-J2 cells were cultured in RPMI 1640-F12 medium supplemented with 10% neonatal bovine serum at 37 °C with 5% CO_2.

In apoptosis-related experiments, TNF-α (Beyotime, Shanghai, China) was added to IPEC-J2 cells as a positive control for induced apoptosis. In the OdDHL group, different concentrations of OdDHL (25 μmol/L, 50 μmol/L, 100 μmol/L, 200 μmol/L) were added to each well and co-incubated with IPEC-J2 for 0.5 h [25]. Furthermore, 200 μmol/L OdDHL was added for apoptosis rate detection in different OdDHL exposure times (0.5 h, 1 h, 2 h, 4 h). In the QSI treatment, a mixture of OdDHL (200 μmol/L) and QSI (10 μg/mL) was added to each well for incubation. PBS was used as a control. The experiments were repeated three times.

In the lipid raft experiments, for the M-β-CD group, 2 mg/mL of M-β-CD was added to remove lipid rafts from the IPEC-J2 surface, followed by 1 h of incubation. In the cholesterol group, cells were incubated with 400 μg/mL cholesterol for 1 h before M-β-CD treatment to reduce potential damage to the lipid rafts. The ganglioside GM1 is a commonly used raft marker. In order to validate the disruption and recovery of the lipid raft structure, the efficiency was determined by detecting fluorescent CTB-FITC binding to GM1 using flow cytometry [54]. Decreased or increased fluorescence intensity indicated disruption or recovery of lipid raft on cells.

4.4. Piglet Intestinal Hematoxylin and Eosin (H&E) Staining

The intestinal tissues were fixed with paraformaldehyde, dehydrated, embedded in paraffin, and sectioned using a microtome (with a section thickness of 5 μm) [57]. After the tissue sections were placed on slides and air-dried, they were stained using a H&E staining kit (Beyotime, Shanghai, China) and observed under an Olympus CK43 microscope (Olympus, Tokyo, Japan). For each intestinal segment, the 10 longest, straight-traced, and well-extended small intestinal villi were selected from each slice. The length of selected small intestinal villi (from the base of the villus to the tip of the villus) and the depth of the crypt (from the base of the crypt to the base of the villus) were measured. The mean values of villi length and crypt depth were calculated.

4.5. Enzyme-Linked Immunosorbent Assay (ELISA)

Serum levels of diamine oxidase (DAO), D-lactate, and cell levels of matrix metalloproteinase (MMP)2, MMP9, caspase 3, caspase 8, and caspase 9 were measured using ELISA kits (Mlbio, Shanghai, China) as specified in the manufacturer's instructions. It mainly consists of steps such as addition of sample, incubation, washing, addition of substrate, and termination. The absorbance at 450 nm of each sample was measured using a microplate reader (Epoch, BioTek, Winooski, VT, USA).

For blood samples (for DAO and D-lactate detection), the precipitated serum was used for subsequent ELISA. For cell samples (for MMP2, MMP9, caspase 3, caspase 8, and caspase 9 measurement), the adherent cells were digested by trypsin. Cells were washed with PBS, then an appropriate amount of lysate solution for resuspension was added. Cells were lysed in an ice bath for 30 min, then centrifuged at 12,000 rpm for 10 min at 4 °C, and the supernatant was aspirated for subsequent experiments.

For DAO, D-lactate, and MMPs, a standard curve is constructed using the standards in the kit so that the content can be calculated. Caspase Activity was determined by calculating the multiplicity of OD_{450} values of samples/OD_{450} values of negative control, in which group PBS was added into cells. The experiments were repeated three times.

4.6. TUNEL Assay for Detecting Piglet Intestinal Apoptosis

The TUNEL assay was performed according to the manufacturer's instructions (Beyotime, Shanghai, China). Briefly, the paraffin-embedded tissue sections were washed twice with phosphate-buffered saline (PBS), and excess water was removed. Each sample was mixed with 50 μL of the TdT enzyme-reaction solution, covered with a cover slip, and incubated at 37 °C in the dark for 60 min. After three washes with PBS, excess solution was removed, and 50 μL of streptavidin-HRP working solution was added, followed by incubation at 37 °C in the dark for 30 min. After three more washes with PBS, 100 μL of DAB (diaminobenzidine)-HRP working solution was added for color development at room temperature for 10 min. The sections were washed with PBS, air-dried, and observed under a microscope. Ten villi were randomly selected from each section and the proportion of TUNEL-positive cells in each villus was counted.

4.7. Annexin-FITC Measurement of Apoptosis Rate

Apoptosis assessment by Annexin-FITC was performed as specified in the manufacturer's instructions (Beyotime, Shanghai, China). TNF-α-induced apoptosis was used as a positive control. Pre-treated cells were washed with pre-chilled PBS, re-suspended, and counted. A total of 50,000 cells were incubated with 195 μL of Annexin-FITC binding solution, followed by 5 μL of Annexin-FITC stain. Then, 5 μL of propidium iodide (PI) was added and mixed, and the cells were incubated at room temperature in the dark for 20 min. The apoptosis rate of each experimental group was measured using a flow cytometer (Cytoflex, Beckman Colter, CA, USA). The experiments were repeated three times.

4.8. Western Blot

Western blotting was conducted to detect the expression levels of CytC, Bid, BAX, Bcl2, and FASCD95, following a method described in a related study [57]. Treated cells were washed three times with PBS. Subsequently, 200 μL of cell lysis buffer was added to the cells. After conducting SDS-PAGE and electro transfer, primary antibodies (Wuhan Sanying Biotechnology Co., Wuhan, Hubei, China) were added (Anti-CytC antibody at 1:1000, Bid at 1:1000, BAX at 1:1000, Bcl2 at 1:500, FASCD95 at 1:500, and GAPDH at 1:2000). After the secondary antibody (goat anti-mouse IgG HRP at 1:2000) was added, detection was carried out with a BeyoECL Plus Kit (Beyotime, Shanghai, China). Protein bands were quantificated by densitometric analysis (ImageJ v.1.49, National Institutes of Health, Bethesda, MD, USA). The experiments were repeated three times.

4.9. OdDHL Adsorption to Lipid Rafts

OdDHL-FITC conjugates (Cayman Chemical, Ann Arbor, MI, USA) were added to treated cells in each group and incubated for 1 h to observe the affinity of OdDHL for lipid rafts. PBS was used to wash away unbound OdDHL-FITC, and fluorescence microscopy was used for qualitative observation.

The cells from each group were digested with trypsin (without EDTA) and washed with PBS to remove unbound OdDHL-FITC. The fluorescence intensity of each group of cells was quantified using flow cytometry to assess the affinity of OdDHL. The experiments were repeated three times.

4.10. OdDHL Lipid Affinity Detection

DOPC (1,2-dioleoyl-sn-glycero-3-phosphocholine), DPPC (1,2-dipalmitoyl-sn-glycero-3-phosphorylcholine), SM (sphingomyelin), Chol (cholesterol), and PE (phosphatidylethanolamine) were prepared at a concentration of 0.01 g/L. Lipids were purchased from Shanghai Aladdin Biochemical Technology Co. (Shanghai, China).

Briefly, 50 µL of each lipid solution was added to a 96-well plate, with four wells containing each lipid solution. For the DOPC/DPPC group, the two lipids were mixed at a 1:1 ratio. For the SM/Chol/DOPC group, the three lipids were mixed at a 6:3:1 ratio to simulate lipid raft formation [42]. After loading the samples, the plate was placed in an incubator at 37 °C for sufficient drying time. Subsequently, 50 µL of OdDHL-FITC was added to each well, incubated at room temperature for 30 min, and washed with PBS. Finally, 20 µL of PBS was added to each well, and the fluorescence intensity was measured using a BioTek Synergy HTX microplate reader (BioTek, Winooski, VT, USA) at an emission wavelength of 485 nm and excitation wavelength of 520 nm. The experiments were repeated three times.

4.11. Statistical Analysis

All experimental data were analyzed by unpaired Student's *t*-test using Prism 5.0 software (GraphPad Inc., San Diego, CA, USA). $p < 0.05$ was considered to be significant statistically. Error bars in the figures represent the standard deviation of the dataset (mean ± standard deviation). * $p < 0.05$, ** $p < 0.01$, *** $p < 0.001$, **** $p < 0.0001$.

5. Conclusions

This study verified that bacterial OdDHL signal may disrupt the host intestinal barrier by OdDHL-induced apoptosis. The mechanism of OdDHL-induced apoptosis in a piglet model was preliminarily explored. Furthermore, QSI effectively antagonized this OdDHL-induced intestinal barrier damage. Lipid rafts are intimately involved in the process of OdDHL adsorption to the cell surface and thus play an important role in OdDHL-mediated apoptosis.

Supplementary Materials: The following are available online at https://www.mdpi.com/article/10.3390/vetsci12030233/s1, Figure S1: the original image of the Western blot from Figure 4D, Figure S2: the original image of the Western blot from Figure 5C, Figure S3: flow cytometry diagram in Figure 5F, Figure S4: flow cytometry diagram in Figure 3B, Figure S5: flow cytometry diagram in Figure 3C, Figure S6: flow cytometry diagram in Figure 5F.

Author Contributions: Conceptualization, Y.Y. and G.Z.; methodology, Y.Y., X.Z., J.Y. and Z.W.; validation, X.Z., J.Y. and Z.W.; formal analysis, Y.Y.; investigation, Y.Y.; data curation, X.Z., J.Y., Z.W., C.M. and J.L.; writing—original draft preparation, Y.Y. and R.S.; writing—review and editing, G.Z.; visualization, G.Z.; supervision, Y.Y. and G.Z.; funding acquisition, Y.Y. and G.Z. All authors have read and agreed to the published version of the manuscript.

Funding: This research was funded by the Chinese National Science Foundation Grant (No. 31972708), the Yangzhou City Science and Technology Bureau Sponsored Program 2022- Social Development (Grant No. YZ2022092), the Open Project Program of Jiangsu Key Laboratory of Zoonosis (R2302), and the Priority Academic Program Development of Jiangsu Higher Education Institutions (PAPD).

Institutional Review Board Statement: The animal study protocol was approved by the Animal Welfare and Ethics Committee of Yangzhou University (Approval Number: 202103003; Approval Data: 20210302) and was conducted in compliance with the ethical standards of animal welfare and the Jiangsu Provincial Animal Management Committee.

Informed Consent Statement: Not applicable. The experimental pigs mentioned in the manuscript were procured by the authors using the funds mentioned in the article and were used exclusively for this study. The authors of the manuscript had ownership of the experimental pigs.

Data Availability Statement: All datasets generated for this study are included in the article.

Acknowledgments: We acknowledge and apologize to the researchers whose outstanding work was not cited due to space constraints.

Conflicts of Interest: The authors declare that they have no conflicts of interest.

References

1. Chelakkot, C.; Ghim, J.; Ryu, S.H. Mechanisms regulating intestinal barrier integrity and its pathological implications. *Exp. Mol. Med.* **2018**, *50*, 1–9. [CrossRef]
2. Ma, J.; Piao, X.; Mahfuz, S.; Long, S.; Wang, J. The interaction among gut microbes, the intestinal barrier and short chain fatty acids. *Anim. Nutr.* **2022**, *9*, 159–174. [CrossRef]
3. Horowitz, A.; Chanez-Paredes, S.D.; Haest, X.; Turner, J.R. Paracellular permeability and tight junction regulation in gut health and disease. *Nat. Rev. Gastroenterol. Hepatol.* **2023**, *20*, 417–432. [CrossRef] [PubMed]
4. Peterson, L.W.; Artis, D. Intestinal epithelial cells: Regulators of barrier function and immune homeostasis. *Nat. Rev. Immunol.* **2014**, *14*, 141–153. [CrossRef] [PubMed]
5. Song, C.; Chai, Z.; Chen, S.; Zhang, H.; Zhang, X.; Zhou, Y. Intestinal mucus components and secretion mechanisms: What we do and do not know. *Exp. Mol. Med.* **2023**, *55*, 681–691. [CrossRef] [PubMed]
6. Turner, J.R. Intestinal mucosal barrier function in health and disease. *Nat. Rev. Immunol.* **2009**, *9*, 799–809. [CrossRef]
7. Gresse, R.; Chaucheyras-Durand, F.; Fleury, M.A.; Van de Wiele, T.; Forano, E.; Blanquet-Diot, S. Gut Microbiota Dysbiosis in Postweaning Piglets: Understanding the Keys to Health. *Trends Microbiol.* **2017**, *25*, 851–873. [CrossRef]
8. Yang, Y.; Shao, J.; Zhou, M.; Duan, Q.; Zhang, X.; Zhu, G. Effects of Quorum Sensing AHL Signaling on the Biological Characteristics of Porcine Derived F4ac+ Enterotoxigenic Escherichia coli. *Pak. J. Zool.* **2022**, *54*, 701–708. [CrossRef]
9. Yang, Y.; Zhang, X.; Zhang, B.; Zhou, M.; Duan, Q.; Li, Z.; Zhang, X.; Zhu, G. Quorum sensing-1 signaling of N-hexanoyl-l-homoserine lactone contributes to virulence in avian pathogenic Escherichia coli. *Arch. Microbiol.* **2021**, *203*, 6079–6089. [CrossRef]
10. Yang, Y.; Zhou, M.; Hardwidge, P.R.; Cui, H.; Zhu, G. Isolation and Characterization of N-acyl Homoserine Lactone-Producing Bacteria From Cattle Rumen and Swine Intestines. *Front. Cell. Infect. Microbiol.* **2018**, *8*, 155. [CrossRef]
11. Yang, Y.; Liu, Y.; Zhou, M.; Zhu, G. Both Quorum Sensing (Qs)-I and Ii Systems Regulate Escherichia coli Flagellin Expression. *Pak. J. Zool.* **2018**, *50*, 1807–1813. [CrossRef]
12. Yang, Y.; Zhou, M.; Hou, H.; Zhu, J.; Yao, F.; Zhang, X.; Zhu, X.; Hardwidge, P.R.; Zhu, G. Quorum-sensing gene luxS regulates flagella expression and Shiga-like toxin production in F18ab Escherichia coli. *Can. J. Microbiol.* **2014**, *60*, 355–361. [CrossRef]
13. Yang, Y.; Yao, F.; Zhou, M.; Zhu, J.; Zhang, X.; Bao, W.; Wu, S.; Hardwidge, P.R.; Zhu, G. F18ab Escherichia coli flagella expression is regulated by acyl-homoserine lactone and contributes to bacterial virulence. *Vet. Microbiol.* **2013**, *165*, 378–383. [CrossRef]

14. Sturbelle, R.T.; Avila, L.F.; Roos, T.B.; Borchardt, J.L.; de Cassia, D.S.D.C.; Dellagostin, O.A.; Leite, F.P. The role of quorum sensing in Escherichia coli (ETEC) virulence factors. *Vet. Microbiol.* **2015**, *180*, 245–252. [CrossRef] [PubMed]
15. Schwarzer, C.; Fu, Z.; Shuai, S.; Babbar, S.; Zhao, G.; Li, C.; Machen, T.E. Pseudomonas aeruginosa homoserine lactone triggers apoptosis and Bak/Bax-independent release of mitochondrial cytochrome C in fibroblasts. *Cell. Microbiol.* **2014**, *16*, 1094–1104. [CrossRef] [PubMed]
16. Carneiro, D.G.; Almeida, F.A.; Aguilar, A.P.; Vieira, N.M.; Pinto, U.M.; Mendes, T.A.O.; Vanetti, M.C.D. Salmonella enterica Optimizes Metabolism After Addition of Acyl-Homoserine Lactone Under Anaerobic Conditions. *Front. Microbiol.* **2020**, *11*, 1459. [CrossRef]
17. Campos-Galvao, M.E.; Ribon, A.O.; Araujo, E.F.; Vanetti, M.C. Changes in the Salmonella enterica Enteritidis phenotypes in presence of acyl homoserine lactone quorum sensing signals. *J. Basic Microbiol.* **2016**, *56*, 493–501. [CrossRef]
18. Nesse, L.L.; Berg, K.; Vestby, L.K.; Olsaker, I.; Djønne, B. Salmonella Typhimurium invasion of HEp-2 epithelial cells in vitro is increased by N-acylhomoserine lactone quorum sensing signals. *Acta Vet. Scand.* **2011**, *53*, 44. [CrossRef]
19. Tao, S.; Xiong, Y.; Wang, Z.; Wu, Y.; Li, N.; Pi, Y.; Han, D.; Zhao, J.; Wang, J. N-Acyl-Homoserine Lactones May Affect the Gut Health of Low-Birth-Weight Piglets by Altering Intestinal Epithelial Cell Barrier Function and Amino Acid Metabolism. *J. Nutr.* **2021**, *151*, 1736–1746. [CrossRef]
20. Tao, S.; Xiong, Y.; Han, D.; Pi, Y.; Zhang, H.; Wang, J. N-(3-oxododecanoyl)-l -homoserine lactone disrupts intestinal epithelial barrier through triggering apoptosis and collapsing extracellular matrix and tight junction. *J. Cell. Physiol.* **2021**, *236*, 5771–5784. [CrossRef]
21. Bertheloot, D.; Latz, E.; Franklin, B.S. Necroptosis, pyroptosis and apoptosis: An intricate game of cell death. *Cell. Mol. Immunol.* **2021**, *18*, 1106–1121. [CrossRef]
22. Ketelut-Carneiro, N.; Fitzgerald, K.A. Apoptosis, Pyroptosis, and Necroptosis—Oh My! The Many Ways a Cell Can Die. *J. Mol. Biol.* **2022**, *434*, 167378. [CrossRef] [PubMed]
23. Wan, Y.; Yang, L.; Jiang, S.; Qian, D.; Duan, J. Excessive Apoptosis in Ulcerative Colitis: Crosstalk Between Apoptosis, ROS, ER Stress, and Intestinal Homeostasis. *Inflamm. Bowel Dis.* **2022**, *28*, 639–648. [CrossRef]
24. Cheng, W.; Wang, Z.; Xiong, Y.; Wu, Z.; Tan, X.; Yang, Y.; Zhang, H.; Zhu, X.; Wei, H.; Tao, S. N-(3-oxododecanoyl)-homoserine lactone disrupts intestinal barrier and induces systemic inflammation through perturbing gut microbiome in mice. *Sci. Total Environ.* **2021**, *778*, 146347. [CrossRef] [PubMed]
25. Yang, Y.; Wang, C.; Liu, Y.; Ma, X.; Xia, P.; Zhou, M.; Duan, Q.; Zhu, G. Impact of Quorum Sensing Signal N-3-Oxododecanoyl Homoserine Lactone on Intestinal Porcine Enterocyte Cell. *J. Biomed. Nanotechnol.* **2023**, *19*, 794–803. [CrossRef]
26. Striednig, B.; Hilbi, H. Bacterial quorum sensing and phenotypic heterogeneity: How the collective shapes the individual. *Trends Microbiol.* **2022**, *30*, 379–389. [CrossRef] [PubMed]
27. Kumar, L.; Patel, S.K.S.; Kharga, K.; Kumar, R.; Kumar, P.; Pandohee, J.; Kulshresha, S.; Harjai, K.; Chhibber, S. Molecular Mechanisms and Applications of N-Acyl Homoserine Lactone-Mediated Quorum Sensing in Bacteria. *Molecules* **2022**, *27*, 7584. [CrossRef]
28. Kondakindi, V.R.; Nalam, R.S.S.; Pabbati, R.; Maddela, N.R. Quorum quenching a sustainable biofilm mitigation strategy. *Res. J. Biotechnol.* **2022**, *17*, 200–202. [CrossRef]
29. Pacheco, T.; Gomes, A.É.I.; Siqueira, N.M.G.; Assoni, L.; Darrieux, M.; Venter, H.; Ferraz, L.F.C. SdiA, a Quorum-Sensing Regulator, Suppresses Fimbriae Expression, Biofilm Formation, and Quorum-Sensing Signaling Molecules Production in Klebsiella pneumoniae. *Front. Microbiol.* **2021**, *12*, 597735. [CrossRef]
30. Hemmati, F.; Salehi, R.; Ghotaslou, R.; Samadi, K.H.; Hasani, A.; Gholizadeh, P.; Nouri, R.; Ahangarzadeh, R.M. Quorum Quenching: A Potential Target for Antipseudomonal Therapy. *Infect. Drug Resist.* **2020**, *13*, 2989–3005. [CrossRef]
31. Adeyemo, R.O.; Famuyide, I.M.; Dzoyem, J.P.; Lyndy, J.M. Anti-Biofilm, Antibacterial, and Anti-Quorum Sensing Activities of Selected South African Plants Traditionally Used to Treat Diarrhoea. *Evid.-Based Complement. Altern. Med.* **2022**, *2022*, 1307801. [CrossRef] [PubMed]
32. Wei, Q.; Bhasme, P.; Wang, Z.; Wang, L.; Wang, S.; Zeng, Y.; Wang, Y.; Ma, L.Z.; Li, Y. Chinese medicinal herb extract inhibits PQS-mediated quorum sensing system in Pseudomonas aeruginosa. *J. Ethnopharmacol.* **2020**, *248*, 112272. [CrossRef]
33. Moradi, F.; Hadi, N.; Bazargani, A. Evaluation of quorum-sensing inhibitory effects of extracts of three traditional medicine plants with known antibacterial properties. *New Microbes New Infect.* **2020**, *38*, 100769. [CrossRef] [PubMed]
34. Subramanian, K.; Selvaraj, H.; K, B.; Sampath Renuga, P.; S, V.; Aruni, W.; Palanivel, V.; Velmurugan, P. Anti-Quorum Sensing in Pathogenic Microbes Using Plant-Based Bioactive Phytochemicals. *Adv. Mater. Sci. Eng.* **2022**, *2022*, 1–7. [CrossRef]
35. Fazal, A.; Yang, M.; Wang, M.; Ali, F.; Wen, Z.; Yin, T.; Zhao, X.; Hua, X.; Han, H.; Lin, H.; et al. Assessment of shikonin and acetyl-shikonin for mitigating quorum sensing potential of C. violaceum. *Plant Growth Regul.* **2021**, *94*, 233–243. [CrossRef]

36. Chong, Y.M.; How, K.Y.; Yin, W.F.; Chan, K.G. The Effects of Chinese Herbal Medicines on the Quorum Sensing-Regulated Virulence in Pseudomonas aeruginosa PAO1. *Molecules* **2018**, *23*, 972. [CrossRef]
37. Husain, F.M.; Ahmad, I.; Al-Thubiani, A.S.; Abulreesh, H.H.; AlHazza, I.M.; Aqil, F. Leaf Extracts of Mangifera indica L. Inhibit Quorum Sensing—Regulated Production of Virulence Factors and Biofilm in Test Bacteria. *Front. Microbiol.* **2017**, *8*, 727. [CrossRef]
38. Eum, S.Y.; Jaraki, D.; Bertrand, L.; András, I.E.; Toborek, M. Disruption of epithelial barrier by quorum-sensing N-3-(oxododecanoyl)-homoserine lactone is mediated by matrix metalloproteinases. *Am. J. Physiol.-Gastroint. Liver Physiol.* **2014**, *306*, G992–G1001. [CrossRef]
39. Vikström, E.; Bui, L.; Konradsson, P.; Magnusson, K. The junctional integrity of epithelial cells is modulated by Pseudomonas aeruginosa quorum sensing molecule through phosphorylation-dependent mechanisms. *Exp. Cell Res.* **2009**, *315*, 313–326. [CrossRef]
40. Pereira, A.M.; Strasberg-Rieber, M.; Rieber, M. Invasion-associated MMP-2 and MMP-9 are up-regulated intracellularly in concert with apoptosis linked to melanoma cell detachment. *Clin. Exp. Metastasis.* **2005**, *22*, 285–295. [CrossRef]
41. Sahoo, G.; Samal, D.; Khandayataray, P.; Murthy, M.K. A Review on Caspases: Key Regulators of Biological Activities and Apoptosis. *Mol. Neurobiol.* **2023**, *60*, 5805–5837. [CrossRef]
42. Zaborowska, M.; Dziubak, D.; Matyszewska, D.; Bilewicz, R. Surface and electrochemical properties of lipid raft model membranes and how they are affected by incorporation of statin. *Electrochim. Acta* **2021**, *386*, 138514. [CrossRef]
43. Zhang, Q.; Gao, X.; Wu, J.; Chen, M. The Correlation between Endotoxin, D-Lactate, and Diamine Oxidase with Endoscopic Activity in Inflammatory Bowel Disease. *Dis. Markers* **2022**, *2022*, 9171436. [CrossRef] [PubMed]
44. Kyrylkova, K.; Kyryachenko, S.; Leid, M.; Kioussi, C. Detection of apoptosis by TUNEL assay. *Methods Mol. Biol.* **2012**, *887*, 41–47. [CrossRef]
45. Yu, H.; Zhang, H.; Zhao, W.; Guo, L.; Li, X.; Li, Y.; Zhang, X.; Sun, Y. Gypenoside Protects against Myocardial Ischemia-Reperfusion Injury by Inhibiting Cardiomyocytes Apoptosis via Inhibition of CHOP Pathway and Activation of PI3K/Akt Pathway In Vivo and In Vitro. *Cell Physiol. Biochem.* **2016**, *39*, 123–136. [CrossRef]
46. Tabas, I.; Ron, D. Integrating the mechanisms of apoptosis induced by endoplasmic reticulum stress. *Nat. Cell Biol.* **2011**, *13*, 184–190. [CrossRef] [PubMed]
47. Singh, R.; Letai, A.; Sarosiek, K. Regulation of apoptosis in health and disease: The balancing act of BCL-2 family proteins. *Nat. Rev. Mol. Cell Biol.* **2019**, *20*, 175–193. [CrossRef]
48. Laulier, C.; Lopez, B.S. The secret life of Bcl-2: Apoptosis-independent inhibition of DNA repair by Bcl-2 family members. *Mutat. Res.* **2012**, *751*, 247–257. [CrossRef] [PubMed]
49. Chen, W.; Huang, C.; Shi, Y.; Li, N.; Wang, E.; Hu, R.; Li, G.; Yang, F.; Zhuang, Y.; Liu, P.; et al. Investigation of the Crosstalk between GRP78/PERK/ATF-4 Signaling Pathway and Renal Apoptosis Induced by Nephropathogenic Infectious Bronchitis Virus Infection. *J. Virol.* **2022**, *96*, e0142921. [CrossRef]
50. Ouweneel, A.B.; Thomas, M.J.; Sorci-Thomas, M.G. The ins and outs of lipid rafts: Functions in intracellular cholesterol homeostasis, microparticles, and cell membranes: Thematic Review Series: Biology of Lipid Rafts. *J. Lipid. Res.* **2020**, *61*, 676. [CrossRef]
51. Sezgin, E.; Levental, I.; Mayor, S.; Eggeling, C. The mystery of membrane organization: Composition, regulation and roles of lipid rafts. *Nat. Rev. Mol. Cell Biol.* **2017**, *18*, 361–374. [CrossRef] [PubMed]
52. Gee, Y.J.; Sea, Y.L.; Lal, S.K. Viral modulation of lipid rafts and their potential as putative antiviral targets. *Rev. Med. Virol.* **2023**, *33*, e2413. [CrossRef] [PubMed]
53. Kulkarni, R.; Wiemer, E.; Chang, W. Role of Lipid Rafts in Pathogen-Host Interaction—A Mini Review. *Front. Immunol.* **2021**, *12*, 815020. [CrossRef]
54. Zhou, M.; Duan, Q.; Li, Y.; Yang, Y.; Hardwidge, P.R.; Zhu, G. Membrane cholesterol plays an important role in enteropathogen adhesion and the activation of innate immunity via flagellin-TLR5 signaling. *Arch. Microbiol.* **2015**, *197*, 797–803. [CrossRef]
55. Truchado, P.; Larrosa, M.; Castro-Ibáñez, I.; Allende, A. Plant food extracts and phytochemicals: Their role as Quorum Sensing Inhibitors. *Trends Food Sci. Technol.* **2015**, *43*, 189–204. [CrossRef]
56. Duan, Q.; Zhou, M.; Zhu, X.; Yang, Y.; Zhu, J.; Bao, W.; Wu, S.; Ruan, X.; Zhang, W.; Zhu, G. Flagella from F18+Escherichia coli play a role in adhesion to pig epithelial cell lines. *Microb. Pathog.* **2013**, *55*, 32–38. [CrossRef]
57. Liu, P.; Lu, Y.; Yang, Y.; Chen, X.; Xiong, W.; Sun, J. Expression of tumour transcription factor GLI1 in canine mammary tumours tissue. *Vet. Med. Sci.* **2022**, *8*, 1451–1457. [CrossRef]

veterinary sciences

MDPI

Article

The Occurrence and Meta-Analysis of Investigations on Intestinal Parasitic Infections Among Captive Wild Mammals in Mainland China

Xueping Zhang [1,2,†], Hongyu Zhou [1,2,†], Lina Ye [1,2], Jiayu Shi [1,2], Huiling Zhang [3,*] and Tangjie Zhang [1,2,*]

1 Institute of Comparative Medicine, College of Veterinary Medicine, Yangzhou University, Yangzhou 225009, China; zhangxp1995@foxmail.com (X.Z.); zhou2096177403@foxmail.com (H.Z.); 201901123@stu.yzu.edu.cn (L.Y.); 18251238559@163.com (J.S.)

2 Jiangsu Co-Innovation Center for Prevention and Control of Important Animal Infectious Diseases and Zoonoses, Yangzhou 225009, China

3 Independent Researcher, New York, NY 11355, USA

* Correspondence: huz21@pitt.edu (H.Z.); slx@yzu.edu.cn (T.Z.)

† These authors contributed equally to this work and should be considered co-first authors.

Simple Summary: To ensure the welfare of captive wild animals and public health, it is essential to assess and monitor the gastrointestinal parasitic infection status of captive wild mammals. This study presents the first meta-analysis and systematic review to evaluate the prevalence of gastrointestinal parasitic infections among captive wild mammals in mainland China and the associated risk factors. A total of 29 eligible studies were included in this comprehensive analysis. The findings revealed a high prevalence of gastrointestinal parasitic infections among captive wild mammals in mainland China, with nematodes identified as the most dominant parasite group. Among taxonomic orders, primates exhibited the highest infection rates, followed by Artiodactyla and Rodentia, whereas Proboscidea showed the lowest prevalence. Seasonally, infection rates peaked during summer and winter, likely due to favorable environmental conditions and behavioral patterns of captive animals. Furthermore, infection prevalence was lower in economically developed regions, suggesting that improved sanitation and management practices play a critical role in mitigating parasite transmission. We emphasize the need for a sanitation management program and regular antiparasitic treatment in captive wild animals.

Abstract: This study investigated the infection rates of intestinal parasites in captive wild animals from 2000 to 2024 and analyzed the associated risk factors using a meta-analysis. We retrieved 29 studies, with a total of 8421 captive wild mammals, published between 2000 and 2024. We used the DerSimonian-Laird model and calculated infection rate estimates with the variance stabilizing double arcsine transformation. The results revealed that the overall prevalence of gastrointestinal parasitic infections in captive wild mammals in mainland China was found to be 53.9%. The highest infection rate was observed for nematodes at 45.1%. Seasonal subgroup analysis revealed the highest incidence in summer at 61.8% and 61.6% in winter. In the class order Mammalia, the highest infection rate was found in the Primates at 66.5% and similarly high values were determined in several other orders like Artiodactyla (59%), Rodentia (57.1%), Carnivora (53.3%) and extremely low in Proboscidea (19.9%). The highest overall infection rate was recorded in summer. The infection rate of gastrointestinal parasites in captive wild mammals in mainland China is notably high. Additionally, parasitic infection rates in captive animals were found to be lower in economically developed regions of mainland China.

Keywords: meta-analysis; captive wild mammals; intestinal parasite infections; mainland China; cross-sectional studies

1. Introduction

China is home to approximately one of tenth of the world's total wild mammal species [1]. However, factors such as environmental changes, poaching, and illegal trade pose significant threats to wildlife, increasing the risk of animal-borne diseases [2,3]. Since the rise of human civilization, 83% of wild mammal species have become extinct. From mice to elephants, only one-sixth of wild mammal species remain [4]. To protect wildlife and maintain biodiversity, some wild mammals are kept within human-controlled environments. Captive wildlife mainly includes animals from zoos, conservation centers, artificial breeding facilities, parks, etc. [5].

Parasites play a critical role in shaping ecological communities. They are important in food webs, competitive interactions, biodiversity patterns, and the regulation of keystone species [6,7]. The gut microbiota of wild mammals, which includes the normal intestinal flora and microorganisms encountered in the environment, is closely associated with the physiology and health of individual animals. Healthy ecosystems often have a high diversity of parasites. Parasitic infections are ubiquitous in wildlife, livestock, and human populations [8]. However, the limited enclosure sizes in captive environments lead to smaller activity spaces compared to the wild, resulting in higher stocking densities. Consequently, they are more susceptible to disease invasion and transmission than their counterparts in the wild. Research indicates that parasitic diseases are among the primary afflictions endangering wild mammals [9,10]. These diseases can lead to malnutrition, tissue and organ damage, and toxic effects. The rise in parasitic diseases in captive wild mammals may also impact on human health and well-being [11,12]. Captive wild mammals are also a source of parasitic diseases for humans and livestock. Zoonotic parasites are likely to be transmitted from animals to human due to direct contact between animal caretakers or tourists and wild animals.

Research on parasitic diseases in wildlife remains underdeveloped in mainland China, with only a few systematic studies conducted [13,14]. This study aimed to assess the prevalence and associated risk factors of intestinal parasites in captive wild mammals in mainland China through a systematic review and meta-analysis.

2. Materials and Methods

2.1. Search Strategy

This study followed the PRISMA guidelines (Preferred Reporting Items for Systematic Reviews and Meta-Analyses) [15]. The PRISMA checklist was used to ensure that all relevant information was included in the analysis (see Supplementary Data S1).

Literature searches were conducted using four databases: China National Knowledge Infrastructure (CNKI), Wanfang Data Knowledge Service Platform, PubMed, and Web of Science. The Chinese search terms were "parasites" or "intestinal parasites", "captive wild animals", "Zoo" or "Park". The English search terms were "parasites" or "intestinal parasites", "captive wild animal", "Zoo" or "Park" and "China". The search covered the period from the inception of the databases to December 2024. Additionally, the reference lists of each retrieved article were manually searched to identify further relevant articles.

2.2. Inclusion and Exclusion Criteria

After removing duplicate and review articles, preliminary screening was conducted based on the title and abstract of each article to further exclude irrelevant studies.

Full texts were obtained and reviewed to determine if they met the following criteria: ① The target subjects were captive wild mammals. ② The detected parasites were intestinal parasites. ③ Clear information on sample collection time and source. ④ Explicit detection methods (molecular or microscopic techniques). ⑤ Cross-sectional studies.

If a disagreement or uncertainty arose about the eligibility of a study, the discrepancies were resolved by a third party or through discussion.

2.3. Data Extraction

Two reviewers independently extracted and recorded specific information from the selected literature. A data collection form was created using Microsoft Excel (2016 for Windows), including: ① Study characteristics: literature title, first author, publication date. ② Study methods: sampling time, sample type, detection methods. ③ Animal classification (host): order. ④ Name and classification of detected parasites (protozoan or nematode or trematode or cestoda). ⑤ Sample size and positive size.

2.4. Literature Quality Assessment

Cross-sectional studies in animals differ from randomized clinical trials; their systematic evaluation methods are not as mature, and they lack control groups. Therefore, the systematic evaluation methods for animal cross-sectional clinical trials were adjusted based on Cochrane quality evaluation criteria, and incorporated the high-quality project assessment scale developed by Munn et al. [16] and the modified quality assessment checklist by Ding et al. [17], we adjusted the systematic evaluation method for this study. The assessment items for evaluating the risk of bias in the included studies are as follows: ① Is the research question/objective clearly described and stated? ② Is the source of wild mammals clearly stated? ③ Is the sampling time clearly stated? ④ Is the parasitic detection method clearly and described in detail? ⑤ Is there detailed information on the detected intestinal parasites, including specific parasite names and classifications? ⑥ Is the classification of animals clearly specified?

2.5. Statistical Analysis

Data collection and organization were conducted using Microsoft Excel (2016 for Windows). Data processing and statistical analysis in this meta-analysis were performed using Stata 15.1 software. Given the presence of extreme values in the included single-group rate data (100% or 0%), forest plots were analyzed using the metaprop package [18], with data transformed using the Freeman-Tukey double arcsine transformation [19,20]. The formula is as follows: $t = \arcsin[\text{sqrt}\ \{r/(n + 1)\}] + \arcsin[\text{sqrt}\ \{(r + 1)/(n + 1)\}]$, where (t) represents the transformed infection rate, (r) represents the number of positive cases, and (n) represents the sample size. The standard error (se) of (t) is calculated as $se(t) = \text{sqrt}\{1/(n + 0.5)\}$ Proportional reverse transformation was performed using the following formula: $p = (\sin(t/2))^2$.

Heterogeneity assessment of infection rates in the included studies was conducted using I^2 and Q tests. If ($p \geq 0.10$) and ($I^2 < 50\%$), it indicates low or no statistical heterogeneity among the effect sizes of the studies, and a fixed-effects model should be used for analysis [21]. If ($p < 0.10$) and ($I^2 \geq 50\%$), it indicates significant statistical heterogeneity among the study effects, and a random-effects model, specifically the Der-simonian-Laird

(D-L) method, should be used to synthesize all estimated values of the combined infection rate and its 95% confidence interval.

Through grouping based on potentially relevant features, further investigation of heterogeneous potential sources can be conducted. In this study, subgroup analysis was performed based on season, year, parasite category, animal order and feeding habits.

Bias and sensitivity analysis were conducted in this study using funnel plots. In asymmetric funnel plots, the trim-and-fill method was used to interpolate potentially missing studies and estimate the corrected prevalence [22,23]. Furthermore, sensitivity analysis was conducted to test the robustness of the results obtained in this study [24], and heterogeneity factors were identified using subgroup analysis.

3. Results

3.1. Literature Retrieval and Basic Characteristics of Included Studies

Following the previously described search strategy, 314 articles published from 1 January 2000, to 1 December 2024, were retrieved from four databases. After excluding 45 duplicated articles from the databases and reviewing the remaining ones based on the inclusion criteria following the Cochrane handbook's literature selection process, a total of 29 eligible studies were considered for meta-analysis. The selection process and results are illustrated in Figure 1.

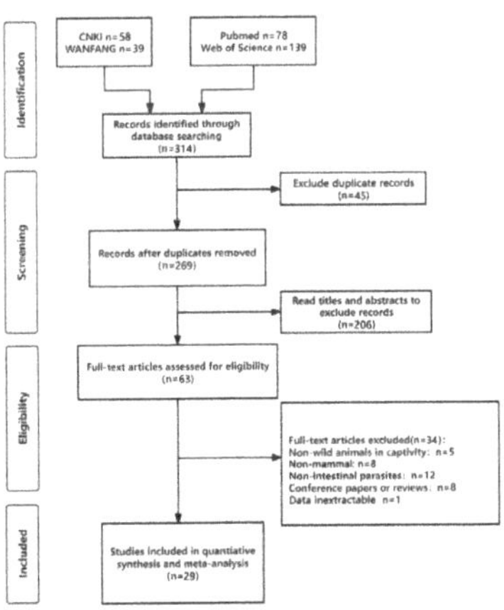

Figure 1. Flow diagram of the selection of eligible studies.

These studies, published between 1 January 2007, and 1 December 2024, were analyzed. A total of 8421 captive wild mammals were included, ranging from 12 to 3349 individuals. Characteristics of the included studies are summarized in Table 1 and Supplementary Data S2.

Table 1. Characteristics of the eligible studies.

No	Authors and Publication Years	Study Years	Diagnos-tictechniques	Location	Sample Type	Sample Size	No. Positive	Parasites
1	Dang et al. [25]	2008	①, ④	Zoo	feces, stomach contents	302	92	Protozoan, Nematode
2	Qi et al. [26]	2009	①, ②	Zoo	feces	88	17	Protozoan, Nematode

Table 1. *Cont.*

No	Authors and Publication Years	Study Years	Diagnos-tictechniques	Location	Sample Type	Sample Size	No. Positive	Parasites
3	Cai et al. [27]	2009	③, ⑤	Zoo	feces	49	41	Protozoan, Nematode, Nematode Cestoda
4	Qi et al. [28]	2009	①	Zoo	feces	35	35	Protozoan, Nematode
5	Lu et al. [29]	2010	③, ⑤	Zoo	feces	71	53	Protozoan, Nematode, Nematode
6	Li et al. [30]	2010	①, ②, ④	Zoo	feces	951	218	Protozoan, Nematode, Nematode
7	Jin et al. [31]	2011	①, ②	Zoo	feces	116	83	Protozoan, Nematode, Nematode
8	Yang et al. [32]	2011	①, ②, ⑤	Zoo, Animal Protection Center	feces	85	48	Protozoan, Nematode
9	Qi et al. [33]	2011	①, ②, ⑤	Zoo	feces	180	134	Protozoan, Nematode, Nematode, Cestoda
10	Zheng et al. [34]	2012	①, ②	Zoo	feces	71	16	Protozoan, Nematode, Cestoda
11	Chen [35]	2013	⑦, ⑨	Zoo	feces	78	45	Protozoan, Nematode, Nematode, Cestoda
12	Zhou et al. [36]	2013	①, ②, ⑤	Zoo	feces	64	39	Protozoan, Nematode
13	Zhao et al. [37]	2014	①, ②	Zoo	feces	94	46	Protozoan, Nematode
14	Luo et al. [38]	2015	①, ⑧	Zoo	feces	500	253	Protozoan, Nematode, Nematode, Cestoda
15	He et al. [39]	2017	①	Zoo	feces	12	8	Protozoan
16	Ji et al. [40]	2017	①, ③, ⑤	Zoo	feces	102	59	Protozoan, Nematode, Nematode
17	Yu [41]	2017	①, ②, ⑤	Zoo	feces	160	103	Protozoan, Nematode
18	Li et al. [42]	2017	①, ⑤	Zoo	feces	3349	1811	Protozoan, Nematode
19	Lu et al. [43]	2018	③, ⑤	Zoo	feces	80	55	Protozoan, Nematode
20	Zheng [44]	2019	①, ②	Zoo	feces	135	74	Protozoan, Nematode, Nematode

Table 1. *Cont.*

No	Authors and Publication Years	Study Years	Diagnos-tictechniques	Location	Sample Type	Sample Size	No. Positive	Parasites
21	Xiao et al. [45]	2019	①, ②	Zoo	feces	79	48	Protozoan, Nematode, Nematode, Cestoda
22	Pei et al. [46]	2021	①, ②	Zoo	feces	328	32	Protozoan, Nematode
23	Peng et al. [47]	2021	①, ②, ⑨	Zoo	feces	120	77	Protozoan, Nematode
24	Zhao [48]	2021	①, ⑧	Siberian Tiger Park	feces	90	64	Protozoan, Nematode
25	Zhang et al. [49]	2021	⑥	Zoo	feces	289	118	Protozoan
26	Zhang et al. [50]	2021	⑥	Zoo	feces	182	72	Protozoan
27	Yan et al. [51]	2022	①, ②, ⑤	Zoo	feces	437	108	Protozoan, Nematode, Cestoda
28	Qi et al. [52]	2023	⑥	Zoo	feces	224	108	Protozoan
29	An et al. [53]	2024	①, ②	Zoo	feces	150	96	Protozoan, Nematode

Legend: ① Flotation method/saturated sugar solution flotation method/saturated salt solution flotation method; ② Sedimentation method/centrifugal sedimentation method/natural sedimentation method/water sedimentation method; ③ Direct smear method/physiological saline smear method; ④ Modified acid-fast staining method; ⑤ Iodine staining method/Lugol's iodine staining method; ⑥ Nested PCR; ⑦ McMaster method; ⑧ Direct observation method; ⑨ Larval culture method.

3.2. Outcomes of Literature Quality Evaluation

Results of study quality are presented in Table 2. Based on six quality assessment criteria, the maximum score was 12 points. The mean $\pm$ standard deviation of the overall quality assessment scores was 10.63 $\pm$ 1.33. The median score was 11 with the score ranging from 6 to 12 points (Supplementary Data S3).

Table 2. Study quality assessment showing the number of the included studies in each category of a simple rating scale based on a checklist of six items.

NO.	Items	No. of Included Studies for Each Category		
		Yes Score 2	Unsure Score 1	No Score 0
1	Is the research question or objective clearly described and stated?	29	0	0
2	Are the sources of wild animals clearly specified?	28	0	1
3	Is the sampling time clearly stated?	25	0	4
4	Is the parasite detection method clearly and detailed described?	21	8	0
5	Is there detailed information on the detected intestinal parasites?	25	4	0
6	Is the classification of animals clearly specified?	15	12	2

3.3. Prevalence Heterogeneity Analysis

Figure 2 presents the forest plot depicting the prevalence of intestinal parasite infection in captive wild mammals in China from 2007 to 2024.

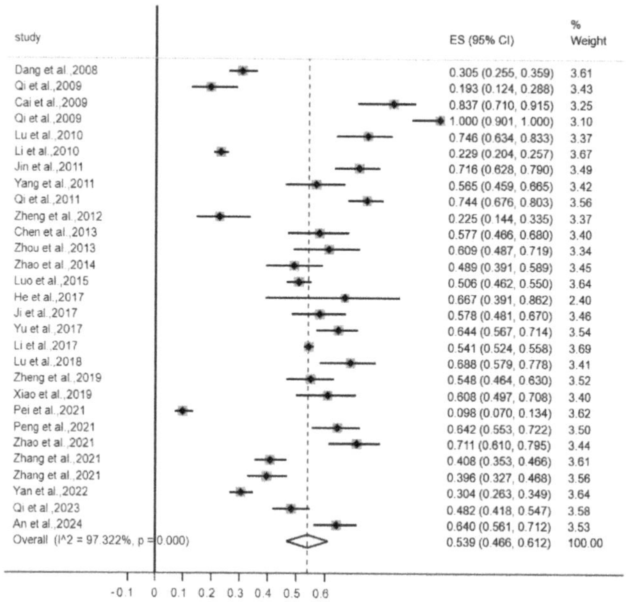

Figure 2. Forest plot depicting intestinal parasite infection among captive wild animals with random-effects analyses (ES, effect size; CI, confidence interval) [25–53].

The overall prevalence of intestinal parasitism in captive wild mammals in mainland China ranged from 9.8% to 100%, as reported in the included studies. Significant heterogeneity was observed among the studies ($\chi^2 = 1045.578$; $p < 0.001$; $I^2 = 97.322\%$). A random-effects model was applied to calculate the pooled prevalence, yielding an estimate of 53.9% (95% CI, 46.6–61.2%).

3.4. Subgroup Analysis

The subgroup analysis based on season, year, parasite category, feeding habits, order, and detection method is shown in Table 3. There was a significant difference in parasitic infection rates across seasons ($p = 0.038$). The parasitic infection rates ranked from highest to lowest were: summer 61.8% (95% CI, 51.2–71.9%),winter 61.6% (95% CI, 46–76.1%), autumn 50% (95% CI, 6.6–93.4%), spring 43% (95% CI, 24.4–62.6%), and year-round 42.6% (95% CI, 32–53.5%).

There is significant heterogeneity in infection rates across regions ($p = 0.009$). The highest infection rate was observed in the Northeast region at 71.2% (95% CI, 61.6–79.9%), followed by the Western region at 56.9% (95% CI, 47.4–87%), the Central region at 51.5% (95% CI, 38.7–64.2%), and the lowest at 45.9% (95% CI, 47.4–87%) in the Eastern region.

The infection rates of different species of parasites showed extremely significant differences ($p = 0.0001$). The highest infection rate was occurred in nematodes at 45.1% (95% CI, 37.2–53.1%), followed by protozoa at 27.4% (95% CI, 18.2–37.8%), trematodes at 5.5% (95% CI, 1.1–12.4%), and the lowest infection rate occurred in cestodes at 2.7% (95% CI, 1.3–4.6%).

Table 3. Pooled estimates of the intestinal parasite infection of captive wild mammals by potential risk factors with meta-analysis.

Intestinal Parasite Infection	No. of Studies	Total No. of Animals	No. of Positive Animals	Infection Rate		Heterogeneity			p
				Estimates	95% CI	x^2	PQ	I^2 (%)	
Overall	29	8421	3978	53.90%	0.466–0.612	1045.578	0.0001	97.32%	
Group									
Season									
Spring	4	578	233	43.00%	0.244–0.626	59.563	0.0001	94.96%	
Summer	2	81	56	61.80%	0.512–0.719	/	0.0001	/	
Autumn	3	589	230	50.00%	0.066–0.934	297.017	0.0001	99.33%	0.038
Winter	3	229	138	61.60%	0.460–0.761	11.385	0.003	82.43%	
Year aroud	8	6010	2758	42.60%	0.320–0.535	377.219	0.0001	98.14%	
Region									
Northeast region	2	102	72	71.20%	0.616–0.799	/	0.0001	/	
Eastern region	11	2876	1096	45.90%	0.338–0.583	422.143	0.0001	97.631	0.009
Western region	8	2663	1458	56.90%	0.474–0.870	114.978	0.0001	93.91%	
Central region	13	2755	1304	51.50%	0.387–0.642	458.96	0.0001	97.39%	
Parasite									
Protozoan	27	7899	2239	27.40%	0.182–0.378	2044.973	0.0001	98.73%	
Nematode	25	7282	2365	45.10%	0.372–0.531	859.641	0.0001	97.21%	0.0001
Trematode	9	1185	65	5.50%	0.011–0.124	119.756	0.0001	93.32%	
Cestoda	9	1958	42	2.70%	0.013–0.046	23.193	0.003	65.51%	
Order Mammalia									
Diprotodontia	5	33	20	43.70%	0.056–0.858	7.737	0.052	61.23%	
Primates	16	710	376	66.50%	0.533–0.787	142.484	0.0001	89.47%	
Rodentia	9	59	45	57.10%	0.088–0.986	60.218	0.0001	86.72%	
Artiodactyla	17	1322	636	59.00%	0.456–0.717	332.356	0.0001	95.19%	0.004
Perissodactyla	10	137	47	32.10%	0.160–0.503	32.235	0.0001	72.08%	
Carnivora	17	774	413	53.30%	0.429–0.636	120.267	0.0001	86.70%	
Proboscidea	6	58	14	19.90%	0.039–0.411	5.086	0.279	21.35%	
Sampling years									
2007–2012	11	2085	793	50.90%	0.358–0.659	403.933	0.0001	97.52%	
2013–2018	7	1052	584	57.70%	0.527–0.627	12.361	0.054	51.46%	0.459
2019–2024	6	1349	510	46.70%	0.265–0.674	290.873	0.0001	98.28%	
Detection methods									
Microscopic techniques	26	7726	3680	55.30%	0.471–0.634	1036.759	0.0001	97.59%	0.012
PCR	3	695	298	42.90%	0.377–0.481	3.872	0.144	48.35%	
Feeding habits									
Carnivores	16	500	301	55.00%	0.436–0.661	75.003	0.0001	80.00%	
Herbivores	18	1606	747	56.60%	0.451–0.678	322.585	0.0001	94.73%	0.971
Omnivores	20	987	503	55.50%	0.444–0.664	204.329	0.0001	90.70%	

A subgroup analysis was performed on animals from various orders. The analysis revealed significant differences in intestinal parasite infection rates among them ($p = 0.004$). Within the class Mammalia, the highest infection rate was observed in Primates at 66.5% (95% CI, 53.3–78.7%), followed by Artiodactyla at 59% (95% CI, 45.6–71.7%), Rodentia at 57.1% (95% CI, 8.8–98.6%), and the lowest occurred in the Proboscidea at 19.9%(95% CI, 3.9–41.4%) (Figure 3). The distribution of parasitic infections among different animal orders is illustrated in Figure 4. Overall, nematodes and protozoa exhibited the highest infection rates. Nematode infections were predominant in Primates, Rodentia, Artiodactyla, Perissodactyla, and Carnivora, whereas protozoan infections were more prevalent in Diprotodontia and Proboscidea. In contrast, trematodes and cestodes infection rates were comparatively low.

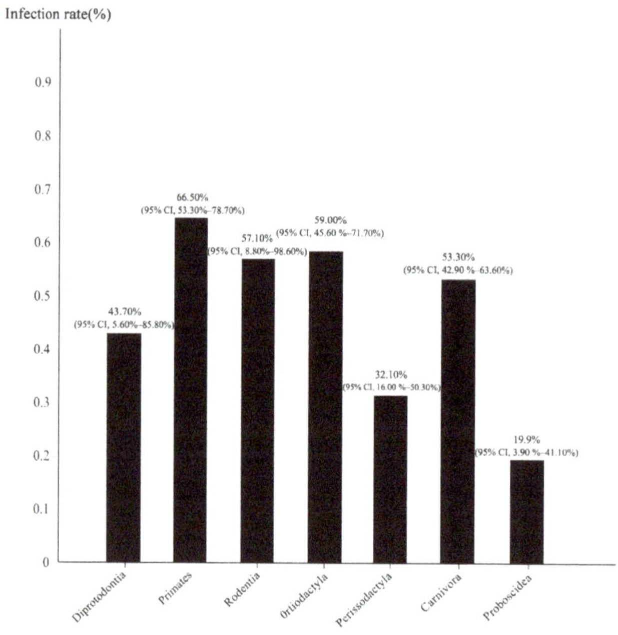

Figure 3. The infection status of intestinal parasites in captive wild mammals from different families.

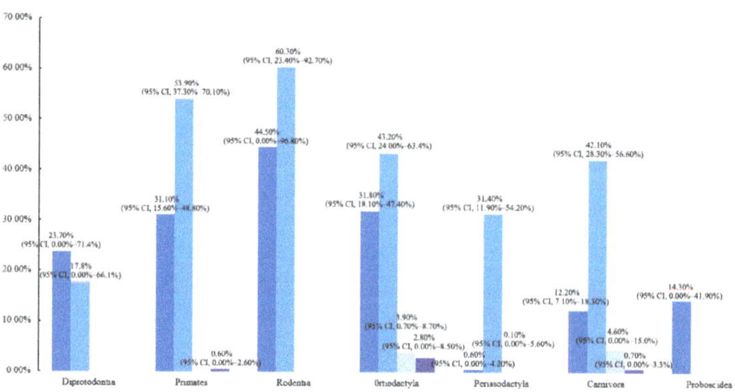

Figure 4. Intestinal parasite infection in captive wild mammals across different orders.

The subgroup analysis of detection methods revealed a significant difference between conventional methods and PCR ($p = 0.012$). The infection rate detected using conventional methods was 55.3% (95% CI, 47.1–63.4%), whereas PCR detected an infection rate of 42.9% (95% CI, 37.7–48.1%).

The feeding habits subgroup analysis showed no significant difference in infection rates among carnivores, herbivores, and omnivores ($p = 0.971$). The infection rates were 55.0% (95% CI, 43.6–66.1%) for carnivores, 56.6% (95% CI, 45.1–66.4%) for herbivores, and 55.5% (95% CI, 44.4–66.4%) for omnivores.

A review of sampling times indicated 11 articles from 2007 to 2012, seven articles from 2013 to 2018, and six articles from 2019 to 2024. Subgroup analysis by sampling time indicated no significant difference ($p = 0.459$).

3.5. Publication Bias and Sensitivity Analysis

A portion of the studies (15) fell outside the confidence interval as shown in the funnel plot (Figure 5), indicating publication bias among the included studies.

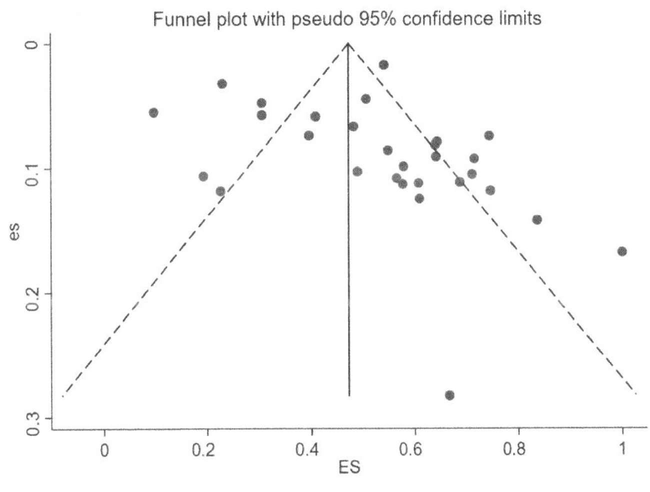

Figure 5. Funnel plot showing publication bias in studies reporting the intestinal parasite infection among captive wild mammals.

Given that publication bias can cause funnel plot asymmetry, an iterative method was used to estimate the number of missing studies. A new meta-analysis was then conducted using the trim-and-fill method (the Metatrim command from Stata 15.1 software) to determine the impact of publication bias on the research results [22,23]. After incorporating seven additional virtual studies, the meta-analysis was performed again, resulting in an adjusted p-value of 0.000 (Figure 6). Heterogeneity remained significant, suggesting that the adjusted results were statistically robust.

The results of the sensitivity analysis indicated that the prevalence of intestinal parasite infection remained within the 95% confidence interval (Figure 7). This suggests that the results of the meta-analysis were robust and not significantly affected by variations in study inclusion.

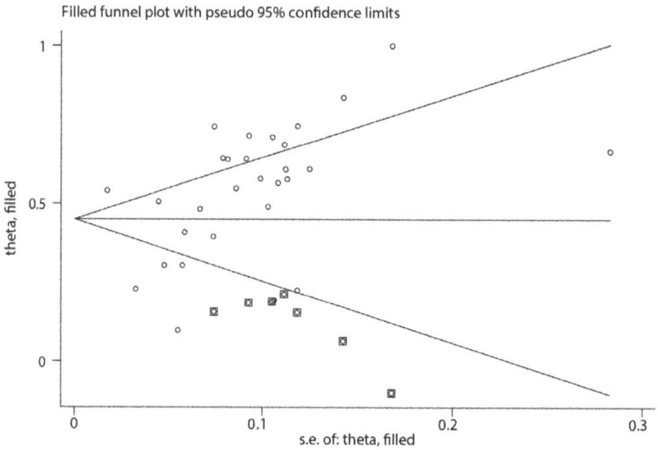

Figure 6. Funnel plot with trim and fill testing for publication bias. (dot: the actual studies; box: the imputed missing studies).

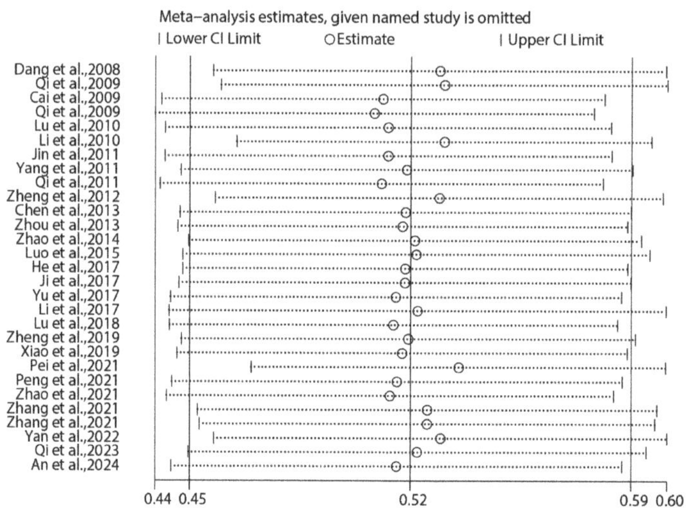

Figure 7. Influence of individual trials on pooled ES. The circles indicate the pooled relative risk estimate when each individual trial is omitted [25–53].

4. Discussion

Globally, 60% of terrestrial mammals are domesticated livestock, predominantly consisting of cattle and pigs. Humans account for 36% of the mammalian population, while wild mammals represent a mere 4%. Humans are highly efficient at exploiting natural resources, and across nearly all continents, hunting has driven many wild mammals to extinction, often for food or recreation [4]. To our knowledge, this is the first meta-analysis and systematic review on the prevalence of intestinal parasite infections in captive wild mammals in China. The combined infection rate in this study was derived from a comprehensive analysis of scientific publications on the prevalence of intestinal parasite infections in captive wild mammals in mainland China from 2007 to 2024. Our study estimated the intestinal parasite infection status of 8421 captive wild animal samples from 29 studies. The combined prevalence of intestinal parasite infection was 53.9% (95% CI, 46.6–61.2%). This was lower than the prevalence reported in the Rio de Janeiro Zoo (68.3%) in southeastern Brazil, Rabat Zoo (70%) in Morocco, and in Bangladesh (65.3%, 95% CI: 53.14–76.12%) [54–56]. However, it was higher than the prevalence reported in zoos in Slovenia (45%) [57]. This suggests that most captive wild mammals worldwide have a high prevalence of gastrointestinal parasitic infections.

Our study found that the infection rate of nematodes in captive mammals in mainland China was the highest, at 45.1% (95% CI, 37.2–53.1%). This high rate may be related to the variety and abundance of nematodes reported in studies conducted in mainland China, warranting further investigation. Mir et al. and Ferdous et al. also reported that in wild mammals, the prevalence of nematode infections was higher than that of other parasites [55,58]. Nematodes can be transmitted through soil, water sources, food, and direct contact, facilitating their spread in captive environments. Among various nematode species, roundworms, whipworms, and hookworms are zoonotic parasites [59–61], posing significant public health risks. Zoos should prioritize enhanced nematode control through effective deworming management.

Seasonality is one of the important factors affecting parasite infection of. Conditions such as warmth and humidity favor the development of eggs and larvae in the external environment, making infections more common in spring and summer [62,63]. The transmission of some parasitic diseases depends on intermediate hosts and vector arthropods. As a result, the transmission and infection seasons coincide align with the emergence of their vector arthropods, influencing the infection rate [64]. Morgan et al. demonstrated that climate strongly influences the free-living stages of gastrointestinal nematodes in sheep [65]. Both extreme heat and cold are detrimental to their development and survival. Chen's research found that the infection rate of gastrointestinal parasites in Ili horses in Xinjiang is typically higher in spring than in autumn [66]. Similarly, Hu et al. found in wild giant pandas that the prevalence of parasite infections correlates with rising temperatures [67].

Our research findings show significant variations in gastrointestinal parasite infection rates across different seasons and regions within the same season. The highest infection rate was observed in summer at 61.8%, while winter also showed a relatively high infection rate of 61.6%. The high temperatures and humid conditions in summer create an optimal environment for the proliferation of mosquitoes, ticks, and other vectors, facilitating the expansion of intermediate hosts and transmission pathways, ultimately contributing to an increased parasite infection rate [68].

In captive environments, animals often exhibit reduced outdoor activities during winter, leading to more congregations within enclosures, reduced indoor ventilation compared to spring and autumn, and hygiene challenges. These factors collectively contribute to higher parasite infection rates during winter [69], a phenomenon potentially common among captive wild mammals in mainland China. Some parasites may also be more easily transmitted in winter [61]. Carlsson et al. found in experiments that, contrary to most parasitic nematodes, Marshallagia marshalli of Svalbard reindeer is transmitted during the Arctic winter [70]. Further research is needed on the impact of temperature on the infection rate of parasites in captive wild mammals.

China's vast landmass exhibits significant geographical and climatic variations. Generally, the warm and humid climate of southern regions provides favorable conditions for parasite reproduction and transmission, leading to relatively high infection rates. In contrast, northern China is characterized by a temperate monsoon climate, while the northwest experiences an arid climate with scarce precipitation, both of which contribute to relatively lower parasite infection rates. However, the level of economic development in mainland

China may exert a more significant influence on parasite infection [71]. Economically developed regions typically possess well-established infrastructure, including access to clean drinking water and improved sanitation. These regions also have better medical resources, such as specialized personnel, equipment, and medications, along with higher public awareness of health and hygiene [72]. Health education programs are more effectively implemented, and substantial investments are made in research related to parasite prevention and control [73]. In this study, the eastern regions of China, which are economically developed, exhibited the lowest parasite infection rate (45.9%). Conversely, the highest infection rate (71.2%) was observed in northeastern China, which may be attributed to lower levels of economic development. However, the relatively small sample size in this study necessitates further research to validate these findings.

This study found significant differences in infection rates among animals from different taxonomic orders. Specifically, the highest infection rate of gastrointestinal parasites occurred in the order Primates, at 66.5%. This rate is significantly higher than the rates reported for non-human primates in zoological institutions in France [74] (53.9%) and for primates at Zoo Negara in Malaysia [75] (54.5%). As social animals, primates frequently interact within and between groups, facilitating parasite transmission. In mainland China, increased interactions between zoo visitors and primates may further contribute to the spread of gastrointestinal parasites.

Additionally, the study found that the infection rate in Artiodactyla (59%) was higher than that in Perissodactyla (32.1%), consistent with findings from the Rio de Janeiro Zoo [54]. Variation in parasitic infection rates across different taxonomic orders may be influenced by differences in digestive physiology, social behaviors, environmental adaptability, and immune responses. For instance, Artiodactyla generally exhibit more complex social structures than Perissodactyla, which tend to have smaller group sizes and simpler social dynamics. Furthermore, Artiodactyla engage in rumination, and their multi-chambered stomachs harbor diverse microbial populations, potentially affecting parasite susceptibility [76]. Therefore, zoos should implement targeted parasite control measures tailored to the specific biological and behavioral characteristics of each animal order. With regard to dietary habits, several studies have demonstrated that herbivorous animals tend to have a higher prevalence of gastrointestinal parasite infections than carnivorous animals [55,77]. However, in our study, no significant difference in infection rates was observed among animals with different feeding habits, with only a marginal difference. This suggests that feeding habits of captive animals in mainland China do not significantly influence parasite infection rates.

The differences observed in gastrointestinal parasite infections among different taxa are likely due to various factors, such as regional differences in husbandry practices and habits, sanitation management programs, and individual variations among animals. It is recommended that zoos implement scientific hygiene management plans, including regular monitoring, and treat gastrointestinal parasites in captive wild mammals.

There are significant differences between traditional detection methods and molecular detection in terms of detection results. Historically, the detection of intestinal parasites in literature mainly relied on microscopic examination of fecal samples. Microscopic parasite examination is highly dependent on the operator's experience and technical skills. Such reliance can lead to false positives due to subjective judgment [78,79]. In the literature we reviewed, PCR has been used to detect parasites since its introduction in 2021. New diagnostic techniques offer more accurate, sensitive, and specific results. As a result, traditional detection methods often identify higher infection levels than PCR methods, which were introduced later. Considering the advancements in molecular diagnostic technology, we recommend prioritizing molecular detection methods or experienced clinical personnel for the diagnosis of gastrointestinal parasites in captive wild mammals, provided that the necessary resources and conditions are available.

This study has several limitations: 1. The current methodology does not comprehensively capture all gastrointestinal parasitic infections in captive wild mammals in mainland China. 2. The literature included in the study lacked comprehensive information on intestinal parasitic infections. This limitation suggests that there may be other factors influencing the results may have been overlooked. 3. Age was not investigated for its impact on gastrointestinal parasite infections in wild mammals. This omission could overlook significant nuances in the dynamics of parasite infections across different age groups. 4. This study covers 15 provincial-level administrative regions in China. Although it represents a wide range of regions in mainland China, its geographical coverage remains incomplete, with less representation from the northeast region in the literature.

5. Conclusions

The overall prevalence of gastrointestinal parasitic infections among captive wild mammals in mainland China is relatively high (53.9%). Significant differences in infection rates are observed across various seasons, regions, diagnostic methods, parasite species, and mammalian orders. Among the parasites, nematodes have the highest infection rate. Within the seven mammalian orders, primates exhibit the highest infection rate. No significant differences in infection rates are found among different dietary types or sampling times. The level of economic development in mainland China also exerts a significant influence on the parasitic infection rates of captive animals.

Supplementary Materials: The following supporting information can be downloaded at: https://www.mdpi.com/article/10.3390/vetsci12020182/s1, Supplementary Data S1; Table Supplementary Data S2; Table Supplementary Data S3.

Author Contributions: Conceptualization, X.Z. and H.Z. (Hongyu Zhou); methodology, X.Z.; software, X.Z. and L.Y.; validation, X.Z., H.Z. (Hongyu Zhou) and T.Z.; formal analysis, H.Z. (Hongyu Zhou) and L.Y.; investigation, H.Z. (Hongyu Zhou) and J.S.; resources, X.Z. and H.Z. (Hongyu Zhou); data curation, X.Z and H.Z. (Huiling Zhang); writing—original draft preparation, X.Z. and H.Z. (Huiling Zhang); writing—review and editing, H.Z. (Huiling Zhang); visualization, X.Z.; supervision, T.Z.; project administration, T.Z. and H.Z. (Huiling Zhang); funding acquisition, T.Z. All authors have read and agreed to the published version of the manuscript.

Funding: Supported by the 111 Project D18007; The Priority Academic Program Development of Jiangsu Higher Education Institutions (PAPD); The Postgraduate Research & Practice Innovation Program of Jiangsu Province (Yangzhou University) [grant number SJXC23_2020].

Institutional Review Board Statement: Not applicable.

Informed Consent Statement: Not applicable.

Data Availability Statement: Data are contained within the article.

Conflicts of Interest: The authors declare no conflicts of interest.

References

1. Gong, S.P.; Shi, H.T.; Jiang, A.W.; Fong, J.J.; Gaillard, D.; Wang, J.C. Integrating and updating wildlife conservation in China. *Curr. Biol.* **2020**, *30*, R915–R919. [CrossRef]
2. Gong, S.P.; Shi, H.T.; Jiang, A.W.; Fong, J.J.; Gaillard, D.; Wang, J.C. Disappearance of endangered turtles within China's nature reserves. *Curr Biol.* **2017**, *27*, R170–R171. [CrossRef]

3. Zhang, L.; Hua, N.; Sun, S. Wildlife trade, consumption and conservation awareness in southwest China. *Biodivers. Conserv.* **2008**, *17*, 1493–1516. [CrossRef] [PubMed]
4. Bar-On, Y.M.; Phillips, R.; Milo, R. The biomass distribution on Earth. *Proc. Natl. Acad. Sci. USA* **2018**, *115*, 6506–6511. [CrossRef]
5. Ryser-Degiorgis, M.P.; Pewsner, M.; Angst, C. Joining the dots—Understanding the complex interplay between the values we place on wildlife, biodiversity conservation, human and animal health: A review. *Schweiz. Arch. Tierheilkd.* **2015**, *157*, 243–253. [CrossRef]
6. Marcogliese, D.J. Parasites of the superorganism: Are they indicators of ecosystem health? *Int. J. Parasitol.* **2005**, *35*, 705–716. [CrossRef] [PubMed]
7. Maizels, R.M.; McSorley, H.J. Regulation of the host immune system by helminth parasites. *J. Allergy Clin. Immunol.* **2016**, *138*, 666–675. [CrossRef]
8. Zhou, B.; Yuan, Y.T.; Zhang, S.S.; Guo, C.; Li, X.L.; Li, G.Y.; Xiong, W.; Zeng, Z.Y. Intestinal Flora and Disease Mutually Shape the Regional Immune System in the Intestinal Tract. *Front. Immunol.* **2020**, *11*, 575. [CrossRef] [PubMed]
9. Bosi, G.; Maynard, B.J.; Pironi, F.; Dezfuli, B.S. Parasites and the neuroendocrine control of fish intestinal function: An ancient struggle between pathogens and host. *Parasitology* **2022**, *149*, 1842–1861. [CrossRef]
10. Sharma, H.P.; Achhami, B. Gastro-intestinal parasites of sympatric red panda and livestock in protected areas of Nepal. *Vet. Med. Sci.* **2022**, *8*, 568–577. [CrossRef] [PubMed]
11. Medina-Vogel, G. Emerging Infectious Diseases of Wildlife and Species Conservation. *Microbiol. Spectr.* **2013**, *1*. [CrossRef]
12. Juncker-Voss, M.; Prosl, H.; Lussy, H.; Enzenberg, U.; Auer, H.; Lassnig, H.; Müller, M.; Nowotny, N. Untersuchungen auf Antikörper egen Zoonoseerreger bei Angestellten des Wiener Tiergartens Schönbrunn [Screening for antibodies against zoonotic agents among employees of the Zoological Garden of Vienna, Schönbrunn, Austria]. *Berl. Munch. Tierarztl. Wochenschr.* **2004**, *117*, 404–409.
13. Hou, Z.J.; Zhang, Z.J.; Liu, W.S.; Tang, G.Y.; Li, H.P.; Li, Y.C.; Hua, Y.P. A case of raccoon fungal infection in a fur animal farm worker. *Heilongjiang Anim. Husb. Vet Med.* **2010**, *20*, 61.
14. Pan, H. Investigation and Analysis on the Infection Status of Leptospira in Export Pig Farms in Fuqing City. *Chin. J. Front. Health Quar.* **2004**, *02*, 72–73.
15. Moher, D.; Liberati, A.; Tetzlaff, J.; Altman, D.G. Preferred reporting items for systematic reviews and meta-analyses: The PRISMA Statement. *Open Med.* **2009**, *89*, 873–880.
16. Munn, Z.; Moola, S.; Riitano, D.; Lisy, K. The development of a critical appraisal tool for use in systematic reviews addressing questions of prevalence. *Int. J. Health Policy Manag.* **2014**, *3*, 123–128. [CrossRef]
17. Ding, H.; Gao, Y.M.; Deng, Y.; Lamberton, P.H.L.; Lu, D.B. A systematic review and meta-analysis of the seroprevalence of Toxoplasma gondii in cats in mainland China. *Parasites Vectors* **2017**, *10*, 27. [CrossRef] [PubMed]
18. Nyaga, V.N.; Arbyn, M.; Aerts, M. Metaprop: A Stata command to perform meta-analysis of binomial data. *Arch. Public Health* **2014**, *72*, 39. [CrossRef]
19. Hu, T.; Zhang, H.; Zhang, X.; Hong, X.; Zhang, T. Prevalence and Risk Factors Associated with Feline Infectious Peritonitis (FIP) in Mainland China between 2008 and 2023: A Systematic Review and Meta-Analysis. *Animals* **2024**, *14*, 1220. [CrossRef] [PubMed]
20. Lin, L.; Xu, C. Arcsine-based transformations for meta-analysis of proportions: Pros, cons, and alternatives. *Health Sci. Rep.* **2020**, *3*, e178. [CrossRef]
21. Jackson, D.; White, I.R. When should meta-analysis avoid making hidden normality assumptions? *Biom. J.* **2018**, *60*, 1040–1058. [CrossRef]
22. Duval, S.; Tweedie, R. Trim and fill: A simple funnel-plot-based method of testing and adjusting for publication bias in me-ta-analysis. *Biometrics* **2000**, *56*, 455–463. [CrossRef]
23. Shi, L.Y.; Lin, L.F. The trim-and-fill method for publication bias: Practical guidelines and recommendations based on a large database of meta-analyses. *Medicine* **2019**, *98*, e15987. [CrossRef] [PubMed]
24. Begg, C.B.; Mazumdar, M. Operating characteristics of a rank correlation test for publication bias. *Biometrics* **1994**, *50*, 1088–1101. [CrossRef]
25. Dang, H.L. Epidemiology Investigation on Intestinal Parasites and Biological Characterization of Cryptosporidium Isolates from Some Wild Animals. Master's Thesis, Henan Agricultural University, Zhengzhou, China, 2008.
26. Qi, M.; Feng, C.; Han, T.G.; Wang, H.l.; Ning, C.S.; Zhang, L.X. Retrospective analysis of intestinal parasitic infections and treatment effects in herbivorous animals in zoos. *Chin. Anim. Husb. Vet. Med.* **2009**, *36*, 172–174.
27. Cai, J.; Qiao, J.Y.; Zhang, X.; Jin, X.L.; Ren, J.S.; Ma, Q.Y.; Wu, X.M. Observation on Intestinal Parasitic Morphology and Infectionin Captive Rare Wildlife in Shaanxi Province. *Chin. J. Zool.* **2009**, *44*, 63–69.
28. Qi, M.; Lv, C.C.; Wang, R.J.; Li, T.Y.; Li, D.Z.; Zhang, L.X.; Jian, F.C.; Ning, C.S. Investigation on Intestinal Parasites of Squirrels in Zhengzhou Zoo and Identification of Coccidia Species. *Heilongjiang Anim. Sci. Vet. Med.* **2009**, *17*, 75–76.
29. Lv, X.H. Intestinal Parasitic Infection and Morphology Observation of Rare Wild Animals. Master's Thesis, Northwest University, Xi'an, China, 2010.

30. Li, B. Study on Gastrointestinal Parasites and *Escherichia coil* in Non-Human Primates of Some Zoological Gardens in China. Master's Thesis, Sichuan Agricultural University, Ya'an, China, 2010.
31. Jin, L.M.; Zhou, J.W.; Gao, X.M.; Xu, M.N.; Cai, Z.H.; Xia, G.Q.; Zhou, L.L.; Shen, S.J. An Investigation of Intestinal Parasitic Infection in Ornamental Carnivorous Animals Raised in Scatters and in Sites. *J. Jinling Inst. Technol.* **2011**, *28*, 88–92.
32. Yang, N.; Qi, M.; Su, F.C.; Zhang, L.X.; Ning, C.S. Survey on the Infection Status of Intestinal Parasites in Captive Wild-animals. *Henan Agric. Sci.* **2011**, *40*, 154–157. (In Chinese)
33. Qi, M.; Dong, H.J.; Hu, L.; Peng, G.N.; Chen, Z.Q.; Zhou, Z.J.; Wan, J.J. Investigation on the Prevalence of Intestinal Parasites in Wild Animal Zoo. *Chin. Anim. Husb. Vet. Med.* **2011**, *38*, 188–191.
34. Zheng, L.; Li, L.; Xu, Q.M.; Zhang, Z.Z.; Wang, G.H.; Li, P.Y. Investigation on Parasitic Infection in Digestive System of Mammals in a Wildlife Park in Anhui Province. *Anim. Husb. Vet. Med.* **2012**, *44*, 82–84.
35. Chen, X.L. Investigation of Intestinal Parasitic Infection of Captive Mammals in the Fuzhou Zoo. Master's Thesis, Fujian Agriculture and Forestry University, Fuzhou, China, 2013.
36. Zhou, Z.J. Discussion on Intestinal Parasite Infection of Ex-situ Wild Animals. *China Sci. Technol. Inf.* **2013**, *11*, 164.
37. Zhao, J.; Zhou, Y.; Wang, S.; Tu, G.; Tang, X.; Wu, X. Preliminary report on the intestinal parasites and their diversity in captive Chinese alligators. *Nutr. Hosp.* **2014**, *31*, 813–819. [PubMed]
38. Luo, Q.; Tan, L.P.; Hu, W.; Zheng, G.C.; Lin, L.Q.; Lu, P.Y.; Li, G.Q. Investigation of parasitic infections in the digestive tract of wild animals in zoos. *Chin. J. Anim. Infect. Dis.* **2015**, *23*, 27–32.
39. He, X.B.; Hao, C.X.; Li, R.N.; Zhang, H.J.; Guan, C.Y. Diagnosis and Treatment of Parasitosis in Captive Primates and Effectiveness Evaluation. *Chin. J. Wildl.* **2017**, *38*, 200–204.
40. Ji, J.X.; Zhang, M.; Zhang, K.X.; Gao, T.T.; Fu, Y.H.; Li, J.N.; Wang, Y.Z.; Zhao, H.T.; Tian, X.F. Investigation on the intestinal parasite infection in 21 species of animals in the wildlife park in Tangshan city. *J. Trop. Dis. Parasitol.* **2017**, *15*, 28–31.
41. Yu, F.C. Epidemiologic Study and Zoonotic Risk Analyze of Three Intestinal Parasitic Protozoa in Captive Golden Snub-Nosed Monkey (*Rhinopithecus roxcllanae*) in China and Chindren Hospitalized in Zhengzhou. Master's Thesis, Henan Agricultural University, Zhengzhou, China, 2017.
42. Li, J.; Dong, H.; Wang, R.; Yu, F.; Wu, Y.; Chang, Y.; Wang, C.; Qi, M.; Zhang, L. An investigation of parasitic infections and review of molecular characterization of the intestinal protozoa in nonhuman primates in China from 2009 to 2015. *Int. J. Parasitol. Parasites Wildl.* **2017**, *6*, 8–15. [CrossRef]
43. Lu, X.J.; Jin, F.Y.; Peng, H.J.; Tao, Y.H.; Li, C.G.; Xu, G.D.; Zhu, G.L.; Wu, Z.M.; Ji, C.H. Investigation on Intestinal Parasite Infection in Wild Animals. *Anim. Breed. Feed* **2018**, *08*, 11–13.
44. Zheng, Y.Y. Investigation on Parasite Infection of Wild Animals in Guiyang Forest Wildlife Park. Master's Thesis, Guizhou University, Guiyang, China, 2019.
45. Xiao, S.H.; Wang, J.; Hu, M.; He, Y.; Xie, Y.Y.; Xiao, Y.Z.; Qiu, Q.G.; Liu, W. A Survey of Parasitic Infection in Changsha Ecological Zoo. *J. Econ. Anim.* **2019**, *23*, 110–114.
46. Pei, Z.Y.; Guo, X.X.; Yang, X.X.; Pu, T.C.; Qiao, F.; Ning, C.S.; Yan, Y.Q.; Qi, M. Investigating the prevalence of intestinal parasites in wildlife in Beijing zoo. *Mod. J. Anim. Husb. Vet. Med.* **2021**, *11*, 71–74.
47. Peng, Y.M.; Zhou, J.; Wang, C.X.; Bao, G.R.; Li, W.C.; Gu, Y.F.; Cheng, W.K.; Liu, X.C. Investigation of Intestinal Parasites Infection in Non-human Primates in a Zoo. *J. Anhui Sci. Technol. Univ.* **2021**, *35*, 25–30.
48. Zhao, Z.Y. Investigation on Intestinal Parasitic Infection of Artificially Reared Amur Tigers and Observation on Insecticidal Effect. Master's Thesis, Northeast Agricultural University, Harbin, China, 2021.
49. Zhang, K.; Zheng, S.; Wang, Y.; Wang, K.; Wang, Y.; Gazizova, A.; Han, K.; Yu, F.; Chen, Y.; Zhang, L. Occurrence and molecular characterization of *Cryptosporidium* spp., *Giardia duodenalis*, *Enterocytozoon bieneusi*, and *Blastocystis* sp. in captive wild animals in zoos in Henan, China. *BMC. Vet. Res.* **2021**, *17*, 332. [CrossRef] [PubMed]
50. Zhang, Y.; Mi, R.; Yang, L.; Gong, H.; Xu, C.; Feng, Y.; Chen, X.; Huang, Y.; Han, X.; Chen, Z. Wildlife Is a Potential Source of Human Infections of *Enterocytozoon bieneusi* and *Giardia duodenalis* in Southeastern China. *Front. Microbiol.* **2021**, *12*, 692837. [CrossRef]
51. Yan, W.L.; Zhang, M.; Zhao, L.Z.; Li, X.J.; Liu, H.Y.; Jiang, L.L.; Lv, C.C.; Wang, T.Q.; Yan, W.C.; Li, J.P. Investigation of intestinal parasitic infections in captive wild animals in some areas of Henan Province. *Chin. J. Anim. Infect. Dis.* **2022**, *32*, 145–152.
52. Qi, T.S.; Zheng, W.L.; Guo, L.T.; Sun, Y.L.; Li, J.X.; Kang, M. First description of *Blastocystis* sp. and *Entamoeba* sp. infecting zoo animals in the Qinghai-Tibetan plateau area, China. *Front. Cell. Infect. Microbiol.* **2023**, *13*, 1212617. [CrossRef]
53. An, D.Y.; Chen, Y.J.; Pei, Y.Q.; Zhu, Z.F.; Feng, Z.X.; Wang, L.F.; Liu, Q.; Jia, T.; Zhang, C.S.; Liu, J. Investigation on Intestinal Parasite Infection of Animals in Zoos. *Chin. J. Med.* **2024**, *60*, 48–53.
54. Barbosa, S.A.D.; Pinheiro, J.L.; Santos, C.R.D.; Lima, C.S.C.C.; Dib, L.V.; Echarte, G.V.; Augusto, A.M.; Bastos, A.C.M.P.; Uchôa, C.M.A.; Bastos, O.M.P.; et al. Gastrointestinal Parasites in Captive Animals at the Rio de Janeiro Zoo. *Acta Parasitol.* **2020**, *65*, 237–249. [CrossRef]

55. Ferdous, S.; Chowdhury, J.; Hasan, T.; Dutta, P.; Rahman, M.M.; Hassan, M.M.; Faruque, M.R.; Alim, M.A. Prevalence of gastrointestinal parasitic infections in wild mammals of a safari park and a zoo in Bangladesh. *Vet. Med. Sci.* **2023**, *9*, 1385–1394. [CrossRef] [PubMed]
56. Taki, Y.; Bourquia, M. Gastrointestinal and respiratory parasites in captive mammals at Rabat zoo, with the first record of *Capillaria* spp. in the Fennec fox (*Vulpes zerda*). *Vet. Ital.* **2023**, *59*, 207–217.
57. Kvapil, P.; Kastelic, M.; Dovc, A.; Bartova, E.; Cizek, P.; Lima, N.; Strus, S. An eight-year survey of the intestinal parasites of carnivores, hoofed mammals, primates, ratites and reptiles in the Ljubljana zoo in Slovenia. *Folia Parasitol.* **2017**, *64*, 013. [CrossRef]
58. Mir, A.Q.; Dua, K.; Singla, L.D.; Sharma, S.; Singh, M.P. Prevalence of parasitic infection in captive wild animals in Bir Moti Bagh mini zoo (Deer Park), Patiala, Punjab. *Vet. World* **2016**, *9*, 540–543. [CrossRef] [PubMed]
59. Fite, G.L. Veterinary Medicine and Human Health. *JAMA* **1969**, *208*, 1910–1911. [CrossRef]
60. Otranto, D.; Strube, C.; Xiao, L. Zoonotic parasites: The One Health challenge. *Parasitol. Res.* **2021**, *120*, 4073–4074. [CrossRef]
61. World Health Organization. WHO Health Topic Page: Zoonoses. Available online: https://www.who.int/topics/zoonoses/en/ (accessed on 20 July 2020).
62. Hudson, P.J.; Cattadori, I.M.; Boag, B.; Dobson, A.P. Climate disruption and parasite-host dynamics: Patterns and processes associated with warming and the frequency of extreme climatic events. *J. Helminthol.* **2006**, *80*, 175–182. [CrossRef]
63. Morgan, E.R.; Wall, R. Climate change and parasitic disease: Farmer mitigation? *Trends Parasitol.* **2009**, *25*, 308–313. [CrossRef]
64. O'Connor, L.J.; Walkden-Brown, S.W.; Kahn, L.P. Ecology of the free-living stages of major trichostrongylid parasites of sheep. *Vet. Parasitol.* **2006**, *142*, 1–15. [CrossRef] [PubMed]
65. Morgan, E.R.; Van Dijk, J. Climate and the epidemiology of gastrointestinal nematode infections of sheep in Europe. *Vet. Parasitol.* **2012**, *189*, 8–14. [CrossRef]
66. Chen, H.; Ma, Y.H.; Ma, C.; Zhao, G.D.; Lv, Y.; Li, X.B.; Turupa, P.; Mukaidai, S.M.; Ayisubiyi, A.; Nabe, D. Investigation and Study on the Infection of Gastrointestinal Parasites in Ili Horses of Different Seasons and Age Groups in Zhaosu County, Xinjiang. *Anim Husb. Feed Sci.* **2019**, *40*, 1–5.
67. Hu, H.; Zhang, X.; Pei, J.F.; Su, L.N.; Zhang, H.F.; Liu, Y.; Wu, X.M. Investigation on the Morphology and Infection Situation of Intestinal Parasites in the Wild Giant Pandas. *J. Econ. Anim.* **2018**, *22*, 106–111+124.
68. Oswaldo, C.V.; Ali, A.; Catherine, A.L.; Sadie, J.R.; Leah, R.J. Influence of environmental, geographic, socio-demographic, and epidemiological factors on presence of malaria at the community level in two continents. *Sci. Rep.* **2024**, *14*, 16734.
69. Gracenea, M.; Gómez, M.S.; Torres, J.; Carné, E.; Fernández-Morán, J. Transmission dynamics of Cryptosporidium in primates and herbivores at the Barcelona zoo: A long-term study. *Vet. Parasitol.* **2002**, *104*, 19–26. [CrossRef]
70. Carlsson, A.M.; Justin, I.R.; Wilson, K. Disease transmission in an extreme environment: Nematode parasites infect reindeer during the Arctic winter. *Int. J. Parasitol.* **2012**, *42*, 789–795. [CrossRef]
71. Singh, G.; Angwafor, S.A.; Njamnshi, A.K.; Fraimow, H.; Sander, J.W. Zoonotic and vector-borne parasites and epilepsy in low-income and middle-income countries. *Nat. Rev. Neurol.* **2020**, *16*, 333–345. [CrossRef] [PubMed]
72. Cai, L. Research on Zoo Habitat Construction and Development of Local Exhibition Area in the Middle and Lower Reaches of the Yangtze River. Master's Thesis, Nanjing Tech University, Nanjing, China, 2023.
73. Liu, Y.; Chen, H.W. Comparison of the Zoo Public Education between China and Abroad. *Sichuan J. Zool.* **2017**, *36*, 100–103.
74. Vonfeld, I.; Prenant, T.; Polack, B.; Guillot, J.; Quintard, B. Gastrointestinal parasites in non-human primates in zoological institutions in France. *Parasite* **2022**, *29*, 43. [CrossRef]
75. Lim, Y.A.L.; Ngui, R.; Shukri, J.; Rohela, M.; Naim, H.R.M. Intestinal parasites in various animals at a zoo in Malaysia. *Vet. Parasitol.* **2008**, *157*, 154–159. [CrossRef]
76. Deusch, S.; Camarinha-Silva, A.; Conrad, J.; Beifuss, U.; Rodehutscord, M.; Seifert, J. A Structural and Functional Elucidation of the Rumen Microbiome Influenced by Various Diets and Microenvironments. *Front. Microbiol.* **2017**, *8*, 1605. [CrossRef]
77. Sangpeng, J.; Eamudomkarn, C.; Hongsrichan, N.; Artchayasawat, A.; Chaisongkram, C.; Ponsrila, K.; Kimkamkaew, S.; Laoprom, N.; Boonmars, T.; Sithithaworn, P.; et al. Prevalence of gastrointestinal parasites in captive mammals at Khon Kaen Zoo. *Vet. World* **2013**, *16*, 2416–2424. [CrossRef]
78. Richard, S.B.; Sarah, G.H.S.; Idzi, P.; Blaine, A.M.; John, F.; Abhishek, M.; Harsha, S.; Francesca, T.; Marc, R.C.; Peter, C.; et al. Where Have All the Diagnostic Morphological Parasitologists Gone? *J. Clin. Microbiol.* **2022**, *60*, e0098622.
79. Cancrini, G.; Iori, A. Traditional and innovative diagnostic tools: When and why they should be applied. *Parassitologia* **2004**, *46*, 173–176. [PubMed]

veterinary sciences

MDPI

Article

Investigation of Off-Season Breeding Effects on Egg-Laying Performance, Serum Biochemical Parameters, and Reproductive Hormones in Zhedong White

Jiaqiao Zhu [1,2,3,†], Yonggang Ma [1,2,3,†], Waseem Ali [1,2,3], Rui Yu [1,2,3], Hui Zou [1,2,3] and Zongping Liu [1,2,3,*]

1 College of Veterinary Medicine, Yangzhou University, Yangzhou 225009, China; jqzhu@yzu.edu.cn (J.Z.); 007854@yzu.edu.cn (Y.M.); 008301@yzu.edu.cn (W.A.); 1759006631@163.com (R.Y.); zouhui@yzu.edu.cn (H.Z.)

2 Jiangsu Co-Innovation Center for Prevention and Control of Important Animal Infectious Diseases and Zoonoses, Yangzhou 225009, China

3 Joint International Research Laboratory of Agriculture and Agri-Product Safety of the Ministry of Education of China, Yangzhou University, Yangzhou 225009, China

* Correspondence: liuzongping@yzu.edu.cn

† These authors contributed equally to this work.

Simple Summary: This study investigates the effects of off-season breeding on Zhedong white geese egg production performance, physiological functions, and reproductive hormones. In this study, off-season breeding significantly changed Zhedong white geese serum biochemical indicators and some key hormones, which effectively increased egg production. These results provide a theoretical basis for the application of this breeding strategy in Zhedong white goose.

Abstract: Off-season breeding is a method of encouraging animals through manual intervention to reproduce during the non-breeding season. The egg-laying period of the Zhedong white goose usually starts from September and ends in April of the following year. During the summer months from May to August, with long daylight hours and high temperatures, the white geese rest and do not lay eggs. This study explored the effects of off-season breeding on egg production performance, serum biochemical indicators, and hormone secretion in Zhedong white goose. The experiment was divided into normal breeding and off-seasonal breeding. The results suggested that the peak laying period lasted 3 months, and egg production was 15.62% higher than that in the control group in the off-season breeding group. The off-season breeding group had a 4.13% higher egg fertilization rate in November (laying period) than the control group, and the hatching rate of hatched eggs and fertilized eggs in the peak laying period was 2.77% and 2.3% higher, respectively, than the control group. The experimental group showed significantly higher serum P, ALB (albumin), TG (triglyceride), and TC (total cholesterol) levels; serum AST (glutamic oxalacetic transaminase) activity; and CREA (creatinine) and BUN (blood urea nitrogen) levels than the control group at different time points. The serum LH (luteinizing hormone), FSH (follicle-stimulating hormone), P4 (progesterone), and E2 (estradiol) levels were significantly higher in the off-season breeding group than in the control group. The serum T3 (triiodothyronine) and T4 (Tetraiodothyronine) levels were significantly higher in the off-season breeding group than in the control group at different points in the breeding cycle. The control group showed a significantly higher gene expression of GnRH (gonadotropin releasing hormone) in the hypothalamus; GnRH, FSH, and LH in the pituitary; and GnRH in the ovary and significantly lower gene expression of VIP (vasoactine intestinal peptide) and PRL (prolactin) in the pituitary than the off-season breeding group. Thus, the off-season breeding of Zhedong white geese may prolong the peak laying period and improve egg production performance, thus enhancing the economic benefits of goose breeding.

Keywords: Zhedong white goose; off-season reproduction; seasonal reproduction; egg-laying performance

1. Introduction

China has the highest number of geese worldwide. In 2021, the number of commercially bred geese in China was approximately 570 million, accounting for more than 90% of the global production [1]. Xiangshan has become the largest breeding base of Zhedong white geese in China. The county has a total of 1.26 million geese in stock and produces and sells 10 million goslings annually. The total output value of the entire industrial chain has reached CNY 560 million. Zhedong white goose, also known as Xiangshan white goose and Zhejiang white goose, is a medium-sized goose bred for their meat and liver. Zhedong white geese are typically seasonal breeders; the breeding cycle begins at the end of autumn as the days gradually shorten (from the end of September to the beginning of October) and ends at the end of spring (from April to June) as the days gradually lengthen, and then enters the rest period in summer [2]. The average reproductive cycle of the Zhedong goose is 150–180 days, which includes three to four laying periods; 8–11 eggs are laid in each laying period, the duration of each laying period is 15–40 min, and the average laying interval is 24–40 h. Thus, typically, 30–45 eggs are produced by a female goose annually; the average egg weight is 160–170 g, and the fertilization rate is 85% [3]. Therefore, how to effectively prolong the breeding cycle of Zhedong white geese is very important.

Off-season breeding aims to use human intervention to make geese produce eggs in the non-breeding season contrary to their physiological characteristics [4]. Previous studies have successfully used light regulation and forced molting to induce geese to produce eggs in the summer of the second year and achieved high-temperature egg production in the summer of the third year [5]. Although the average number of eggs produced in the off-season breeding cycle was lower than that in the normal breeding cycle, the price of goslings produced in the summer was much higher than the price of goslings raised in the normal breeding cycle, and the production cost was divided between the two seasons. The implementation of feeding technology in Landes geese was reported to increase the average profit by 20% compared with that generated using the normal breeding cycle, which affirms the potential of developing off-season breeding of geese in China. With the economic development and the consequent increase in food consumption, the market demand for goose products is gradually growing. The application of off-season breeding can effectively induce Zhedong white geese to produce eggs in the non-breeding season, which has economic benefits and ensures a stable supply to the market.

It is not difficult to find that counter-seasonal breeding can effectively increase economic benefits. It is widely acknowledged that animals, particularly birds, mainly regulate the seasonality of reproduction and other physiological activities by depending on the annual variations of the daily light cycle [6]. Egg production requires the synthesis of large amounts of lipids, proteins, and minerals. When laying starts, plasma triglycerides increase substantially and lipoproteins are needed to transport them from liver to ovary [7]. In a previous study, the elevation of TG and ALB indicated that the lipid and protein synthesis capacity of liver is greatly improved during egg laying [8]. The study also found that egg production requires large amounts of trace elements, especially Ca and P, to form eggshell, and the transport of Ca requires participation of Na and Cl [9]. It is well known that hormonal changes are closely related to the reproductive performance of animals. Bédécarrats et al. found that the photoperiod signals by the pineal gland, acting on photoreceptors located in the retina and the deep brain, directly regulate GnRH, and indirectly affect GnIH secretions through melatonin secretions [10]. GnRH and GnIH, respectively, stimulate and inhibit secretions of luteinizing hormone (LH) and follicle-stimulating hormone (FSH) through the pituitary gland so as to activate or inhibit reproductive activities. In addition, photo-signals trigger the release of thyroid-stimulating hormone (TSH) by stimulating the photoreceptor cells. TSH stimulates the local conversion of thyroid hormone total thyroxin (T4) into triiodothyronine (T3), which may in turn stimulate the release of GnRH-I in the median eminence, which subsequently induces reproductive activation [11]. Therefore, hormone regulation plays an essential role in the application of off-season breeding. A growing body of evidence suggests that serum biomarkers have been developed as indicators for clinical and breeding purposes in humans and animals [12,13]. In animal breeding, serum parameters serve as indirect indicators for both animal health and economic traits [14]. Such as high levels of creatine kinase (CK) and lactate dehydrogenase (LDH) are associated with pale, soft, exudative meat. Qi et al. found that total protein (TP), albumin (ALB), globulins (GLB), total cholesterol (CHOL), total triglyceride (TG) content, and glutamic-pyruvic transaminase (GPT) are related to egg production [13].

There is no doubt that the application of off-season breeding can effectively induce Zhedong white geese to produce eggs in the non-breeding season, which has economic benefits and ensures a stable supply to the market. However, because off-season breeding is a production strategy that goes against physiological laws, it remains unknown whether it affects the egg laying performance, physiological functions, and reproductive endocrine regulation of Zhedong white goose. Therefore, this study aims to investigate the effects of off-season breeding on the egg-laying performance, physiological functions, and reproductive hormones of Zhedong white geese.

2. Materials and Methods

2.1. Experimental Design

This study was conducted at an eastern Zhejiang white goose breeding farm located in Xiangshan County, Ningbo City, Zhejiang Province, China. Experiments were approved by the Animal Ethics Committee of the university (Yangzhou, China; approval no. SYXK (Su) 2022-0044). A total of 2000, 120 days, female Zhedong white geese of the same batch were selected; the number of male geese was 4000, 120 days. An amount of 1000 each were allocated to the normal seasonal breeding (control) group and off-season breeding group, respectively. Male geese were housed with female geese in a ratio of 2:1 to facilitate mating in each group. In the control group, geese were housed in a facility equipped with automatic feeders and egg laying houses furnished with appropriate bedding materials, and they were provided with ample space for activity along with an artificial pool outdoors. In contrast, geese in the off-season breeding group were accommodated in a light-protected breeding shed with controlled lighting conditions. The method was as follows: 1. Lighting Control. Implement artificial lighting: from 6 a.m. to 6 p.m., provide artificial supplementary lighting to the goose house for 10 to 12 h each day, which can increase the duration of light in the goose house to over 16 h; Adjust the intensity of light: farmers can control the number and intensity of external light sources to adjust the intensity of light. Generally, an intensity of 1500 to 2000 lux is appropriate. 2. Environmental Regulation. 1. Adjust the temperature in the goose house: a room temperature of 15 to 20 degrees Celsius is suitable for the egg-laying and reproduction of geese. During off-season reproduction in winter, heaters or radiators can be used to maintain the temperature in the goose house, allowing geese to live in a suitable temperature to increase the reproduction rate. 2. Maintain the humidity in the goose house: appropriate humidity can help geese maintain their body's water balance and normal metabolism. During off-season reproduction in winter, the humidity in the goose house should be maintained at 60% to 70%. Spraying water or adding humidity equipment can achieve this goal.

2.2. Sample Selection

A total of 100 Zhejiang white geese were randomly selected from the off-seasonal breeding group and the control group for blood collection in 7, 8, 9 months. Fasting was ensured for 12 h before blood collection. In total, 5 mL of whole blood was collected from the wing vein using vacuum blood collection vessels, and the whole blood samples were kept at 4 °C for 30–60 min. The whole blood was centrifuged in a refrigerated centrifuge for 10 min, after which, the supernatant was drawn in a 5 mL centrifuge tube and stored at −20 °C.

2.3. Experimental Period

The control group (normal seasonal breeding) underwent an experimental period spanning 11 months, from 1 August 2018 to 1 June 2019. This period encompassed both the rest phase (August to September 2018 and May to June 2019) and the egg laying phase (October 2018 to March 2019), which included the peak egg laying periods in December 2018 and January and March 2019.

For off-season breeding group, the experimental period lasted a total of 11 months—from 1 April 2018 to 1 February 2019—including the body production period (April to May 2018, December 2018, and January and February 2019), egg laying period (June to November 2018), and peak egg laying period (August to October 2018).

2.4. Implementation of Off-Season Breeding Strategy

Throughout the entire experimental period, the control group was afforded opportunities for physical activity and provided with ample water and feed in accordance with the standard seasonal breeding protocol to uphold a regular laying cycle and rest period. In August 2018, the control group entered the laying phase. From 1 August to 1 June 2019, eggs were collected every morning and evening, and the total egg number and number of fertilized eggs, hatched goslings, and lost young geese (dead-in-shell) were determined.

The feed intake in the off-season breeding group was restricted from 1 to 20 February 2018; the total intake of fine and coarse feed was maintained at 150 g per animal, and green feed (spartina anglica) was provided ad libitum. From 1 February, the geese were driven into the shed every day after sunset and continued to be given artificial light, and the long light treatment was maintained for 18 h a day. During this period, the nutritional status of the breeding geese in the off-season breeding group was poor, so they gradually stopped egg production. From 21 February to 1 March, the feed intake for geese in the off-season breeding group was gradually increased to 200–250 g per animal, and the light exposure period was maintained at 18 h per day. During this period, the geese in this group began to molt. From 1 March to 31 March, normal feeding was resumed, with 200–250 g feed concentrate per animal, and green feed was provided ad libitum. From March 1, the geese in the off-season breeding group shed their feathers. From 1 to 30 April, 200–250 g feed concentrate was administered per animal, and light exposure was increased by 1 h every week from 15 April to 1 June. After 1 month of molt recovery, feathers grew back and egg production started. From June 1, a shorter light exposure period of 12 h a day was used to promote egg production; the temperature of the goose house was maintained below 25 °C. Other management methods and feeding materials were the same as those in the control group, and each house was swept once to ensure clean conditions. From 1 April 2018 to 1 February 2019, eggs were collected once every morning and evening and screened to identify fertilized eggs. The total egg number and numbers of fertilized eggs, goslings born, and young geese lost in the 11-month production cycle were determined. Table 1 shows the composition of the feed provided to Zhedong white geese during the laying period.

Table 1. Feed composition of Zhedong white geese during laying period.

Type of Raw Material	Concentrated Feed	Roughage Feed
Corn (%)	48	24
Barley (%)	30	15
Bran (%)	5	8.5
Soybean meal (%)	12	6
Rice chaff (%)	2	45
Premix feed (%)	3	1.5
Salt (%)	0.3	0.3

The nesting behavior of geese in the control and off-season breeding groups was checked once every 3–4 d. If nesting behavior was observed, the feed intake of the geese was restricted for 7–10 d to promote the termination of nesting, and the birds were returned to the flock.

2.5. Detection of Serum Biochemical Indicators

By employing an automatic blood biochemical analyzer, the indices related to protein metabolism, fat metabolism, mineral metabolism, liver function, and kidney function of geese were measured. The determined indicators encompass the following: protein metabolism indicators such as total protein (TP), albumin (ALB), and globulin (GLOB); mineral metabolism indicators like calcium (Ca) and phosphorus (P); fat metabolism indicators including total cholesterol (TC) and triglycerides (TG); liver function indicators, namely aspartate aminotransferase (AST), alanine aminotransferase (ALT), total bilirubin (TBU), gamma-glutamyl transferase (GGT), creatine kinase (CK), and lactate dehydrogenase (LDH); and kidney function indicators, creatinine (CREA) and blood urea nitrogen (BUN).

2.6. Examination of the Content of Serum Hormone E2, P4, FSH, LH, PRL, and T4

After serum sample preparation, ELISA was performed according to the kit protocol. Finally, with the optical density (OD) value of the measured standard as the abscissa and the concentration value of the standard as the ordinate, a standard curve was drawn using Excel, and a linear regression equation was obtained. The OD value of the sample was substituted into the equation to calculate the concentration of the sample.

2.7. Examination of the Gene Expression of Hypothalamus, Ovary Tissues FSH, LH, PRL, GnRH, and VIP

According to the mRNA sequences of actin, FSH, LH, PRL, gonadotropin-releasing hormone (GnRH), and VIP of geese in GenBank, qRT-PCR primers were designed by Oligo 7.6 software, and the primer sequences were finally obtained by NCBI Primer Blast for comparison. All primers were synthesized by BGI Ltd. (Shenzheng, China). The primer sequences of target genes are listed in Table 2.

The cDNA template was processed in steps according to the reference instructions provided by the kit. After qRT-PCR, the tissue specificity of the primers was determined according to the dissolution curve. β-actin was used as the internal reference, and the gene expression was calculated using the $2^{-\Delta\Delta Ct}$ method.

2.8. Data Statistics and Analysis

All data are expressed as the as mean ± SEM. Statistical analyses were performed using GraphPad Prism version 8.02. and the differences between the two groups were analyzed using the *t*-test. Statistical significance was declared. "*" denotes a significant difference compared with the control group ($p < 0.05$); "ns" indicates no significant difference compared with the control group ($p > 0.05$).

Table 2. Primer sequences of target genes.

Gene Name	Primer Sequence
β-actin	F: TGACGCAGATCATGTTTGAGA R: GCAGAGCGTAGCCCTCATAG
GnRH	F: CTGGGACCCTTGCTGTTTTG R: AGGGGACTTCCAACCATCAC
VIP	F: ACCAGTGTCTACAGCCATCTTTTG R: AGGTGGCTCAGCAGTTCATCTACA
FSH	F: GTGGTGCTCAGGATACTGCTTCA R: GTGCAGTTCAGTGCTATCAGTGTCA
LH	F: GACCCGGGAACCGGTGTA R: AGCAGCCACCGCTCGTAG
PRL	F: TGCTCAGGGTCGGGGTTTCA R: GCTTGGAGTCCTCATCGGCAAGTT

3. Results

3.1. Effect of Breeding Season on the Reproductive Performance of Eastern Zhejiang White Geese

In this study, the reproductive performance during the whole breeding cycle (11 months) of the off-season breeding and control group of eastern Zhejiang white geese was examined (Table 3). The hatching rate of eggs in the off-season breeding group was 1.58% higher than that in the control group, and the hatching rate (numbers of hatching geese/number of eggs) was 0.95% higher than that in the control group. The average economic benefit per goose was 35.5 RMB higher in the off-season breeding group than in the control group in the breeding cycle from April 2018 to June 2019.

Table 3. Effect of breeding season on production performance of Zhedong white geese in a breeding cycle.

Project Types	Control	Out of Season Breeding Group
Egg Production (Pieces)	31,537	30,496
Average Egg Production (Pieces)	31.54	30.50
Fertilized Egg (Pieces)	24,942	24,600
Number of spawn (Number)	21,788	21,473
Stillborn Egg (Pieces)	2065	2053
Loss of Goose Seeding (Number)	1089	1074
Hatching Rate of Eggs (%)	79.08	80.66
Hatching Rate of Fertilized Eggs (%)	69.09	70.04
Still Birth Rate (%)	87.35	87.28
Total Sales Amount of Goose Seeding (%)	8.28	8.35
Total Sales Amount of Goose Seeding (RMB)	491,170	526,702
The Economic Benefic of The Average Goose (RMB)	491.17	526.70

3.2. Effect of Breeding Season on Production Performance of Zhedong White Geese

As shown in Figure 1A, the off-season breeding group maintained the peak period of laying (more than 4000 eggs per month) for three consecutive months (August to October) during the laying period, whereas the control group exhibited peak laying only in January. The egg production in the off-season breeding group was 15.62% higher than that in the control group. In the off-season breeding group, egg production decreased in December and entered the perinatal period, which lasted for 6 months. As shown in Figure 1B, the number of chicks in the control group reached more than 3000 only in January during the peak laying period, whereas the number of chicks in the off-season breeding group reached more than 3000 for three consecutive months during the peak laying period, which was 20.60% higher than that in the control group. As shown in Figure 1C, the fertilization rate of the off-season breeding group was 1.58% higher than that of the control group. In November, the egg fertilization rate in the off-season breeding group was 4.13% higher than that in the control group. The egg fertilization rate in the off-season breeding group in December, January, and February was 3.81%, 9.3%, and 9.57% higher, respectively, than that in the control group. As shown in Figure 1D, the hatching rate of the control group was 0.95% lower than that in the off-season group during the whole breeding cycle. The egg hatching rate during the peak laying period and rest period (January to February) in the off-season breeding group was 2.77% and 6.24% higher, respectively, than that in the control group. As shown in Figure 1E, there was no significant difference in the hatching rate of fertilized eggs between the two groups during the whole breeding period, but the hatching rate of fertilized eggs in the off-season breeding group was 2.3% higher than that in the control group at the peak of laying.

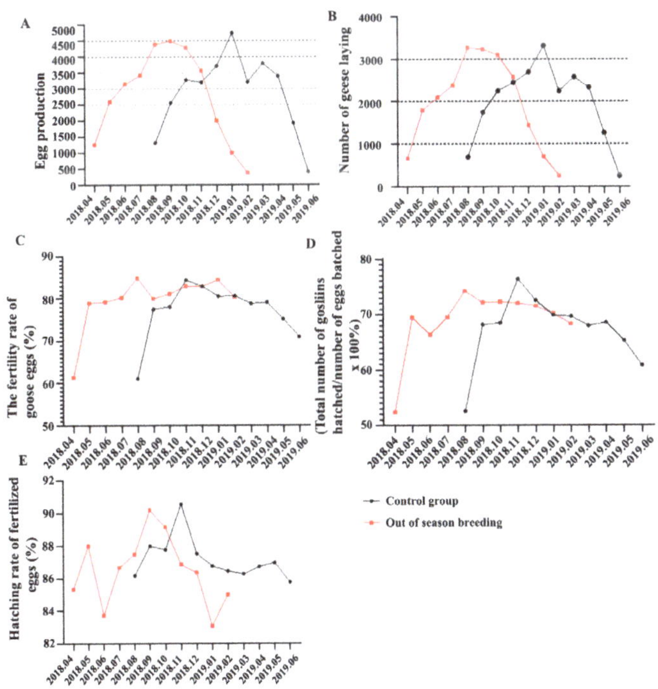

Figure 1. Effect of off-season breeding on the production performance of eastern Zhejiang white geese. (**A**) Egg production of Zhedong white geese; (**B**) Nestlings of Zhedong white geese; (**C**) Egg fertilization rate of Zhedong white goose; (**D**) Hatching rate of Zhedong white geese; (**E**) The hatchability of fertilized eggs of Zhedong white geese.

3.3. Effects of Breeding Season on Serum Biochemical Indexes of Zhedong White Geese

As shown in Figure 2, compared with the control group, there was no significant difference in serum TP and GLOB levels in the control and off-season breeding groups ($p > 0.05$), and serum ALB content was significantly higher in the off-season breeding group than in the control group ($p < 0.05$ or $p < 0.01$). Next, we examined the effect of off-season breeding on serum ion levels in eastern Zhejiang white geese. The results showed that the serum P content of the off-season breeding group was significantly higher than that of the control group ($p < 0.05$), but the serum Ca content was not significantly different ($p > 0.05$). The results of serum lipid metabolism showed that the serum triglyceride (TG) levels in the off-season breeding group were significantly higher than those in the control group ($p < 0.05$). Analysis of serum liver function showed that the serum TBil content in the off-season breeding group was significantly higher than that in the control group in July ($p < 0.05$). The activity of CK, LDH, and AST in the serum were significantly different from that in the control group at various time points ($p < 0.05$). The serum CREA level of the off-season breeding group was significantly higher than that of the control group in August and September ($p < 0.05$), and the serum BUN content was extremely significantly higher than that of the control group in August ($p < 0.01$). There was no significant difference at other time points ($p > 0.05$). The above results showed that the off-season breeding of eastern Zhejiang white geese increased the serum ALB, TG, and P levels and AST activity, promoted the metabolism of proteins, fats, and minerals, and helped to maintain egg production performance at the peak of laying and showed that off-season breeding had no adverse effects on serum biochemical indexes of eastern Zhejiang white geese.

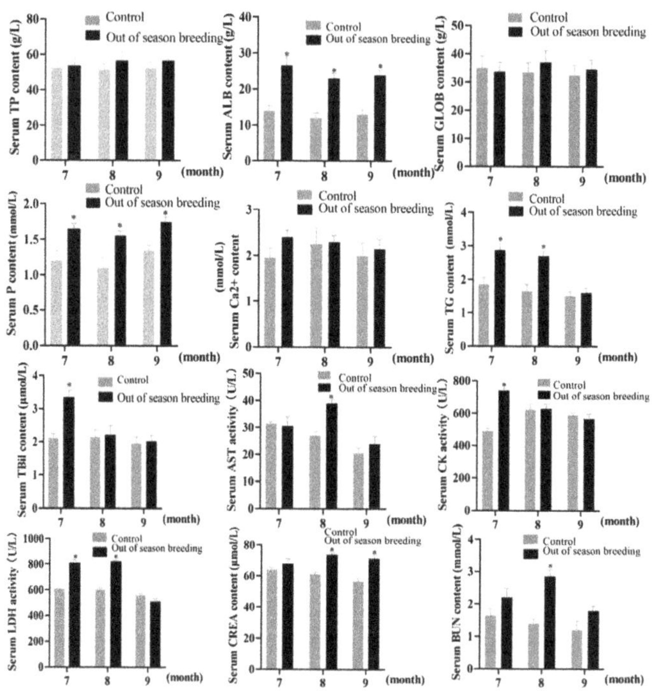

Figure 2. Effects of off-season breeding on serum biochemical indexes of Zhedong white geese. ELISA kit was used to analyze the serum TP, ALB, GLOB, P, Ca^{2+}, TG, TBil, AST, CK, LDH, CREA, and BUN content. * $p < 0.05$ indicates a significant difference in comparison with the control.

3.4. Effects of Breeding Season on Serum Levels of Reproductive Hormones in Zhedong White Geese

As shown in Figure 3A,B, the changes in serum E2 and P4 levels in the two groups were basically the same. The serum E2 content in the off-season breeding group was significantly higher than that in the control group at the corresponding time points in April and July ($p < 0.05$). The serum E2 content in the off-season breeding group was significantly lower than that in the control group at the 12th month ($p < 0.05$), and there was no significant difference between the two groups at other time points ($p > 0.05$). The serum P4 content in the off-season breeding group was significantly higher than that in the control group at corresponding time points from May to July ($p < 0.05$), and there was no significant difference between the two groups at other time points ($p > 0.05$).

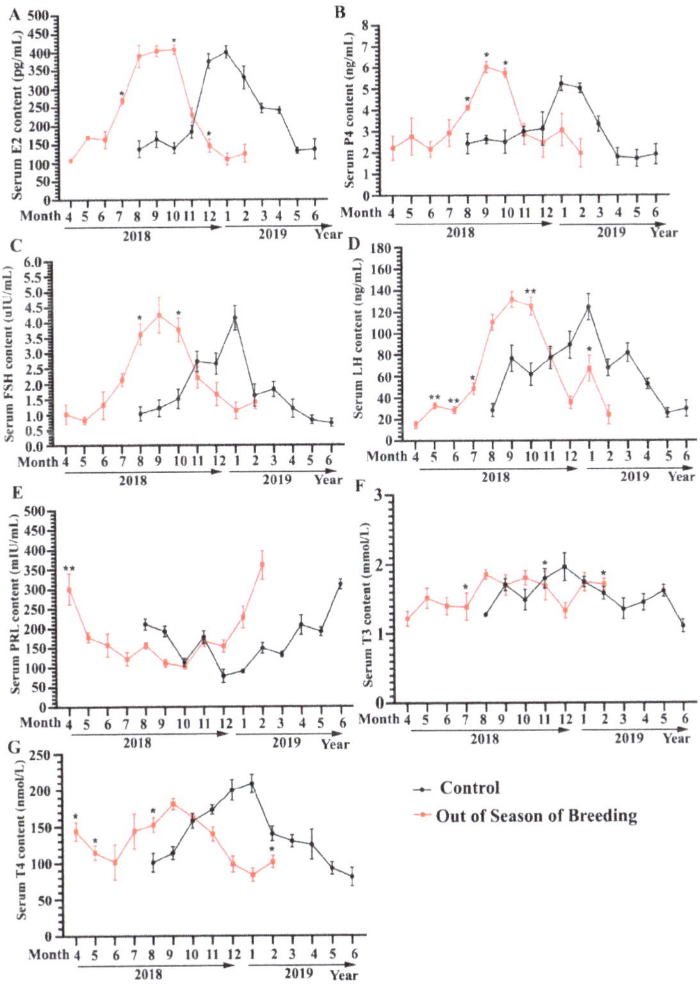

Figure 3. Effects of off-season breeding on serum reproductive hormone levels of Zhedong white geese. (**A–G**) ELISA kit was used to analyze the serum E2, P4, FSH, LH, PRL, T3, and T4 content; * $p < 0.05$ indicates a significant difference, and ** $p < 0.01$ indicates a very significant difference.

As shown in Figure 3C,D, the trend of change in serum FSH and LH levels was similar between the two groups, and the highest levels of these hormones were consistent with the peak egg laying period. The serum FSH content in the off-season breeding group was significantly higher than that in the control group at the corresponding time points in August and October ($p < 0.05$), and there was no significant difference between the two groups at the other time points ($p > 0.05$). The serum LH content of the off-season breeding group was significantly or extremely significantly lower than that of the control group at corresponding time points from February to April ($p < 0.05$ or $p < 0.01$), whereas the serum LH content of the off-season breeding group was significantly higher than that of the control group at the 7th and 10th months ($p < 0.05$); there was no significant difference between the two groups at other time points ($p > 0.05$).

As shown in Figure 3E, the trend of change in serum PRL content was similar in the two groups, and the highest hormone level was consistent with the rest period. The serum PRL content of the off-season breeding group was significantly higher than that of the control group at the corresponding time points in the first month after the start of breeding ($p < 0.05$), and there was no significant difference between the two groups at the other time points ($p > 0.05$). PRL promotes the reproductive behavior of some animals and enhances the maternal nature of female animals, especially clinginess in birds.

As shown in Figure 3F,G, the changes in serum T3 and T4 levels were consistent between the two groups. The serum T3 levels in the two groups fluctuated, and the overall level was higher at the peak of the laying period. The trend of change in serum T4 content in the two groups was consistent with that in the peak laying period, with a lower level in the rest period and a higher level at the peak of laying. The serum T3 content in the off-season breeding group was significantly higher than that of the control group at the corresponding time points in the 8th and 11th month ($p < 0.05$), and was significantly lower than that of the control group in the 4th month ($p < 0.05$). There was no significant difference between the two groups at other time points ($p > 0.05$). The serum T4 content in the off-season breeding group was significantly higher than that in the control group at the corresponding time points in January to February and November ($p < 0.05$), and was significantly lower than that of the control group in May ($p < 0.05$). There was no significant difference between the two groups at other time points ($p > 0.05$). The above results suggest that off-season breeding affected only the time and season of laying, and the change trend of the contents of reproduction-related regulatory hormones detected throughout the off-season breeding cycle was consistent with those in the normal cycle.

3.5. *Effects of Breeding Season on the Expression of Hormone-Related Genes in Eastern Zhejiang White Geese*

In December, the tissues of hypothalamus and ovary of the two groups of Zhedong white geese were collected. At this time, the off-season breeding group began to enter the fallow period, and the control group was in the high-yield period. As shown in Figure 4, the expression of GnRH in the hypothalamus was significantly higher in the control group than in the off-season breeding group ($p < 0.01$), but there was no significant difference in the expression of the VIP gene between the two groups ($p > 0.05$). The expression levels of the FSH, GnRH, and LH genes in the control group were significantly higher than those in the off-season breeding group ($p < 0.05$ or $p < 0.01$), and the expression levels of the PRL and VIP genes in the control group were significantly lower than those in the off-season breeding group ($p < 0.05$ or $p < 0.01$). The expression of the GnRH gene in the control group was significantly higher than that in the off-season breeding group ($p < 0.01$), and the expression of the VIP gene in the control group was significantly lower than that in the off-season breeding group ($p < 0.01$).

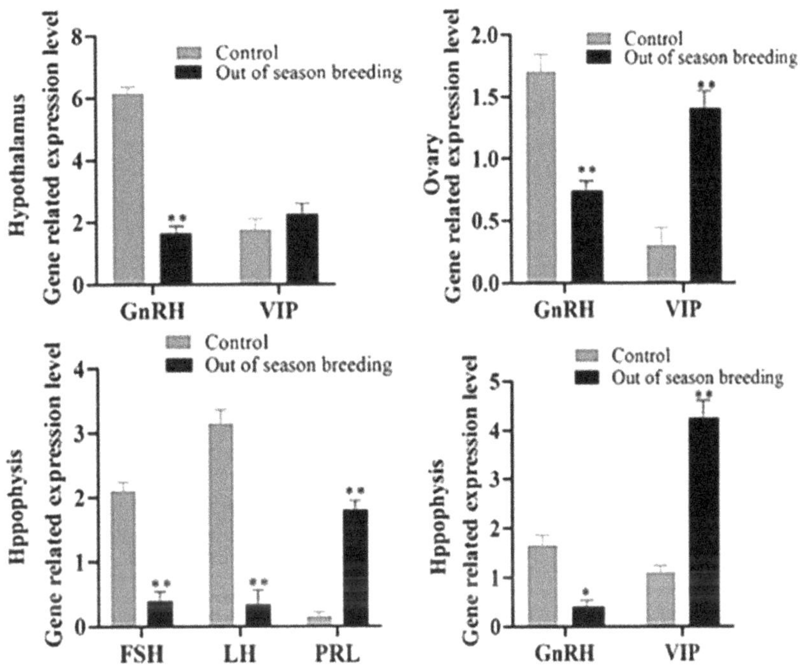

Figure 4. Effects of off-season breeding on the expression of reproductive hormone-related genes in Zhedong white geese. q-RT-PCR was used to analyze the expression of hormone-related genes. * $p < 0.05$ indicates a significant difference; ** $p < 0.01$ indicates a very significant difference.

4. Discussion

The technology of goose off-season reproduction is designed to carry out egg production in the non-reproductive season through artificial intervention against physiological characteristics. Natural light variation is the main factor for the emergence of seasonal reproductive trait characteristics in many animals. The implementation of goose out-of-season breeding production technology is a systematic engineering, the key of which is light and forced moult. Many studies have found that changing the light can affect the reproductive performance of poultry [15,16]. In this study, we successfully established an off-season breeding production technology for Zhedong white geese by limiting feed-forced molting combined with light and temperature regulation. We found that the egg laying peak of the off-season breeding group increased by 15.62% compared with the control group.

Serum biochemical indexes serve as vital parameters for evaluating the nutritional status, physiological function, and production performance of animals in commercial production [17]. The higher the serum content of TP and ALB, the better the absorption and deposition of the protein provided by the feed, the higher the feed utilization rate, and the higher the protein metabolism [18]. The present study found that there was no significant difference in serum TP concentration between the off-season breeding group and the control group at the peak of the laying period, but the serum ALB content was significantly higher than that of the control group, indicating that the off-season breeding of eastern

Zhejiang white geese did not affect protein absorption and metabolism and facilitated protein deposition at the peak of the laying period, which is conducive to maintaining their egg production performance.

Serum biochemistry considered to be a good indicator of health status. Ca and P, the most abundant mineral elements in the body, are crucial for bone growth, muscle development, and the maintenance of egg production performance and physiological function homeostasis during the breeding period [19]. Micaela Sinalair-Black et al. found that serum Ca and P ion levels were correlated with egg laying performance, and blood calcium concentration was related to eggshell formation [20]. Evandro Ferreira Cardoso et al. found that when laying hens suffered from heat stress, their serum Ca and P levels decreased significantly [21]. Alagawany et al. found that different P levels in feed materials had no significant impact on the TP, AST, ALT, and BUN levels [22]. In the present study, it was found that the serum P and Ca levels at the peak of egg production were significantly higher in the off-season breeding group than in the control group, but there was no significant difference, indicating that the off-season breeding of eastern Zhejiang white geese did not affect the absorption and metabolism of mineral Ca and P, and the off-season breeding group increased the utilization efficiency of Ca and P in the feed during egg production, which was conducive to maintaining a high-yield performance. In short, Ca and P content is crucial for maintaining egg production performance.

Hen egg yolk is a natural supramolecular assembly of lipids and proteins with different organization levels [23]. The determination of serum lipid metabolism indexes is crucial for evaluating lipid metabolism. The liver synthesizes majority of the body's cholesterol, while exogenous cholesterol obtained from dietary intake is absorbed through the intestinal tract. Kuzmenko et al. found that serum TG levels fluctuate with season, typically being higher in summer than in winter [24]. This is one of the reasons why poultry produce fewer eggs in winter. Lipids such as TG and cholesterol in the body enter the oocytes through specific receptors under the regulation of reproductive hormones, which promotes the formation of egg yolk. Maintaining normal levels of serum TG and TC provides nutritional guarantee and improves the hatching rate of fertilized eggs [25]. Mourot et al. [26] found that the likelihood of fatty liver was twice as high in Landes geese than in Polish geese, and the serum TG levels in Landes raised for foie gras (foie gras of animals, especially geese, usually eaten as foie gras) were higher and they exhibit higher fat metabolism and deposition in the liver. Tanabe et al. reported that changes in serum TG levels were significantly positively correlated with laying performance in laying hens [27]. When the serum TG levels decreased, the laying performance decreased, and when the serum TG levels recovered and increased, the laying performance gradually increased. In the present study, it was found that the serum TG levels in the off-season breeding group were significantly higher than those of the control group at the peak of the laying period, and the serum TC content also showed an increasing trend, indicating that the off-season breeding of Zhedong white geese did not affect the absorption and metabolism of lipids in the feed, and the higher serum TG and TC levels would help to maintain the energy demand and the nutritional composition of egg yolk at the peak of the laying period.

The concentrations of serum AST and ALT are related to the metabolism and function of the liver in farm animals. After liver cell damage, the concentrations of aspartate aminotransferase and alanine aminotransferase in the blood are usually higher [28]. The liver plays an important role in lipid metabolism, promoting the absorption of fat-soluble vitamins and synthesis and the decomposition of proteins, regulating blood glucose, and participating in hormonal regulation. Jiang Tao [29] reported that seasons significantly affect the liver indexes in different goose breeds; for example, the serum AST activity is higher in summer than in winter, which is more conducive to protein absorption. In the present study, the serum AST activity of the off-season breeding group was slightly higher than that of the control group at the peak of egg production, and significantly higher than that of the control group in August, which was consistent with the study results of Jiang Tao. In conclusion, feeding Zhedong white geese in off-season not only increased egg production but also promoted the increase in serum AST activity during the laying peak, thus improving protein deposition and maintaining high egg-production performance. Kang Bo et al. conducted blood biochemical analysis in white geese and seed geese in northeast China and found significant differences in ALT, AST, and GGT activity and other indicators but no significant differences in TBil and Dbil between the goose breeds [30]. Previous studies indicate that increasing the proportion of protein in the feed of lion head geese increased the egg production and hatching rate, serum ALT activity, and expression of pituitary FSHP and ovarian FSH receptor, which improved the production performance of lion head geese [31,32]. In the present study, there was no significant difference in serum Tbil levels and ALT and GGT activity between the experimental and control groups, indicating that off-season breeding did not affect the liver function of eastern Zhejiang white geese.

It is well known that poultry hormone regulation plays an important role in reproduction, which is conducive to maintaining the physiological pressure of poultry and improving economic benefits. Birds are mostly seasonal and their reproductive process is initiated by environmental queues [33]. In poultry, changes in the environment can regulate the secretion of hormones [34]. However, off-season breeding is a typical method of achieving economic growth by adjusting the rearing environment and feed management. GnRH is one of the key hormones involved in the regulation of seasonal reproduction in geese. Its secretion and synthesis are regulated by photoreceptors in the hypothalamus, its own reproductive function, and the endocrine system [35]. GnRH produced by the hypothalamus regulates LH and FSH production by the pituitary gland, and the continuous secretion of GnRH is regulated by other nerve centers, especially photoreceptors. In birds exposed to long days, the secretion of GnRH increases, causing an increase in plasma gonadotropin concentration and maintaining the ovarian production of somatotropin and follicle development. According to a study, the rhythmic secretion of GnRH activates the adenylate cyclase/cAMP/protein kinase system by binding to specific receptors on gonadotropin-secreting cells in the pituitary gland and induces the secretion of LH and FSH [36]. Changes in FSH and LH levels stimulate the secretion of E2 and P4 in follicular cells to regulate mating and other reproductive behaviors during the breeding period. Yang Haiming et al. [37] studied the expression of GnRH in Yangzhou geese under different light and temperature conditions and found that GnRH gene expression during the laying period was significantly higher than that during the rest period. The present study showed that GnRH and VIP were expressed in the hypothalamus, pituitary, and ovary of eastern Zhejiang white geese. GnRH gene expression in the three tissues in the off-season breeding group during the rest period was significantly lower than that in the control group during the peak laying period; specifically, GnRH expression in the hypothalamus showed a 10-fold difference, which is consistent with the previous findings, indicating that GnRH regulates the laying performance in eastern Zhejiang white geese. Moreover, the expression of FSH and LH genes in the pituitary tissue in the off-season breeding group during the rest period was significantly lower than that in the control group during the peak laying period, consistent with the higher serum FSH and LH levels during the peak period of egg production. The serum FSH and LH levels in the off-season breeding group were higher than those in the control group at various time points, indicating that off-season breeding in eastern

Zhejiang white geese was also strictly regulated by the hypothalamic–pituitary–gonadal axis to maintain the peak laying period.

Estradiol (E2) is predominantly synthesized by follicular cells, ovarian tissue, and the placenta. Avian progesterone (P4) is primarily produced by ovarian granulosa cells and is regulated by follicle-stimulating hormone (FSH) and luteinizing hormone (LH). Gumulka et al. observed significantly elevated serum levels of E2 and P4 in Anding geese during the laying period compared with the rest period [38]. Zhao et al. [39] found that the serum E2 and P4 levels of female Wanxi white geese were low before the breeding season, gradually increased as the laying period began, reached a peak in the middle and late laying periods, and reached the lowest level at the end of the laying period. In the present study, the change trend of serum E2 and P4 levels in the off-season breeding group and the control group was similar, and the highest hormone levels were consistent with the peak of the laying period. The change trend was that the serum E2 and P4 levels peaked at the peak of the laying period, gradually decreased after the peak laying period, and gradually increased during the early laying period after recovery during the rest period. However, the serum E2 and P4 levels in the off-season breeding group were significantly higher than those in the control group at some time points; these differences were consistent with the finding that the off-season breeding group maintained a longer peak laying period.

T3 and T4 are important endocrine hormones that participate in nutrient absorption and metabolism, bone growth and development, and gonadal development. Pond et al. reported that thyroid gland removal in turkeys led to a failure of egg production in antenatal turkeys and the cessation of egg production in turkeys during the laying period [40]. Liu et al. used the estrogen-like effect of daidzein to promote an increase in thyroxine levels in Hy-Line Brown laying hens, which improved their laying performance [41]. Previous studies have shown that T3, T4, and GH are important regulators of animal reproduction [42,43]. T3 and T4 are essential for the initiation and maintenance of egg production and for molting and regulate reproduction via the hypothalamic–pituitary–gonadal axis in poultry [44]. They regulate reproductive function directly and indirectly by regulating PRL secretion. The study showed that the change trend of serum T3 and T4 levels in eastern Zhejiang white geese was similar between the off-season breeding group and the control group; the serum T3 level was generally higher in the peak laying period, whereas the serum T4 level was lower in the rest period and higher in the peak laying period. The results suggest that the changes in serum T3 and T4 levels are closely associated with the energy expenditure and maintenance of laying performance during the laying period.

Off-season breeding prolonged the duration of the peak laying period of eastern Zhejiang white geese, and the egg production in the off-season breeding group was 15.62% higher than that in the control group. The average economic benefit per goose in the off-season breeding group was 7.2% higher than that in the control group. Off-season breeding increased the serum ALB, TG, and P levels and AST activity and promoted the metabolism of proteins, lipids, and minerals, which is conducive to maintaining the laying performance during the peak laying period. The change trends of serum FSH, LH, E2, P4, PRL, T3, and T4 levels in the off-season breeding group were similar to those in the control group. The serum E2, FSH, LH, P4, and T3 levels were higher in the off-season breeding group than in the control group at some time points during the peak laying period, and the highest levels of serum PRL were consistent with the rest period. Reproduction in avian species is controlled by the hypothalamic–pituitary–gonadal axis (HPGA), and the HPGA is a comprehensive feedback system formed under central nervous regulation [45]. HPGA is regulated by multiple hormones, including GnRH from the hypothalamus, prolactin (PRL), follicle-stimulating hormone (FSH), and luteinizing hormone (LH) from the pituitary gland, as well as E2 and P4 from the gonads [46], with PRL regulating nesting behavior through the HPGA. During oviposition, the E2 from the ovary acts on the hypothalamus, enhancing the activity of dopamine and vasoactive intestinal peptide (VIP), which in turn promotes the synthesis and release of VIP. Through a feedback effect, this stimulates the secretion of PRL, thereby driving the nesting behavior [47]. In this study, the gene expression levels of GnRH in the hypothalamus, pituitary, and ovary and FSH and LH in the pituitary were significantly lower in the off-season breeding group than in the control group during the laying period, but the gene expression levels of VIP and PRL were significantly higher than those in the control group. It is not difficult to observe that the reduction in hormone content is conducive to the secretion of VIP and PRL, which increases the nesting behavior of poultry and promotes egg-laying.

Author Contributions: J.Z. and Y.M. designed the experiments, performed the experiments, analyzed and interpreted data, wrote, and edited the manuscript. R.Y. and W.A. performed the experiments, analyzed, and interpreted data in the manuscript. H.Z. and Z.L. contributed reagents/materials/analysis tools for manuscript. All authors have read and agreed to the published version of the manuscript.

Funding: This work was supported by the National Natural Science Foundation of China [grant number 32072933, 32072923].

Institutional Review Board Statement: The study was conducted in accordance with the Declaration of Helsinki, and approved by the Institutional Review Board of Yangzhou University (SYXK (Su) 2022-0044).

Informed Consent Statement: Not applicable.

Data Availability Statement: The datasets used and/or analyzed during the current study are available from the corresponding author on reasonable request.

Conflicts of Interest: The authors declare no conflicts of interest.

Abbreviations

ALB (albumin), AST (glutamic oxalacetic transaminase), BUN (blood urea nitrogen), CREA (creatinine), E2 (estradiol), FSH (follicle-stimulating hormone), GnRH (gonadotropin releasing hormone), PRL (prolactin), P4 (progesterone), TG (triglyceride), TC (total cholesterol), T3 (triiodothyronine), T4 (Tetraiodothyronine), LH (luteinizing hormone), VIP (vasoactine intestinal peptide).

References

1. Yan, X.; Xu, Y.; Zhen, Z.; Li, J.; Zheng, H.; Li, S.; Hu, Q.; Ye, P. Slaughter performance of the main goose breeds raised commercially in China and nutritional value of the meats of the goose breeds: A systematic review. *J. Sci. Food Agric.* **2023**, *103*, 3748–3760. [CrossRef] [PubMed]
2. Wang, Z.; Wang, L.; Zhang, Y.; Yao, Y.; Zhao, W.; Xu, Q.; Chen, G. Characterization of ovarian morphology and reproductive hormones in Zhedong white geese (*Anser cygnoides domesticus*) during the reproductive cycle. *J. Anim. Physiol. Anim. Nutr.* **2021**, *105*, 938–945. [CrossRef] [PubMed]
3. Liu, G.; Guo, Z.; Zhao, X.; Sun, J.; Yue, S.; Li, M.; Chen, Z.; Ma, Z.; Zhao, H. Whole Genome Resequencing Identifies Single-Nucleotide Polymorphism Markers of Growth and Reproduction Traits in Zhedong and Zi Crossbred Geese. *Genes* **2023**, *14*, 487. [CrossRef]
4. Chen, L.; Liu, K.; Zhao, Z.; Blair, H.T.; Zhang, P.; Li, D.; Ma, R.Z. Identification of sheep ovary genes potentially associated with off-season reproduction. *J. Genet. Genom. = Yi Chuan Xue Bao* **2012**, *39*, 181–190. [CrossRef]
5. Hu, M.; Jin, H.; Wu, J.; Zhou, X.; Yang, S.; Zhao, A.; Wang, H. Identification of the differentially expressed genes in the leg muscles of Zhedong white geese (*Anser cygnoides*) reared under different photoperiods. *Poult. Sci.* **2022**, *101*, 102193. [CrossRef]

6. Zhu, H.; Shao, X.; Chen, Z.; Wei, C.; Lei, M.; Ying, S.; Yu, J.; Shi, Z. Induction of out-of-season egg laying by artificial photoperiod in Yangzhou geese and the associated endocrine and molecular regulation mechanisms. *Anim. Reprod. Sci.* **2017**, *180*, 127–136. [CrossRef] [PubMed]
7. Bacon, W.L. The effect of heparin injection on laying turkey hen very-low density lipoprotein metabolism in sexually immature and laying turkey hens. *Comp. Biochem. Physiol. Part A Physiol.* **1994**, *109*, 391–402. [CrossRef]
8. Yang, W.; Lang, X.; Song, D.; Xu, H.; Zhang, C.; Guo, L.; Chen, X. Comparative analysis of reproductive hormones, serum biochemical indexes and ovarian metabolites in Muscovy breeder duck at different laying stages. *Poult. Sci.* **2024**, *103*, 104370. [CrossRef]
9. Han, G.P.; Kim, J.H.; Kim, J.M.; Kil, D.Y. Transcriptomic analysis of the liver in aged laying hens with different eggshell strength. *Poult. Sci.* **2023**, *102*, 102217. [CrossRef] [PubMed]
10. Bédécarrats, G.Y.; Baxter, M.; Sparling, B. An updated model to describe the neuroendocrine control of reproduction in chickens. *Gen. Comp. Endocrinol.* **2016**, *227*, 58–63. [CrossRef] [PubMed]
11. Nakane, Y.; Yoshimura, T. Deep brain photoreceptors and a seasonal signal transduction cascade in birds. *Cell Tissue Res* **2010**, *342*, 341–344. [CrossRef]
12. Martinez, E.; Martorell, J.; Riambau, V. Review of serum biomarkers in carotid atherosclerosis. *J. Vasc. Surg* **2020**, *71*, 329–341. [CrossRef]
13. Qi, J.; Liu, H.; Zhou, Z.; Jiang, Y.; Fan, W.; Hu, J.; Li, J.; Guo, Z.; Xie, M.; Huang, W.; et al. Genome-wide association study identifies multiple loci influencing duck serum biochemical indicators in the laying period. *Br. Poult. Sci.* **2024**, *65*, 8–18. [CrossRef] [PubMed]
14. Qu, D.; Zhou, X.; Yang, F.; Tian, S.; Zhang, X.; Ma, L.; Han, J. Development of class model based on blood biochemical parameters as a diagnostic tool of PSE meat. *Meat Sci.* **2017**, *128*, 24–29. [CrossRef] [PubMed]
15. Liu, G.J.; Chen, Z.F.; Zhao, X.H.; Li, M.Y.; Guo, Z.H. Meta-analysis: Supplementary artificial light and goose reproduction. *Anim. Reprod. Sci.* **2020**, *214*, 106278. [CrossRef] [PubMed]
16. Wang, C.M.; Chen, L.R.; Lee, S.R.; Jea, Y.S.; Kao, J.Y. Supplementary artificial light to increase egg production of geese under natural lighting conditions. *Anim. Reprod. Sci.* **2009**, *113*, 317–321. [CrossRef]
17. El-Naggar, K.; El-Kassas, S.; Abdo, S.E.; Kirrella, A.A.K.; Al Wakeel, R.A. Role of gamma-aminobutyric acid in regulating feed intake in commercial broilers reared under normal and heat stress conditions. *J. Therm. Biol.* **2019**, *84*, 164–175. [CrossRef] [PubMed]
18. Liu, M.; Chen, R.; Wang, T.; Ding, Y.; Zhang, Y.; Huang, G.; Huang, J.; Qu, Q.; Lv, W.; Guo, S. Dietary Chinese herbal mixture supplementation improves production performance by regulating reproductive hormones, antioxidant capacity, immunity, and intestinal health of broiler breeders. *Poult. Sci.* **2024**, *103*, 103201. [CrossRef] [PubMed]
19. Fleming, R.H. Nutritional factors affecting poultry bone health. *Proc. Nutr. Soc.* **2008**, *67*, 177–183. [CrossRef]
20. Sinclair-Black, M.; Garcia, R.A.; Ellestad, L.E. Physiological regulation of calcium and phosphorus utilization in laying hens. *Front. Physiol.* **2023**, *14*, 1112499. [CrossRef]
21. Cardoso, E.F.; Donzele, J.L.; de Oliveira Donzele, R.F.M.; Sufiate, B.L.; Silva, A.D.; Tizziani, T. Non-phytate phosphorus requirement for broilers from 8 to 21 days of age under heat stress conditions. *Trop. Anim. Health Prod.* **2018**, *50*, 317–325. [CrossRef] [PubMed]
22. Alagawany, M.; Ashour, E.A.; El-Kholy, M.S.; Mohamed, L.A.; Abd El-Hack, M.E. Effect of dietary calcium and phosphorus levels on growth, carcass characteristics and liver and kidney functions of growing Egyptian geese. *Poult. Sci.* **2021**, *100*, 101244. [CrossRef]
23. Anton, M. Egg yolk: Structures, functionalities and processes. *J. Sci. Food Agric.* **2013**, *93*, 2871–2880. [CrossRef]
24. Kuzmenko, N.V.; Shchegolev, B.F. Dependence of Seasonal Dynamics in Healthy People's Circulating Lipids and Carbohydrates on Regional Climate: Meta-Analysis. *Indian J. Clin. Biochem. IJCB* **2022**, *37*, 381–398. [CrossRef] [PubMed]
25. Nematbakhsh, S.; Pei Pei, C.; Selamat, J.; Nordin, N.; Idris, L.H.; Abdull Razis, A.F. Molecular Regulation of Lipogenesis, Adipogenesis and Fat Deposition in Chicken. *Genes* **2021**, *12*, 414. [CrossRef] [PubMed]
26. Mourot, J.; Guy, G.; Lagarrigue, S.; Peiniau, P.; Hermier, D. Role of hepatic lipogenesis in the susceptibility to fatty liver in the goose (*Anser anser*). *Comp. Biochem. Physiol. Part B Biochem. Mol. Biol.* **2000**, *126*, 81–87. [CrossRef] [PubMed]
27. Tanabe, Y.; Ogawa, T.; Nakamura, T. The effect of short-term starvation on pituitary and plasma LH, plasma estradiol and progesterone, and on pituitary response to LH-RH in the laying hen (*Gallus domesticus*). *Gen. Comp. Endocrinol.* **1981**, *43*, 392–398. [CrossRef]
28. Enemor, V.H.; Anosike, J.C.; Nwoke, B.E.; Chikezie, P.C. Serum aminotransferases and bilirubin levels in malaria patients. *Int. J. Nat. Appl. Sci.* **2005**, *1*, 85–89. [CrossRef]

29. Tao, J. Comparative Analysis of Nutrient Digestibility and Serum Biochemical Indexes among Different Goose Breeds. *Life Sci* **2015**, *35*, 178–186.
30. Kang, B. Blood biochemical indexes of northeast white goose and seed goose. *Chin. Vet. J.* **2006**, *79*, 649–652.
31. Lin, B.; Zhou, X.; Jiang, D.; Shen, X.; Ouyang, H.; Li, W.; Xu, D.; Fang, L.; Tian, Y.; Li, X.; et al. Comparative transcriptomic analysis reveals candidate genes for seasonal breeding in the male Lion-Head goose. *Br. Poult. Sci.* **2023**, *64*, 157–163. [CrossRef] [PubMed]
32. Mansour, A.H.; Rabie, M.H.; El-Said, E.A.; Abo El-Maaty, H.M. Interactive effects of dietary protein and nano-chitosan on growth performance, immune response, and histological aspects of lymphoid organs in broiler chickens. *Trop. Anim. Health Prod.* **2024**, *56*, 62. [CrossRef]
33. Sharp, P.J. Strategies in avian breeding cycles. *Anim. Reprod. Sci.* **1996**, *42*, 505–513. [CrossRef]
34. Ono, H.; Nakao, N.; Yamamura, T.; Kinoshita, K.; Mizutani, M.; Namikawa, T.; Iigo, M.; Ebihara, S.; Yoshimura, T. Red jungle fowl (*Gallus gallus*) as a model for studying the molecular mechanism of seasonal reproduction. *Anim. Sci. J. = Nihon Chikusan Gakkaiho* **2009**, *80*, 328–332. [CrossRef] [PubMed]
35. Dunn, I.C.; Beattie, K.K.; Maney, D.; Sang, H.M.; Talbot, R.T.; Wilson, P.W.; Sharp, P.J. Regulation of chicken gonadotropin-releasing hormone-I mRNA in incubating, nest-deprived and laying bantam hens. *Neuroendocrinology* **1996**, *63*, 504–513. [CrossRef]
36. Bédécarrats, G.Y.; McFarlane, H.; Maddineni, S.R.; Ramachandran, R. Gonadotropin-inhibitory hormone receptor signaling and its impact on reproduction in chickens. *Gen. Comp. Endocrinol.* **2009**, *163*, 7–11. [CrossRef]
37. Yang, H.M.; Wang, Y.; Wang, Z.Y.; Wang, X.X. Seasonal and photoperiodic regulation of reproductive hormones and related genes in Yangzhou geese. *Poult. Sci.* **2017**, *96*, 486–490. [CrossRef] [PubMed]
38. Gumułka, M.; Avital-Cohen, N.; Rozenboim, I. Determination of Annual Plasma Hormone Levels Associated with Reproduction in Long-Day Breeding Domestic Geese. *Animals* **2021**, *11*, 2363. [CrossRef] [PubMed]
39. Zhao, C.; Jin, T.; Yang, K.; Liu, X.; Ren, M.; She, D.; Hu, Q.; Li, S. The hematopoietic function, histological characteristics, and transcriptome profiling of Wanxi white geese ovary during nesting and late-laying stages. *Poult. Sci.* **2025**, *104*, 104764. [CrossRef]
40. Pond, W.G.; Yen, J.T.; Mersmann, H.J. Effect of severe dietary protein, nonprotein calories or feed restriction during gestation on postnatal growth of progeny in swine. *Growth* **1987**, *51*, 355–371. [PubMed]
41. Liu, J.; Fu, Y.; Zhou, S.; Zhao, P.; Zhao, J.; Yang, Q.; Wu, H.; Ding, M.; Li, Y. Comparison of the effect of quercetin and daidzein on production performance, anti-oxidation, hormones, and cecal microflora in laying hens during the late laying period. *Poult. Sci.* **2023**, *102*, 102674. [CrossRef] [PubMed]
42. Hull, K.L.; Harvey, S. Growth hormone: Roles in female reproduction. *J. Endocrinol.* **2001**, *168*, 1–23. [CrossRef] [PubMed]
43. Hull, K.L.; Harvey, S. Growth hormone: A reproductive endocrine-paracrine regulator? *Rev. Reprod.* **2000**, *5*, 175–182. [CrossRef]
44. Lien, R.J.; Siopes, T.D. Effects of thyroidectomy on egg production, molt, and plasma thyroid hormone concentrations of turkey hens. *Poult. Sci.* **1989**, *68*, 1126–1132. [CrossRef]
45. Rose, E.M.; Haakenson, C.M.; Ball, G.F. Sex differences in seasonal brain plasticity and the neuroendocrine regulation of vocal behavior in songbirds. *Horm. Behav.* **2022**, *142*, 105160. [CrossRef]
46. Gumułka, M.; Hrabia, A.; Avital-Cohen, N.; Andres, K.; Rozenboim, I. The effect of parachlorophenylalanine treatment on the activity of gonadal and lactotrophic axes in native Polish crested chickens stimulated to broodiness. *Poult. Sci.* **2020**, *99*, 2708–2717. [CrossRef] [PubMed]
47. Avital-Cohen, N.; Heiblum, R.; Argov, N.; Rosenstrauch, A.; Chaiseha, Y.; Mobarkey, N.; Rozenboim, I. The effect of active immunization against vasoactive intestinal peptide (VIP) and inhibin on reproductive performance of aging White Leghorn roosters. *Poult. Sci.* **2012**, *91*, 161–174. [CrossRef]

veterinary sciences

Article

Helicobacter hepaticus CdtB Triggers Colonic Mucosal Barrier Disruption in Mice via Epithelial Tight Junction Impairment Mediated by MLCK/pMLC2 Signaling Pathway

Tao Wang [1,†], Xiao Meng [1,†], Miao Qian [1], Shanhao Jin [1], Ruoyu Bao [1], Liqi Zhu [1,*] and Quan Zhang [1,2,3,*]

1 Institute of Comparative Medicine, College of Veterinary Medicine, Yangzhou University, Yangzhou 225009, China
2 Co-Innovation Center for Prevention and Control of Important Animal Infectious Diseases and Zoonoses, Yangzhou University, Yangzhou 225009, China
3 Jiangsu Transgenic Animal Pharmaceutical Engineering Research Center, Yangzhou University, Yangzhou 225009, China
* Correspondence: zhuliqi@yzu.edu.cn (L.Z.); zquan@yzu.edu.cn (Q.Z.); Tel.: +86-514-87979337 (Q.Z.)
† These authors contributed equally to this work.

Simple Summary: Helicobacter hepaticus (*H. hepaticus*) has been demonstrated to have clinical relevance to the development of colitis in rodents. The aim of the present study was to ascertain the impact of CdtB, the active subunit of CDT, on the colonic mucosal barrier during *H. hepaticus* infection. The results demonstrate that the presence of ΔCdtB led to a mitigation of the symptoms associated with *H. hepaticus*-induced colitis. In vitro, CdtB mutation reversed the reduction in tight junction (TJ) marker (ZO-1 and occludin) expression and the impairment of the F-actin structure induced by *H. hepaticus* in either IEC6 cells or intestinal organoids. ROS levels were found to be significantly reduced in cells treated with the ΔCdtB strain. Furthermore, myosin light chain kinase (MLCK) activation was observed in the *H. hepaticus*-infected group in vivo, whereas the MLCK inhibitor ML-7 was found to reverse the CdtB-induced alterations in TJ proteins in IEC6 cells. CdtB plays a pivotal role in the *H. hepaticus*-induced colonic mucosal barrier. This is achieved through the regulation of TJs via the MLCK/pMLC2 signaling pathway, which is linked to elevations in oxidative stress and inflammation within intestinal epithelial cells.

Abstract: Background: *Helicobacter hepaticus* (*H. hepaticus*) has been demonstrated to have clinical relevance to the development of colitis in rodents. *H. hepaticus* produces cytolethal distending toxins (CDTs), which are identified as the most important virulence factors to the pathogenicity of CDT-producing bacteria in animals. However, the precise relationship between CDTs of *H. hepaticus* and intestinal barrier dysfunction remains unclear. The objective of the present study was to ascertain the impact of CdtB, the active subunit of CDTs, on the colonic mucosal barrier during *H. hepaticus* infection. **Materials and Methods**: We investigated the infection of male BALB/c mice, intestinal organoids, and IEC-6 cell monolayers by *H. hepaticus* or CdtB-deficient *H. hepaticus* (ΔCdtB). A comprehensive histopathological examination was conducted, encompassing the assessment of *H. hepaticus* colonization, the levels of mRNA expression for inflammatory cytokines, the expression levels of tight junction proteins, and the related signaling pathways. **Results**: The results demonstrate that the presence of ΔCdtB led to a mitigation of the symptoms associated with *H. hepaticus*-induced colitis, as evidenced by colon length shortening and the colon histological inflammation score. In addition, the levels of pro-inflammatory cytokines were reduced in the ΔCdtB group. Moreover, a downward trend was observed in the phosphorylation levels of STAT3 and nuclear factor-κB (p65). In vitro, the presence of *H.*

hepaticus resulted in a reduction in the expression of tight junction (TJ) markers (ZO-1 and occludin) and an impairment of the F-actin structure in either the intestinal epithelium or intestinal organoids. However, these effects were reversed by CdtB deletion. Concurrently, both ROS levels and apoptosis levels were found to be significantly reduced in cells treated with the ΔCdtB strain. Mechanistically, myosin light chain kinase (MLCK) activation was observed in the *H. hepaticus*-infected group in vivo, whereas the MLCK inhibitor ML-7 was found to reverse the CdtB-induced alterations in TJ proteins in IEC6 cells. **Conclusions**: The collective findings demonstrate that CdtB plays a pivotal role in the *H. hepaticus*-induced colonic mucosal barrier. This is achieved through the regulation of TJs via the MLCK/pMLC2 signaling pathway, which is linked to elevations in oxidative stress and inflammation within intestinal epithelial cells.

Keywords: *Helicobacter hepaticus*; CdtB; colonic mucosal barrier; tight junction; MLCK; pMLC2

1. Introduction

Members of the genus *Helicobacter* have been reported in a wide range of animals, and they have been identified to infect numerous sites, including the stomach, small intestine, large intestine, liver, and bile duct, and they have even been reported in the hearts of chickens and in the milk of sheep and cows [1,2]. *H. hepaticus*, a close relative of *H. pylori* and *C. jejuni*, is a member of the enterohepatic *Helicobacter* species (EHS) [3]. It colonizes the colons, ceca, and hepatobiliary tracts of rodents [4–6] and has been reported to cause colitis and hepatitis in susceptible mouse strains in past decades [7–10]. Pathobionts are defined as members of the gut microbiota that possess the capacity to induce diseases when intestinal homeostasis is disrupted. The hypothesis that these organisms play a significant role in the etiology of enteritis and inflammatory bowel disease (IBD) is currently under investigation. IBD is a group of chronic inflammatory disorders characterized by an aberrant immune response to microbiota. Gut microbiota comprise both beneficial and harmful microorganisms, and enterohepatic *Helicobacter* spp. infection is considered a naturally acquired condition [11]. Intestinal bacterial dysbiosis and *H. hepaticus* infection exert a synergistic effect on the development of IBD and neoplastic progression. The recent detection of *H. hepaticus* in bile samples from patients with pancreatic and biliary tract cancers indicates a potential role for this organism in the progression of gastrointestinal diseases in humans [12,13]. Therefore, a deeper comprehension of the underlying mechanism of *H. hepaticus*-induced colonic mucosal barrier damage is crucial for the advancement of human and animal health.

In a manner analogous to other pathogenic bacteria, *H. hepaticus* produces and secretes cytolethal distending toxin (CDT), which is a critical factor in the development of colitis. CDT is classified as an AB_2 toxin, a category of toxins produced by Gram-negative bacteria [3,14], and consists of CdtA, CdtB, and CdtC [15,16]. The genome of *H. hepaticus* consists of 1,799,146 base pairs (bps), and the CDT gene is located from 1,395,180 bp to 1,398,116 bp. The length of CDT coding DNA sequence is 2070 bp, consisting of CdtA (696 bp), CdtB (822 bp), and CdtC (552 bp) in order. The virulence of CDT is dependent on CdtB, which has DNase activity [16]. As CdtB contributes to the pathogenicity of CDT-producing bacteria in vivo, it is of great significance to explore the pathogenic mechanism of CdtB to ensure the health of humans and animals. The biological functions of CdtB have been documented to include G2/M cell cycle arrest, STAT3 activation, DNA double-strand breaks (DSBs), and apoptosis in intestinal epithelial cells (IECs), which can lead to the development of epithelial dysplasia [16–18].

The surface of the intestinal mucosa is lined with epithelial cells that provide an essential barrier to the diffusion of pathogens, toxins, and allergens from the lumen to the mucosal tissues [19]. Tight junctions represent a crucial element of the intestinal barrier, comprising the assembly of multiple proteins situated at the apical aspect of the outer membrane of epithelial cells, including ZO-1, occludin, and claudin. The aberrant expression and distribution of tight junction proteins in intestinal tissue result in increased intestinal permeability and dysfunction of the intestinal epithelial barrier [20,21]. Our previous work has demonstrated that *H. hepaticus* CdtB is capable of causing disruption to the intestinal mucosa in IL10$^{-/-}$ mice [17] and can induce intestinal epithelial cell death via the mitochondrial apoptotic pathway in vitro [22]. However, it remains unknown whether *H. hepaticus* CdtB affects the intestinal barrier and the pathogenesis of colitis.

The objective of this study was to investigate the effect of *H. hepaticus* CdtB on the intestinal barrier. The data presented herein suggest a potential effect of CdtB on colitis, which is associated with tight junction dysfunction and cytoskeletal disorders. Our study can provide better insights regarding pathogenic mechanisms and treatment strategies for EHS-triggered enteritis.

2. Materials and Methods

2.1. Animals

Male BALB/c mice (6∼8 weeks old) were purchased from the Comparative Medicine Center of Yangzhou University (male BALB/c mice are more sensitive to *H. hepaticus*). The animal experiments were conducted in accordance with the project license [SYXK (Su) 2022-0044] approved by Jiangsu Provincial Science. The ethics of the use of laboratory animals (202203022) was approved by the Institutional Animal Care and Use Committee (IACUC) of Yangzhou University. The mice utilized in this study were free of *Helicobacter* species, as determined by PCR. The mice were randomly divided into three groups, with 30 mice in each group, as the euthanasia was performed in two batches. The *H. hepaticus* and *H. hepaticus* ΔCdtB groups were administered the microorganism (1×10^8 CFU) in 200 μL PBS solution by gavage, while the control group was given 200 μL PBS by gavage every 2 days for a week. Fecal samples were collected on a weekly basis and subjected to PCR analysis to ascertain the presence of infection. Mice from each group were euthanized at the 4th and 9th weeks post-infection (WPI), respectively. Colonic tissues were collected for subsequent experiments.

2.2. Intestinal Organoid Models

The methodology for separating colon organoids is based on the reports of Hiroyuki Miyoshi et al. [23]. In brief, the mice were fasted and provided with water containing a mixture of penicillin, streptomycin, and gentamicin one day prior to sacrifice. Subsequently, the colon samples were collected, longitudinally dissected, and washed repeatedly in PBS containing penicillin, streptomycin, and amphotericin B. The samples were then gently shaken with digestion solution (20 mM EDTA in Ca^{2+}/Mg^{2+}-free PBS) at 37 °C in order to dissociate the epithelial layer. Subsequently, low-speed centrifugation was employed to separate the supernatant, which contains impurities and individual cells. The crypts were present in the precipitate and were diluted with the organoid culture medium to a concentration of 1×10^4 crypts/mL. The crypt suspension was combined with Matrigel in a 1:1 ratio and inoculated into a 24-well plate at a volume of 50 μL per well. The organoid culture medium was formulated with $1 \times N_2$, $1 \times$ B27, 1% GlutaMAX, 12.5 mM NAC,

50 ng/mL EGF, 15% L-WRN medium, 10 μM Y27632, and 10 μM SB-431542 in Advance DMEM/F12.

2.3. Cell Culture and Treatment

IEC-6 cells (ATCC® CRL-1592™) were cultivated in Dulbecco's modified Eagle's medium (DMEM, ATCC® 30-2002™), with the addition of 10% fetal bovine serum (FBS, Gibco), within a humidified atmosphere containing 95% air and 5% CO_2 at a temperature of 37 °C. The cells were seeded in 24-well plates and subsequently exposed to *H. hepaticus* and recombinant *H. hepaticus* CdtB protein once confluent monolayers had formed.

2.4. Bacterial Culture

In our previous study, a mutant strain of *H. hepaticus* (*H. hepaticus* ΔCdtB) that lacks the CdtB gene was constructed. The *H. hepaticus* 3B1 (ATCC 51449) and *H. hepaticus* ΔCdtB strains were cultured on Brucella agar plates supplemented with 5% sheep blood and antibiotics (amphotericin B, vancomycin, and cefoperazone) for 4–5 days under microaerobic conditions (85% N_2, 10% CO_2, and 5% O_2) at 37 °C. *H. hepaticus* was harvested in PBS and employed for research purposes until the optical density (OD) 600 reading of the bacterial solution reached 1.

2.5. DNA Extraction and PCR Analysis

The colons were collected in the same position, and the tissue and bacterial DNA were extracted according to the manufacturer's instructions using the Bacteria DNA Extraction Kit (Vazyme, Nanjing, China). The quantification of *H. hepaticus* in the colon was performed using quantitative polymerase chain reaction (qPCR) with the Applied Biosystems Step One Real-Time PCR System (ABI) and Universal SYBR Green Master (ABI). The level of colonization by *H. hepaticus* was estimated by normalization to micrograms of mouse chromosomal DNA.

2.6. Histopathology

During the necropsy, the colon tissues of each animal were fixed in 4% paraformaldehyde for a period of 24 h. Thereafter, the proximal, middle, and distal parts were embedded in paraffin for microscopic examination. Tissue sections (5 μm) were subjected to staining with hematoxylin–eosin (H&E) and alcian blue. The histological scores were evaluated in accordance with the previously described methodology, taking into account the extent of the inflammatory infiltrate in the mucosa, submucosa, and muscularis/serosa; the severity of epithelial damage; the degree of crypt atrophy; and the presence of submucosal hyperplasia.

2.7. Immunohistochemistry Analysis

Prior to immunostaining, the paraffin-embedded sections were dewaxed and rehydrated in an alcohol series. Subsequently, the sections were treated with 3% hydrogen peroxide, after which antigen repair was performed with EDTA–citrate solution. Following this, bovine serum albumin was incubated with the sections. Subsequently, the sections were incubated overnight at 4 °C with diluted primary antibodies, including ZO-1 (sc-33725, 1:1000, Santa Cruz Biotechnology, Santa Cruz, CA, USA) and occludin (1:1000, Cambridge, MA, USA), for 16 h at 4 °C. The sections were developed using 3,3′-diaminobenzidine tetrahydrochloride (DAB) (VECTOR Laboratories, Burlingame, CA, USA). The nuclei were then counterstained with H&E for 1 min. The sections were then observed under a light microscope.

2.8. Cell Viability Assay

Cells were seeded into 96-well plates at a density of 1×10^4 cells per well and incubated for 24 h. C-terminal octapeptide of cholecystokinin (CCK-8) assay (Beyotime, Shanghai, China) was used to assess cell viability. The OD at 450 nm was detected according to instructions via Synergy H1 Microplate Reader.

2.9. ROS Assessment

After treated with indicated conditions, cells were harvested and stained with DCFH-DA (Beyotime, Shanghai, China) according to the manufacturer's instructions. The ROS were then determined by flow cytometry (excitation at 488 nm and emission at 525 nm).

2.10. Apoptosis Assessment

Following the completion of the treatment regimen, the cells were harvested and subjected to staining with Annexin V-FITC and PI in accordance with the instructions provided by the manufacturer (Solarbio, Beijing, China). Subsequently, the percentage of apoptotic cells was determined by flow cytometry.

2.11. Real-Time RT-PCR

Total RNA from the IEC-6 cells and colon tissues were extracted using Trizol reagent (Vazyme, Nanjing, China). cDNA synthesis was performed using PrimeScript™ RT reagent kit with gDNA Eraser (TaKaRa, Dalian, China). The mRNA levels of the inflammatory factors and the housekeeping gene *GAPDH* were detected using Universal qPCR SYBR Green Master Mix (Vazyme, Nanjing, China) performed on Step One Real-Time PCR System. Relative expression of target gene mRNA was calculated using the $2^{-\Delta\Delta Ct}$ method. Primers (5′-3″) used in this study are listed as follows:

ZO-1 TTGCCACACTGTGACCCTA (forward), GGGGCATGCTCACTAACCTT (reverse); *occludin* GACCTTGTCCGTGGATGACTTCAG (forward), ATCAGCAGCAGCCATGTACTCTTC (reverse); *IL-6* GATACCACTCCCAACAGAC (forward), CTTTTCTCATTTCCACGAT (reverse); *TNF-α* TCCCAGAAAAGCAAGCAACC (forward), TAGACAGAAGAGCGTGGTGG (reverse); *IL-1β* GTGGTGTGTGACGTTCCCATTA (forward), CCGACAGCACGAGGCTTT (reverse); *iNOS* GTTGAAGACTGAGACTCTGG (forward), GACTAGGCTACTCCGTGGA (reverse); *GAPDH* AGGTCGGTGTGAACGGATTTG (forward), TGTAGACCATGTAGTTGAGGTCA (reverse).

2.12. Western Blot Assay

Samples from tissues or cells were collected, and the proteins were extracted from using cold 1 × RIPA buffer (Solarbio, China) and phenylmethylsulfonyl fluoride (PMSF) (1:1000). Protein concentration was measured with a bicinchoninic acid (BCA) protein assay kit (Solarbio) and then separated by 8–15% SDS-PAGE and transferred to PVDF membranes. The membrane was blocked with 5% skimmed milk for 1 h and incubated with the primary antibodies ZO-1 (sc-33725, 1:1000, Santa Cruz Biotechnology), occludin (1:1000, Abcam, Cambridge, UK), MLCK (sc-365352, 1:1000, Santa Cruz Biotechnology), p-MLC antibody (1:1000, Cell Signaling Technology, Danvers, MA, USA), MLC antibody (1:1000, Cell Signaling Technology, USA), and GAPDH (1:1000, Cell Signaling Technology, USA) and then incubated with secondary antibodies (1:5000, AB clonal Technology, Wuhan, China) labeled with horseradish peroxidase (HRP) at room temperature for 1 h. The bands were visualized using Amer sham ECL Select Western blotting detection reagent (Millipore, Miamisburg, OH, USA).

2.13. Fluorescence Assay

The colon tissues or cells were fixed with 4% paraformaldehyde for 15 min, after which they were blocked with 5% BSA (Beyotime, Shanghai, China). The samples were incubated with the indicated primary antibody and secondary antibodies labeled with FITC or rhodamine, respectively. The nuclei were stained with 4′,6-diamidino-2-phenylindole (DAPI; Beyotime, Shanghai, China). The cytoskeleton was stained with phalloidin (Invitrogen, Waltham, MA, USA). The samples were observed using a confocal microscope (Leica, Wetzlar, Germany). Excitation was achieved at 488 nm, with emission occurring at 530 nm.

2.14. Statistical Analysis

All studies were performed as three independent experiments. The data are expressed as the mean ± SD. Significant variance between groups was determined using one-way ANOVA following by Student's *t*-test. $p < 0.05$ was considered statistically significant. Statistical analyses were performed by using Prism 7 software (GraphPad Software, Inc., La Jolla, CA, USA) and SPSS version 18.0 (SPSS, Inc., Chicago, IL, USA).

3. Results

3.1. CdtB Deficiency Alleviates H. hepaticus-Triggered Intestinal Inflammation

The colonization of *H. hepaticus* in the colons of mice was detected at 4 or 9 MPI by using qPCR. As qPCR data lack an RNA integrity assessment and do not report primer efficiencies, colonization was identified by using an indirect immunofluorescence assay. The polyclonal *H. hepaticus* antibody, prepared in our previous study, was used to detect the *H. hepaticus*-specific antigen in the colons of both the WT and ΔCdtB-infected groups. As illustrated in Figure S1A,B, there is no difference in bacterial colonization between the two groups at either 4 MPI or 9 MPI ($p > 0.05$). Moreover, the colons of mice infected with WT *H. hepaticus* were significantly shorter than those of the control group, which was alleviated by the absence of CdtB (Figure 1A). Next, the colonic histopathology was evaluated. As illustrated in Figure 1B, there was a notable increase in inflammatory cell infiltration within the colons and altered crypt morphology in the WT *H. hepaticus*-infected mice at 4 MPI in comparison to the control mice. At 9 MPI, a more pronounced infiltration of inflammatory cells was evident, accompanied by the formation of lymphocyte aggregates and hyperplasia, within the submucosa and mucosal epithelium. Additionally, a markedly altered crypt morphology, characterized by either atrophy or dilatation, was observed. However, all of these pathological changes were observed to be impaired in mice infected with the ΔCdtB strain. Furthermore, the transcription of inflammatory cytokines, including IL-6, IL-1β, TNF-α, and iNOS, driven by *H. hepaticus*, was inhibited by CdtB deletion at both 4 MPI and 9 MPI (Figure 1C,D). Similarly, the elevated levels of p65 and STAT3 phosphorylation were also inhibited by the CdtB deficiency. In conclusion, these findings suggest that CdtB plays a pivotal role in *H. hepaticus*-induced intestinal inflammation, which involves the STAT3 and NF-κB signaling pathways.

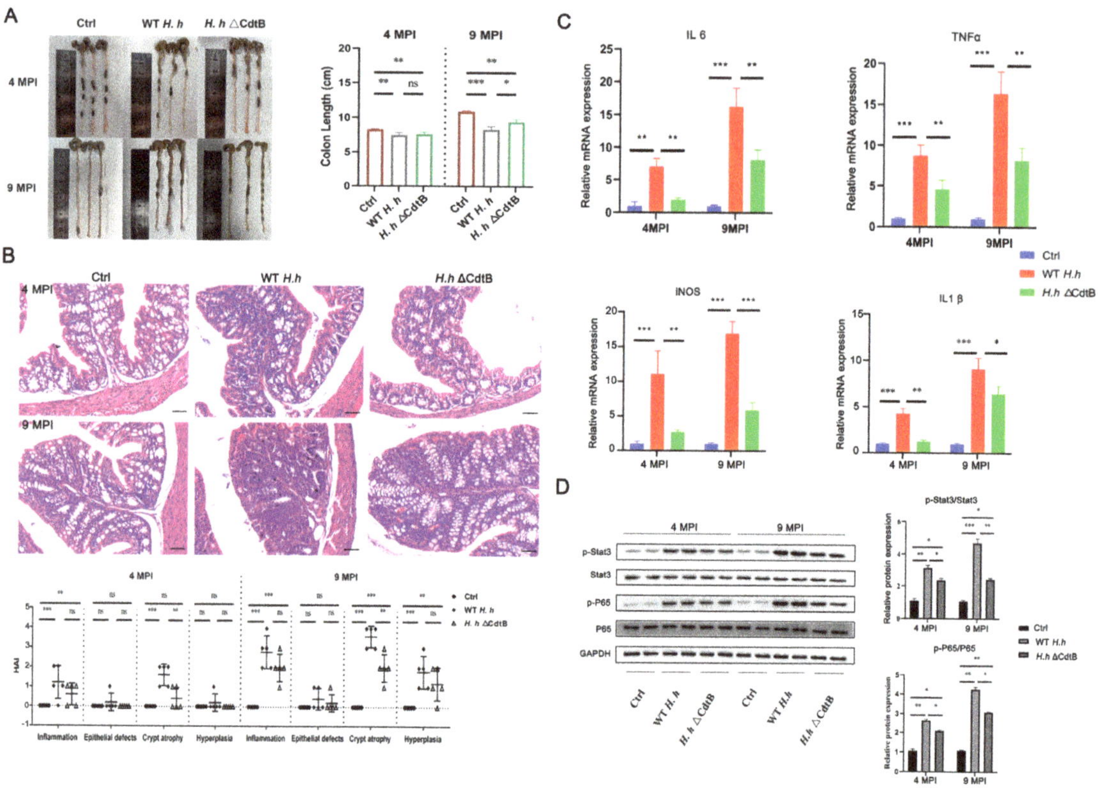

Figure 1. The etiology of colitis induced by *H. hepaticus* is dependent on the presence of CdtB. The mice were treated with *H. hepaticus*, with or without a CdtB mutation, via gavage for either four or nine months. (**A**) Representative colons were photographed, and the lengths were subsequently analyzed. (**B**) Representative images of colon tissue stained with H&E. The histology score was evaluated in accordance with the following criteria: inflammation, damage to the epithelial tissue, atrophy of the crypts, and hyperplasia. The scale bar represents a length of 50 μm. (**C**) The relative mRNA expression levels of IL-6, TNF-α, iNOS, and IL-1β in the mouse colon tissues were quantified by qPCR. (**D**) The relative expression levels of Stat3, p-Stat3, P65, and p-P65 proteins were quantified by a Western blot analysis. GAPDH was employed as a loading control. * $p < 0.05$, ** $p < 0.01$, and *** $p < 0.001$. ns: not significant.

3.2. CdtB Contributes to H. hepaticus-Induced Intestinal Barrier Damage

To investigate whether CdtB is involved in *H. hepaticus*-induced mucin damage, colon sections were stained with alcian blue. As shown in Figure 2A, mice infected with WT *H. hepaticus* showed a significant decrease in the area of acidic mucin in the blue region compared to the control mice, which is reversed by CdtB deletion. The mucus barrier mucin2 (Muc2) acts as the first barrier preventing direct contact between intestinal bacteria and colonic epithelial cells [24]. Similarly, the reduction in Muc2 transcription resulting from *H. hepaticus* infection was also prevented by the CdtB mutation (Figure 2B). Moreover, a reduction in zonula ZO-1 and claudin-1 protein distribution was evident in the intestinal epithelium of mice treated with *H. hepaticus*, whereas no such effect was observed in the mice treated with the ΔCdtB strain (Figure 2C). Furthermore, both the protein and mRNA levels of ZO-1 were increased in the ΔCdtB-infected group in comparison to the WT *H. hepaticus*-infected group. Consistently, comparable results were obtained in the intestinal organoid model (Figure S2). These findings indicate that CdtB is a key factor in the disruption of colonic tight junctions by *H. hepaticus*.

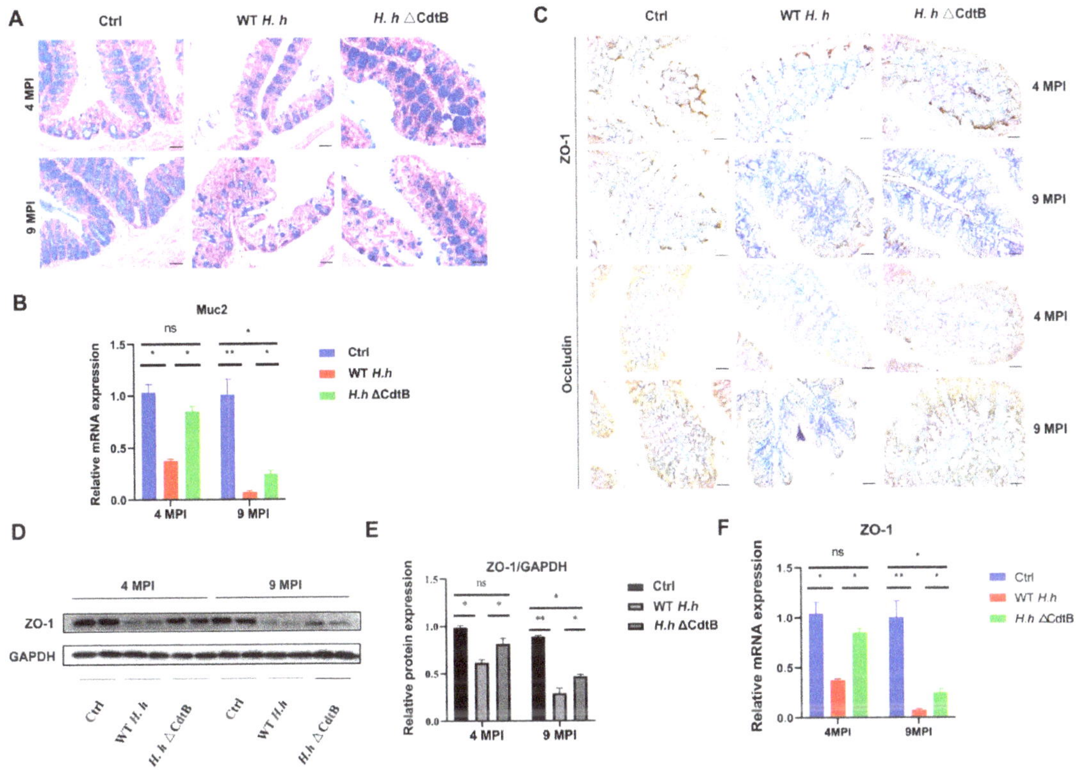

Figure 2. CdtB deficiency alleviates *H. hepaticus*-induced colonic mucosal barrier. Mice were treated with WT *H. hepaticus* or ΔCdtB strain by gavage for 4 and 9 months. (**A**) Representative images of alcian blue-stained colon tissues. (**B**) The relative mRNA expression levels of Muc2 in the mouse colon tissues were tested by q-PCR. (**C**) Representative immunohistochemistry of ZO-1 and occludin staining for colon sections are shown. Scale bar: 50 μm. (**D**,**E**) The relative expression levels of ZO-1 protein were measured by Western blotting. GAPDH was used as a loading control. (**F**) The relative mRNA expression level of ZO-1 was tested by q-PCR. * $p < 0.05$; ** $p < 0.01$. ns not significant.

3.3. CdtB Is Dominant in H. hepaticus-Induced Intestinal Epithelial Cell Damage

To investigate the effects of *H. hepaticus* CdtB on the viability of intestinal epithelial cells, IEC6 cells were treated with different concentrations of *H. hepaticus* or CdtB recombinant protein for 24 h and 48 h. Cell viability was assessed using a CCK-8 assay. As shown in Figure 3A, a significant reduction in cell viability was observed in cells infected with *H. hepaticus* (MOI = 200) for 48 h, while the CdtB recombinant protein (2 μg/mL) demonstrated a notable decline in cell viability at 24 h (Figure 3B). Subsequently, flow cytometry (FCM) was employed to ascertain whether these alterations were linked to CdtB-induced apoptosis. As illustrated in Figure 3C, the CdtB mutation resulted in a reduction in the apoptotic rate of IEC-6 cells relative to the WT *H. hepaticus*-infected group. Moreover, infection with *H. hepaticus* ΔCdtB was found to result in a notable reduction in the mRNA transcription levels of inflammatory cytokines in comparison to those observed in the WT strains. Moreover, comparable effects of CdtB on ROS were discerned in cells utilizing

FCM (Figure 4B). These findings suggest that *H. hepaticus* CdtB reduces the viability of intestinal epithelial cells by triggering an inflammatory and oxidative stress response.

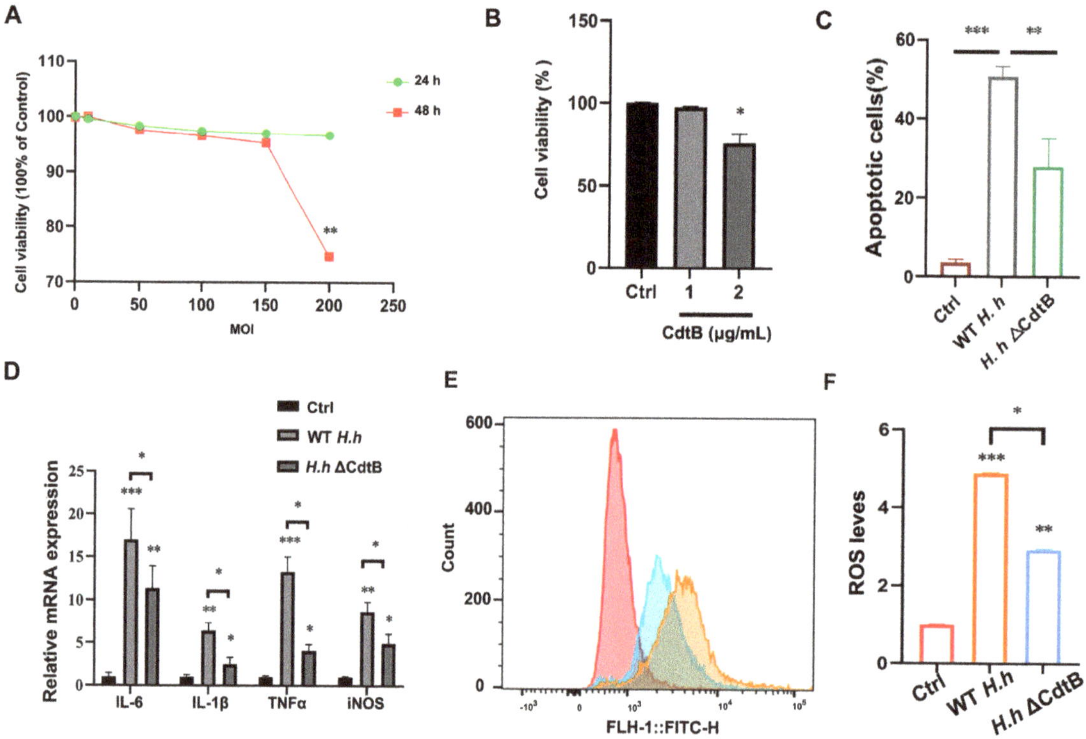

Figure 3. CdtB is attributed to *H. hepaticus*-induced intestinal epithelial cell damage. (**A**) IEC 6 cells were treated with 0–200 MOI *H. hepaticus* for 24 h and 48 h, respectively; (**B**) alternatively, they were treated with different dozens of CdtB protein, prepared in our previous studies, and a CCK-8 assay was performed to evaluated the cell viability. IEC 6 cells were infected with WT *H. hepaticus* or the ΔCdtB strain for 24 h. (**C**) Apoptotic cells were analyzed by using FCM. (**D**) The relative mRNA expression levels of *IL-6*, *IL1β*, *TNF-α*, and *iNOS* were tested by q-PCR. (**E**,**F**) The cellular ROS content was measured and analyzed by using FCM. * $p < 0.05$, ** $p < 0.01$, and *** $p < 0.001$.

3.4. CdtB Causes Intestinal Barrier Disruption via MLCK/pMLC2 Signaling Pathway

To assess the effect of CdtB on tight junctions in vitro, IEC-6 cells were infected with WT *H. hepaticus* or *H. hepaticus* ΔCdtB for 48 h. Both protein expression and mRNA levels of ZO-1 and occludin were decreased by ΔCdtB in cells infected with *H. hepaticus* (Figure 4A,B). Furthermore, exposure to *H. hepaticus* resulted in a marked disruption of ZO-1 distribution, which was impaired by ΔCdtB (Figure 4C). In addition, *H. hepaticus* was detected to significantly reduce the fluorescence intensity of occludin, which was impeded by ΔCdtB as well. It has been demonstrated that MLCK/pMLC2 signaling plays a pivotal role in maintaining the regulation of the intestinal barrier [25]. To ascertain whether MLCK, pMLC2, and MCL2 were involved in the disruption of the intestinal barrier caused by *H. hepaticus* CdtB, their protein expressions were detected by Western blotting. As shown in Figure 5A, the expression of MLCK and the phosphorylation of MCL2 were promoted by *H. hepaticus*, and they were compromised by ΔCdtB (Figure 5A). To further investigate the potential involvement of MLCK/pMLC2 in CdtB-induced intestinal barrier disruption, ML-7, an MLCK inhibitor, was employed to assess the integrity of tight junctions in cells impaired by CdtB. As illustrated in Figure 5B,C, ML-7 was observed to inhibit the reduction in protein and mRNA levels of ZO-1 and occludin, respectively, induced by CdtB. Furthermore, the inhibition of cell apoptosis induced by CdtB was observed with the addition of ML-7 (Figure 5D,E). Furthermore, the redistribution of ZO-1 and occludin was restored by ML-7 (Figure 5F,G). Collectively, these findings suggest that CdtB causes intestinal barrier injury via the MLCK/pMLC2 pathway.

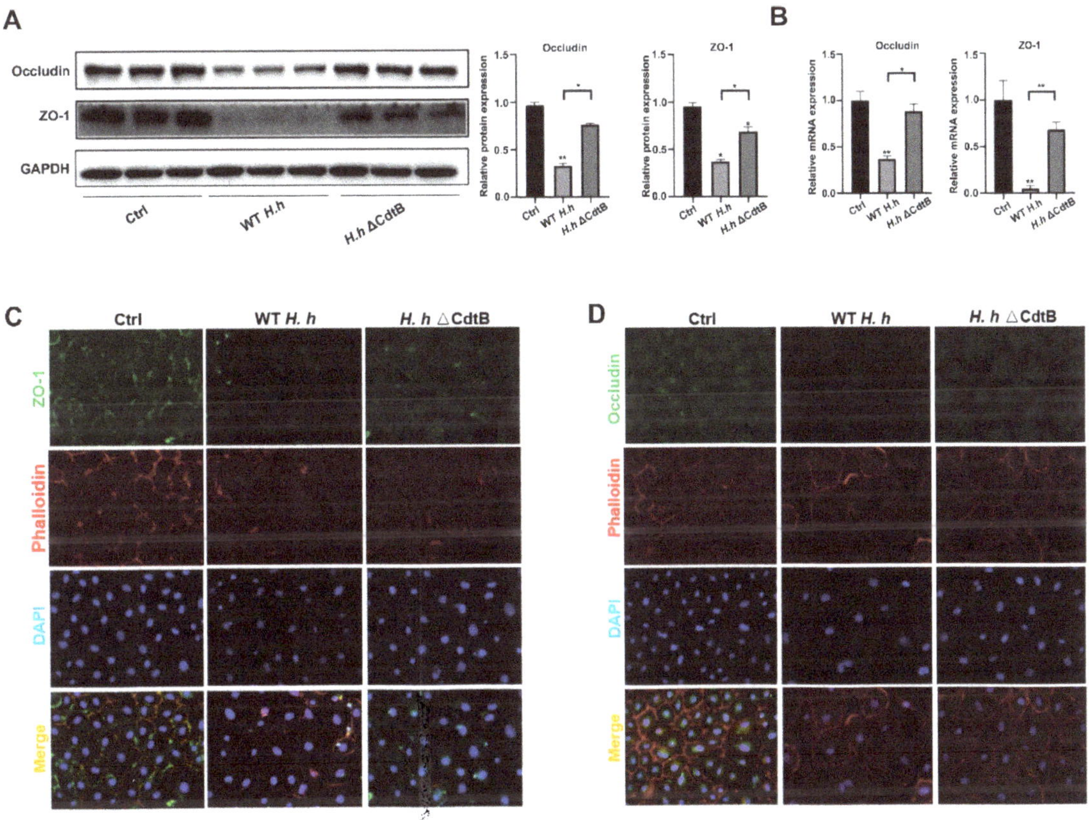

Figure 4. *H. hepaticus* CdtB disturbs the expression and distribution of tight junction proteins in intestinal epithelial cells. IEC 6 cells were treated with WT *H. hepaticus* or the ΔCdtB strain for 24 h. (**A**) Representative occludin and ZO-1 bands from Western blot analyses are shown. Densitometric values were normalized to the GAPDH value. (**B**) The relative mRNA expression levels of occludin and ZO-1 by q-PCR. Representative slides from the control, WT *H. hepaticus*, and *H. hepaticus* ΔCdtB groups stained for (**C**) ZO-1 (green) or (**D**) occludin (green) colocalized with phalloidin (red) were observed using confocal laser scanning microscopy. * $p < 0.05$; ** $p < 0.01$.

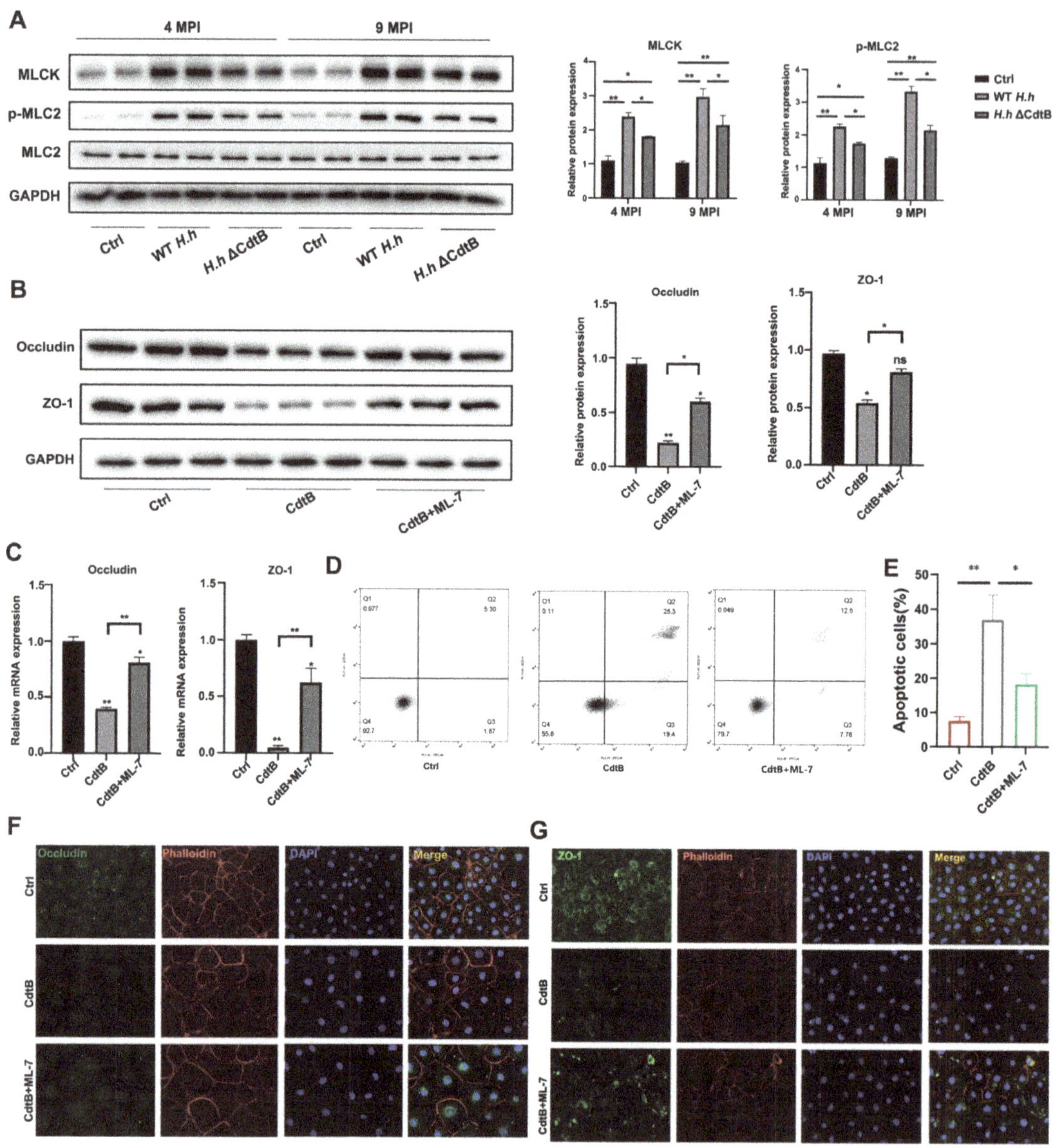

Figure 5. MLCK inhibition suppresses the impairment of tight junctions in intestinal epithelial cells. IEC 6 cells were treated with WT *H. Hepaticus* or the ΔCdtB strain for 24 h. (**A**) Representative MLCK, p-MLC2, and p-MLC2 bands from Western blot analyses are shown. Densitometric values were normalized to the GAPDH value. Cells were treated with CdtB protein in the presence or absence of ML-7. (**B**) Representative occluding and ZO-1 bands from Western blot analyses are shown. (**C**) The relative mRNA expression levels of occluding and ZO-1 obtained by q-PCR. (**D**) Apoptotic cells were analyzed by using FCM. Representative pictures of cells stained for (**E**,**F**) ZO-1 (green) or (**G**) occludin (green) colocalized with phalloidin (red) were observed using confocal laser scanning microscopy. * $p < 0.05$; ** $p < 0.01$. ns: not significant.

4. Discussion

In recent decades, CDTs have been identified as a potential contributor to inflammatory diseases induced by CDT-producing bacteria [26–28]. Multiple pathogenic Gram-negative bacteria, including *Aggregatibacter* (formerly *Actinobacillus*) *actinomycetemcomitans*, *Campylobacter* spp., *Escherichia coli*, *Haemophilus ducreyi*, *Helicobacter* species, *Salmonella enterica* serovar Typhi (*S.* Typhi), and *Shigella* species, have been found to produce CDTs, which pose a threat to chickens, rabbits, canines, and humans [29–31]. Thus, a suitable model should be established for their exploration. CDTs are essential for the persistent colonization of toxin-producing *H. hepaticus* in mice [32]. In this study, *H. hepaticus* was employed to increase the understanding of the virulent properties of CDTs, which can help in creating prophylactic and therapeutic strategies to combat infectious diseases such as diarrhea in animals and inflammatory bowel disease in humans. The findings of this study demonstrate that *H. hepaticus* infection is associated with the promotion of intestinal inflammation and damage to the intestinal barrier in BALB/c male mice, with CdtB emerging as a key factor in this process. Particularly, CdtB disrupts intercellular tight junction structures and induces apoptosis in IEC-6 cells via the MLCK/pMLC2 signaling pathway.

H. hepaticus has been demonstrated to induce experimental colitis in rodents with a fully developed gastrointestinal tract or a developing intestine [33]. In this study, the presence of typical colitis lesions was observed in mice infected with *H. hepaticus*. In particular, the *H. hepaticus* ΔCdtB strain was employed to assess the impact of CdtB on *H. hepaticus*-induced colitis. The restoration of intestinal pathological changes and inflammatory factors induced by *H. hepaticus* by CdtB deletion indicates that CdtB plays a pivotal role in *H. hepaticus*-induced colitis. These observations are in accordance with previous reports on the role of CdtB in *H. hepaticus*-induced colitis [18,34]. The disruption of the intestinal barrier represents a pivotal mechanism in the pathogenesis of intestinal pathogen-induced diarrhea. The function of the cell–cell tight junction in epithelial barrier dysfunction has been documented, yet it has never been examined in the context of *H. hepaticus* infection. Barrier-relevant tight junction proteins, such as ZO-1, occludin, and claudin, are predominantly multiprotein complexes anchored to the cytoskeleton [34,35]. Their impairment causes the leak flux mechanism by increased paracellular permeability [36]. The expression and distribution of tight junction proteins were investigated. To illustrate an intact tight junction network, we demonstrated the colocalization of occludin with ZO-1 in the control samples. CdtB is indispensable for the *H. hepaticus*-induced disruption of barrier-relevant TJ proteins. It is noteworthy that our findings reveal that CdtB induced the redistribution of ZO-1 yet had no effect on occludin. Similarly, no alteration in the distribution of occludin to intracellular compartments was evident in the context of *C. jejuni*-induced disruption of the intestinal epithelial barrier [37]. We propose ongoing changes in tight junction protein expression and distribution with reported effects on barrier function. Further studies are needed to confirm the non-essential role of CdtA or CdtC in virulence by using CdtA or CdtC mutants.

Both inflammation and epithelial cell death are inducing factors of barrier dysfunction induced by bacterial pathogens [37]. Treatment with inflammatory factors was shown to cause barrier dysfunction by tight junction alterations [38]. TNFα-induced barrier dysfunction mostly happens during apoptosis. However, the immune response to the interaction with *C. jejuni* caused barrier impairment and induced overlapping effects of apoptosis and tight junction changes [39]. In terms of inflammatory induction, IL6, TNFα, and IL1β are the major cytokines altered by CdtB mutation during *H. hepaticus* infection. *H. hepaticus* CdtB has been reported to promote IL-6, TNFα, and caspase-3 activity in liver cells [29]. In addition, it was indicated to increase host DNA damage by inducing DNA double-strand break that can upregulate the expression of IL6 and TNFα [18]. Zhu et al. found that

CdtB promotes colitis development in IL10$^{-/-}$ mice by the induction of an inflammatory response [40]. In these models, higher levels of pro-inflammatory cytokines were observed in infections with *H. hepaticus*. Our findings consistently demonstrate that CdtB deficiency resulted in a reduction in *H. hepaticus*-induced apoptosis levels in addition to a decrease in pro-inflammatory cytokines. Therefore, it is possible that there are indirect effects of inflammatory factors on *H. hepaticus* CdtB-induced apoptosis, and further investigations in this area would be beneficial.

The cytoskeleton is essential for the regulation of TJ proteins. In this study, we found that CdtB of *H. hepaticus* disrupts the cytoskeleton of the monolayer intestinal barrier in IEC-6 cells. This finding is consistent with a previous study which showed that both Helicobacter pullorum and *H. hepaticus* CdtB induce nuclear and cytoskeletal remodeling [41,42]. Moreover, CdtB raises ROS levels in IEC-6 cell monolayers, which contributes to intestinal inflammation and intestinal barrier disruption [43,44]. MLCK-driven MLC phosphorylation has been found to be involved in signal transduction pathways employed by diverse extracellular stimuli to influence barrier function and alter tight junction protein expression and localization [25,45]. Su et al. found that MLCK$^{-/-}$ mice effectively inhibited apoptosis intestinal epithelial cells and the disruption of TJs caused by DSS [46]. In comparison, it was observed that apoptosis and tight junction disruption were markedly reduced in IEC-6 cells following the administration of an MLCK inhibitor. This finding suggests that ROS-mediated MLCK activation plays a role in the CdtB-induced colonic mucosal barrier [47]. All of these results suggest that ROS and inflammatory cytokines are involved in the activation of MLCK/pMLC2 signaling induced by CdtB, leading to cytoskeletal disruption and the redistribution of TJ proteins. Since CdtB is already known to be essential for virulence, this study primarily extends existing knowledge laterally. Further research is necessary to determine new functions of CdtB and how it contributes to the *H. hepaticus*-induced intestinal pathogenesis.

5. Conclusions

In conclusion, *H. hepaticus* CdtB can destabilize and down-regulate tight junction proteins by activating the MLCK-MLC signaling pathway, thus ameliorating the tight junctions of intestinal epithelial cells. CDT plays an important role in *H. hepaticus* infection and the virulence of microbially induced gastrointestinal diseases. Our findings can help in the development of new prevention strategies for intestinal infectious diseases in animals infected by CDT-producing bacteria.

Supplementary Materials: The following supporting information can be downloaded at: https://www.mdpi.com/article/10.3390/vetsci12020174/s1, Figure S1: Intestinal colonization of *H. hepaticus* in the is independent of CdtB; Figure S2. *H. hepaticus* CdtB disturbs the expression and distribution of the tight junction proteins in intestinal organoids; Figure S3. The original images of Figure 1D in the article; Figure S4. The original images of Figure 2D in the article; Figure S5. The original images of Figure 4A in the article; Figure S6. The original images of Figure 5A in the article; Figure S7. The original images of Figure 5B in the article.

Author Contributions: Conceptualization, T.W.; methodology, X.M.; formal analysis, S.J. and R.B.; investigation, X.M and M.Q.; resources, L.Z.; writing—original draft preparation, T.W.; writing—review and editing, all authors; supervision, Q.Z.; funding acquisition, Q.Z. All authors have read and agreed to the published version of the manuscript.

Funding: This work was supported by the National Natural Science Foundation of China (32273004), the Key R & D plan of Jiangsu Province (BE2020674), the Qing Lan project of Jiangsu province, Priority Academic Program Development of Jiangsu Higher Education Institutions (PAPD), the Postgraduate

Research and Practice Innovation Program of Jiangsu Province (KYCX21_3209), and the Top-level Talents Support Program of Yangzhou University.

Institutional Review Board Statement: The animal study protocol was approved by the Jiangsu Provincial Science and Technology Department and the Institutional Animal Care and Use Committee (IACUC) of Yangzhou University (code: 202402138 on 26 February 2024).

Informed Consent Statement: Not applicable.

Data Availability Statement: The data that support the findings of this study are available from the corresponding author upon reasonable request.

Conflicts of Interest: The authors declare no conflicts of interest.

References

1. Neubauer, C.; Hess, M. Tissue tropism of in specific pathogen-free chickens determined by culture and nucleic acid detection. *Avian Dis.* **2006**, *50*, 620–623. [CrossRef] [PubMed]
2. Harbour, S.; Sutton, P. Immunogenicity and pathogenicity of infections of veterinary animals. *Vet. Immunol. Immunopthol.* **2008**, *122*, 191–203. [CrossRef] [PubMed]
3. Fox, J.G.; Ge, Z.; Whary, M.T.; Erdman, S.E.; Horwitz, B.H. *Helicobacter hepaticus* infection in mice: Models for understanding lower bowel inflammation and cancer. *Mucosal Immunol.* **2011**, *4*, 22–30. [CrossRef] [PubMed]
4. Kaewpitoon, S.J.; Loyd, R.A.; Rujirakul, R.; Panpimanmas, S.; Matrakool, L.; Tongtawee, T.; Kootanavanichpong, N.; Pengsaa, P.; Kompor, P.; Chavengkun, W.; et al. Helicobacter Species are Possible Risk Factors of Cholangiocarcinoma. *Asian Pac. J. Cancer Prev. APJCP* **2016**, *17*, 37–44. [CrossRef] [PubMed]
5. Zhu, L.; Wu, Z.; Zhu, C.; Yin, J.; Huang, Y.; Feng, J.; Zhang, Q. The Deletion of IL-17A Enhances *Helicobacter hepaticus* Colonization and Triggers Colitis. *J. Inflamm. Res.* **2022**, *15*, 2761–2773. [CrossRef]
6. Falsafi, T.; Mahboubi, M. *Helicobacter hepaticus*, a new pathogenic species of the Helicobacter genus: Similarities and differences with H. pylori. *Iran. J. Microbiol.* **2013**, *5*, 185–194. [PubMed]
7. Jeffery, R.; Ilott, N.E.; Powrie, F. Genetic and environmental factors shape the host response to insights into IBD pathogenesis. *Curr. Opin. Microbiol.* **2022**, *65*, 145–155. [CrossRef]
8. Camilleri, M.; Madsen, K.; Spiller, R.; Greenwood-Van Meerveld, B.; Verne, G.N. Intestinal barrier function in health and gastrointestinal disease. *Neurogastroenterol. Motil.* **2012**, *24*, 503–512. [CrossRef]
9. Parikh, K.; Antanaviciute, A.; Fawkner-Corbett, D.; Jagielowicz, M.; Aulicino, A.; Lagerholm, C.; Davis, S.; Kinchen, J.; Chen, H.H.; Alham, N.K.; et al. Colonic epithelial cell diversity in health and inflammatory bowel disease. *Nature* **2019**, *567*, 49–55. [CrossRef]
10. Qiu, P.; Ishimoto, T.; Fu, L.; Zhang, J.; Zhang, Z.; Liu, Y. The Gut Microbiota in Inflammatory Bowel Disease. *Front. Cell. Infect. Microbiol.* **2022**, *12*, 733992. [CrossRef]
11. Taylor, N.S.; Xu, S.; Nambiar, P.; Dewhirst, F.E.; Fox, J.G. Enterohepatic species are prevalent in mice from commercial and academic institutions in Asia, Europe, and North America. *J. Clin. Microbiol.* **2007**, *45*, 2166–2172. [CrossRef]
12. Kleine, M.; Worbs, T.; Schrem, H.; Vondran, F.W.; Kaltenborn, A.; Klempnauer, J.; Förster, R.; Josenhans, C.; Suerbaum, S.; Bektas, H. *Helicobacter hepaticus* induces an inflammatory response in primary human hepatocytes. *PLoS ONE* **2014**, *9*, e99713. [CrossRef] [PubMed]
13. Yang, J.; Ji, S.; Zhang, Y.; Wang, J. *Helicobacter hepaticus* infection in primary hepatocellular carcinoma tissue. *Singap. Med. J.* **2013**, *54*, 451–457. [CrossRef] [PubMed]
14. Suerbaum, S.; Josenhans, C.; Sterzenbach, T.; Drescher, B.; Brandt, P.; Bell, M.; Droge, M.; Fartmann, B.; Fischer, H.P.; Ge, Z.; et al. The complete genome sequence of the carcinogenic bacterium *Helicobacter hepaticus*. *Proc. Natl. Acad. Sci. USA* **2003**, *100*, 7901–7906. [CrossRef]
15. Haghjoo, E.; Galán, J.E. Salmonella typhi encodes a functional cytolethal distending toxin that is delivered into host cells by a bacterial-internalization pathway. *Proc. Natl. Acad. Sci. USA* **2004**, *101*, 4614–4619. [CrossRef]
16. Guerra, L.; Cortes-Bratti, X.; Guidi, R.; Frisan, T. The biology of the cytolethal distending toxins. *Toxins* **2011**, *3*, 172–190. [CrossRef]
17. Zhu, L.; Zhu, C.; Cao, S.; Zhang, Q. *Helicobacter hepaticus* Induce Colitis in Male IL-$10^{-/-}$ Mice Dependent by Cytolethal Distending Toxin B and via the Activation of Jak/Stat Signaling Pathway. *Front. Cell. Infect. Microbiol.* **2021**, *11*, 616218. [CrossRef]
18. Ge, Z.M.; Feng, Y.; Ge, L.L.; Parry, N.; Muthupalani, S.; Fox, J.G. *Helicobacter hepaticus* cytolethal distending toxin promotes intestinal carcinogenesis in 129-deficient mice. *Cell Microbiol.* **2017**, *19*, e12728. [CrossRef] [PubMed]

19. Martens, E.C.; Neumann, M.; Desai, M.S. Interactions of commensal and pathogenic microorganisms with the intestinal mucosal barrier. *Nat. Rev. Microbiol.* **2018**, *16*, 457–470. [CrossRef] [PubMed]
20. Suzuki, T. Regulation of the intestinal barrier by nutrients: The role of tight junctions. *Anim. Sci. J. Nihon Chikusan Gakkaiho* **2020**, *91*, e13357. [CrossRef]
21. Horowitz, A.; Chanez-Paredes, S.D.; Haest, X.; Turner, J.R. Paracellular permeability and tight junction regulation in gut health and disease. *Nat. Rev. Gastroenterol. Hepatol.* **2023**, *20*, 417–432. [CrossRef]
22. Liyanage, N.P.; Manthey, K.C.; Dassanayake, R.P.; Kuszynski, C.A.; Oakley, G.G.; Duhamel, G.E. *Helicobacter hepaticus* cytolethal distending toxin causes cell death in intestinal epithelial cells via mitochondrial apoptotic pathway. *Helicobacter* **2010**, *15*, 98–107. [CrossRef] [PubMed]
23. Miyoshi, H.; Stappenbeck, T.S. *In vitro* expansion and genetic modification of gastrointestinal stem cells in spheroid culture. *Nat. Protoc.* **2013**, *8*, 2471–2482. [CrossRef] [PubMed]
24. Yao, D.B.; Dai, W.L.; Dong, M.; Dai, C.L.; Wu, S.D. MUC2 and related bacterial factors: Therapeutic targets for ulcerative colitis. *Ebiomedicine* **2021**, *74*, 103751. [CrossRef] [PubMed]
25. Huang, S.W.; Fu, Y.J.; Xu, B.; Liu, C.; Wang, Q.; Luo, S.; Nong, F.F.; Wang, X.J.; Huang, S.Y.; Chen, J.Y.; et al. Wogonoside alleviates colitis by improving intestinal epithelial barrier function via the MLCK/pMLC2 pathway. *Phytomedicine* **2020**, *68*, 153179. [CrossRef]
26. Péré-Védrenne, C.; Cardinaud, B.; Varon, C.; Mocan, I.; Buissonnière, A.; Izotte, J.; Mégraud, F.; Ménard, A. The Cytolethal Distending Toxin Subunit CdtB of Induces a Th17-related and Antimicrobial Signature in Intestinal and Hepatic Cells In Vitro. *J. Infect. Dis.* **2016**, *213*, 1979–1989. [CrossRef]
27. Cao, S.; Zhu, C.; Feng, J.; Zhu, L.; Yin, J.; Xu, Y.; Yang, H.; Huang, Y.; Zhang, Q. *Helicobacter hepaticus* infection induces chronic hepatitis and fibrosis in male BALB/c mice via the activation of NF-κB, Stat3, and MAPK signaling pathways. *Helicobacter* **2020**, *25*, e12677. [CrossRef] [PubMed]
28. Qian, M.; Cao, S.; Wang, T.; Xu, X.; Zhang, Q. Apoptosis triggered by cytolethal distending toxin B subunit of *Helicobacter hepaticus* is aggravated by autophagy inhibition in mouse hepatocytes. *Biochem. Biophys. Res. Commun.* **2022**, *598*, 40–46. [CrossRef] [PubMed]
29. Ge, Z.M.; Schauer, D.B.; Fox, J.G. virulence properties of bacterial cytolethal-distending toxin. *Cell Microbiol.* **2008**, *10*, 1599–1607. [CrossRef] [PubMed]
30. Wising, C.; Mölne, L.; Jonsson, I.M.; Ahiman, K.; Lagergård, T. The cytolethal distending toxin of aggravates dermal lesions in a rabbit model of chancroid. *Microbes Infect.* **2005**, *7*, 867–874. [CrossRef]
31. AbuOun, M.; Manning, G.; Cawthraw, S.A.; Ridley, A.; Ahmed, I.H.; Wassenaar, T.M.; Newell, D.G. Cytolethal distending toxin (CDT)-negative strains and anti-CDT neutralizing antibodies are induced during human infection but not during colonization in chickens. *Infect. Immun.* **2005**, *73*, 3053–3062. [CrossRef] [PubMed]
32. Pons, B.J.; Vignard, J.; Mirey, G. Cytolethal Distending Toxin Subunit B: A Review of Structure-Function Relationship. *Toxins* **2019**, *11*, 595. [CrossRef] [PubMed]
33. Dvorak, K.; Coursodon-Boyiddle, C.F.; Snarrenberg, C.L.; Kananurak, A.; Underwood, M.A.; Dvorak, B. *Helicobacter hepaticus* increases intestinal injury in a rat model of necrotizing enterocolitis. *Am. J. Physiol-Gastr L* **2013**, *305*, G585–G592.
34. Chanez-Paredes, S.D.; Abtahi, S.; Kuo, W.T.; Turner, J.R. Differentiating Between Tight Junction-Dependent and Tight Junction-Independent Intestinal Barrier Loss In Vivo. *Methods Mol. Biol.* **2021**, *2367*, 249–271. [PubMed]
35. Wu, H.; Chen, Q.Y.; Wang, W.Z.; Chu, S.; Liu, X.X.; Liu, Y.J.; Tan, C.; Zhu, F.; Deng, S.J.; Dong, Y.L.; et al. Compound sophorae decoction enhances intestinal barrier function of dextran sodium sulfate induced colitis via regulating notch signaling pathway in mice. *Biomed. Pharmacother. Biomed. Pharmacother.* **2021**, *133*, 110937. [CrossRef]
36. Barmeyer, C.; Fromm, M.; Schulzke, J.D. Active and passive involvement of claudins in the pathophysiology of intestinal inflammatory diseases. *Pflügers Arch. Eur. J. Physiol.* **2017**, *469*, 15–26. [CrossRef] [PubMed]
37. Butkevych, E.; de Sá, F.D.L.; Nattramilarasu, P.K.; Bücker, R. Contribution of Epithelial Apoptosis and Subepithelial Immune Responses in *Campylobacter jejuni*-Induced Barrier Disruption. *Front. Microbiol.* **2020**, *11*, 344. [CrossRef] [PubMed]
38. Bücker, R.; Krug, S.M.; Moos, V.; Bojarski, C.; Schweiger, M.R.; Kerick, M.; Fromm, A.; Janssen, S.; Fromm, M.; Hering, N.A.; et al. *Campylobacter jejuni* impairs sodium transport and epithelial barrier function via cytokine release in human colon. *Mucosal Immunol.* **2018**, *11*, 474–485. [CrossRef]
39. de Sá, F.D.L.; Butkevych, E.; Nattramilarasu, P.K.; Fromm, A.; Mousavi, S.; Moos, V.; Golz, J.C.; Stingl, K.; Kittler, S.; Seinige, D.; et al. Curcumin Mitigates Immune-Induced Epithelial Barrier Dysfunction by *Campylobacter jejuni*. *Int. J. Mol. Sci.* **2019**, *20*, 4830. [CrossRef]

40. Zhu, M.K.; Miao, S.S.; Zhou, W.T.; Elnesr, S.S.; Dong, X.Y.; Zou, X.T. MAPK, AKT/FoxO3a and mTOR pathways are involved in cadmium regulating the cell cycle, proliferation and apoptosis of chicken follicular granulosa cells. *Ecotox Environ. Saf.* **2021**, *214*, 112091. [CrossRef]
41. Ernst, K.; Landenberger, M.; Nieland, J.; Nørgaard, K.; Frick, M.; Fois, G.; Benz, R.; Barth, H. Characterization and Pharmacological Inhibition of the Pore-Forming Clostridioides difficile CDTb Toxin. *Toxins* **2021**, *13*, 390. [CrossRef]
42. Péré-Védrenne, C.; He, W.; Azzi-Martin, L.; Prouzet-Mauléon, V.; Buissonnière, A.; Cardinaud, B.; Lehours, P.; Mégraud, F.; Grosset, C.F.; Ménard, A. The Nuclear Remodeling Induced by Helicobacter Cytolethal Distending Toxin Involves MAFB Oncoprotein. *Toxins* **2020**, *12*, 174. [CrossRef] [PubMed]
43. Liang, B.; Zhong, Y.; Huang, Y.; Lin, X.; Liu, J.; Lin, L.; Hu, M.; Jiang, J.; Dai, M.; Wang, B.; et al. Underestimated health risks: Polystyrene micro- and nanoplastics jointly induce intestinal barrier dysfunction by ROS-mediated epithelial cell apoptosis. *Part. Fibre Toxicol.* **2021**, *18*, 20. [CrossRef]
44. Deng, Y.; Zhang, Z.; Yang, H.; Wang, J.; Feng, L.; Su, Y.; Xu, D. Mycophenolic Acid Induces the Intestinal Epithelial Barrier Damage through Mitochondrial ROS. *Oxidative Med. Cell. Longev.* **2022**, *2022*, 4195699. [CrossRef] [PubMed]
45. Xie, Y.; Zhan, X.; Tu, J.; Xu, K.; Sun, X.; Liu, C.; Ke, C.; Cao, G.; Zhou, Z.; Liu, Y. Atractylodes oil alleviates diarrhea-predominant irritable bowel syndrome by regulating intestinal inflammation and intestinal barrier via SCF/c-kit and MLCK/MLC2 pathways. *J. Ethnopharmacol.* **2021**, *272*, 113925. [CrossRef] [PubMed]
46. Li, X.; Li, Q.; Xiong, B.; Chen, H.; Wang, X.; Zhang, D. Discoidin domain receptor 1(DDR1) promote intestinal barrier disruption in Ulcerative Colitis through tight junction proteins degradation and epithelium apoptosis. *Pharmacol. Res.* **2022**, *183*, 106368. [CrossRef] [PubMed]
47. Wang, H.; Zhai, N.; Chen, Y.; Fu, C.; Huang, K. OTA induces intestinal epithelial barrier dysfunction and tight junction disruption in IPEC-J2 cells through ROS/Ca^{2+}-mediated MLCK activation. *Environ. Pollut.* **2018**, *242 Pt A*, 106–112. [CrossRef]

veterinary sciences

MDPI

Article

Effect of Ascites Syndrome on Diversity of Cecal Microbiota of Broiler Chickens

Aikebaier Reheman [1,†], Zhichao Wang [2,†], Ruihuan Gao [2], Jiang He [2], Juncheng Huang [3], Changqing Shi [3], Meng Qi [1] and Xinwei Feng [1,*]

1 College of Animal Science and Technology, Tarim University, Alar 843300, China; akpar0902@163.com (A.R.); qimengdz@163.com (M.Q.)
2 Engineering Laboratory of Tarim Animal Diseases Diagnosis and Control, Xinjiang Production & Construction Corps, Alar 843300, China; 13177890078@163.com (Z.W.); gaoruihuan0@163.com (R.G.); 17591556804@163.com (J.H.)
3 Key Laboratory of Livestock and Forage Resources Utilization around Tarim, Ministry of Agriculture and Rural Affairs, Tarim University, Alar 843300, China; hjcll20001020@163.com (J.H.); scq926@163.com (C.S.)
* Correspondence: fxwdky@126.com
† These authors contributed equally to this work.

Simple Summary: Ascites syndrome (AS) is a metabolic disease that impacts the growth and development of broiler chickens. The intestinal microbiota is closely associated with the growth and development of broiler chickens. However, the effects of AS on the intestinal microbiota of broiler chickens have not yet been reported. The purpose of this study was to demonstrate whether AS affects the intestinal microbiota of broiler chickens. This study showed that AS significantly increased the abundance of *Bacteroidetes* while reducing the abundance of *Actinobacteria*. This study provides data for the in-depth study of the relationship between AS and intestinal microbiota.

Abstract: Ascites syndrome (AS) is a metabolic disease that seriously affects the growth and development of broiler chickens. Intestinal microbiota play a significant role in the growth of broiler chickens. Therefore, further research on the relationship between AS and intestinal microbiota will help to better understand the impact of AS on broiler growth. In this study, 0.2% sodium chloride was added to the drinking water, which induced AS in broiler chickens, and we detected the influence of AS on the growth performance and cecal microbiota of broiler chickens. The results showed that AS significantly reduced the cecal microbial diversity of broiler chickens and affected the cecal microbial composition at the phylum and genus levels (p = 0.05). Further, LEfSe analysis revealed that AS significantly increased the abundance of *Bacteroidetes* (p = 0.035) while simultaneously reducing the abundance of *Actinobacteria* (p = 0.031) in the cecum. Additionally, the differential metabolites associated with polycyclic aromatic hydrocarbon degradation were significantly diminished. The findings suggest that AS may further impact the growth rate of broiler chickens by altering cecal microorganisms.

Keywords: ascites syndrome; cecal microorganisms; 16S rDNA; broilers

1. Introduction

Broiler ascites syndrome (AS), also known as broiler pulmonary hypertension syndrome (PHS), is a common nutritional and metabolic disease in rapidly growing commercial broilers [1]. Many studies have shown that diet, management, environment, genetics, and other factors are related to the occurrence of AS; among them, low-temperature, high-energy, high-protein, and high-salt diets are the main feeding and management factors leading to AS [2]. The above factors lead to a series of pathophysiological changes in chickens due to hypoxia, such as increased pulmonary artery blood pressure, pulmonary artery remodeling, right ventricular hypertrophy and failure, free radical production, and increased lipid peroxidation [3]. The above shows that hypoxia is the main cause of AS in broilers.

A hypoxic environment can lead to a significant reduction in the intestinal microbial diversity of broiler chickens. Research indicates that the proportion of certain beneficial bacteria, such as lactic acid bacteria, increases under anoxic conditions, while the abundance of other bacteria, including specific anaerobic species, decreases [4,5]. This shift in microbial composition may adversely affect the health and growth of broilers, impairing their immune function and digestive capacity. Additionally, studies have demonstrated that hypoxia diminishes the growth and feed conversion rates of broiler chickens by impairing the utilization efficiency of specific nutrients by microorganisms [6]. This, in turn, adversely impacts the lipid metabolism, growth performance, and overall health status of broiler chickens [7]. Research shows a potential relationship between the occurrence of PHS and eight specific gut microbiota [8]. There is evidence that PHS is associated with increased gut–hypothalamic paraventricular nucleus transport, pathological changes in the gut wall, and gut microbiota imbalance [9]. Based on the above, we speculate that AS may affect the composition of intestinal microbiota.

Gut microbiota refers to the diverse community of microorganisms, including bacteria, fungi, viruses, and protozoa, that inhabit the intestines of animals. These microorganisms play a crucial role in digestion, immune function, nutrient absorption, and overall health. Many studies have shown that the composition and diversity of microorganisms in the intestines of broiler chickens directly affect their growth rate, feed conversion rate, and disease resistance [10–12]. Numerous studies have demonstrated that intestinal microorganisms generate a diverse array of metabolites that are advantageous to the host during metabolic processes, including short-chain fatty acids (SCFAs), vitamins, and other bioactive compounds [13,14]. However, metabolic diseases may alter the gut microbiome and thus negatively impact broiler chickens and disrupt the production of these beneficial metabolites [15,16].

In summary, AS, as a common metabolic disease in broiler chickens, may also affect the composition of the intestinal microbiota of broiler chickens. Therefore, to study the impact of AS on the intestinal flora of broiler chickens, this article analyzed the intestinal microbial diversity of AS broiler chickens by constructing a broiler ascites syndrome model and analyzing intestinal microbial flora.

2. Materials and Methods

2.1. Ethical Statement

The laboratory animals were housed and treated following the Regulations for the Administration of Affairs Concerning Experimental Animals of the People's Republic of China and following the Research Ethics Committee of Tarim University (2023025).

2.2. Study Design and Treatments

A group of 260 healthy broiler chickens were selected at 0 days old, male in gender and Arbor Acres in breed, and raised intensively in single-layer cages in Alar, China, at an altitude of about 1011 m. Daily management adhered to the routine feeding and management technical specifications for broiler chicks, providing free access to feed and water, artificial lighting throughout the day, and daily cleaning of feces trays to maintain a hygienic environment within the chicken house, and weighing occurred on days 0, 18, and 28. The diet ingredients and nutrients provided during the experiment are shown in Table 1. After 7 days of normal feeding, the feed was withheld for 12 h, and the chicks were subsequently weighed. A total of 240 chicks, all of equal weight and in good mental condition, were selected and randomly divided into four groups, each containing three replicates with 20 chicks per replicate. The mental condition of broiler chickens is primarily assessed by evaluating their levels of exercise, food intake, and water consumption. Signs of a good mental state in broiler chickens include quick movement, acute hearing, bright and lively eyes, long and loud vocalization, a strong appetite, and responsiveness to external stimuli. The experimental group received drinking water with sodium chloride concentrations of 0.1%, 0.2%, and 0.3%, respectively. In contrast, the control group was provided with drinking water that contained no sodium chloride, while both groups were maintained under identical feeding conditions. The broiler chickens were raised until 28 days of age, and the ascites rate, body weight, and ascites heart index (AHI) were calculated. The formula for calculating the AHI is as follows: AHI = right ventricular mass/total ventricular mass. If the peritoneal effusion exceeds 20 mL and the ascites heart index is greater than 0.25, the presence of either of the aforementioned conditions can be classified as broiler ascites syndrome.

Table 1. Basic diet composition and nutrient level.

Ingredients	Content (%)	Nutrient Level	
Maize	59.00	Metabolizable energy/(MJ/kg)	13.56
Soybean meal	27.00	Dry matter (%)	86.00
Wheat bran	5.00	Crude protein (%)	19.58
Rapeseed meal	3.00	Crude fiber (%)	6.00
Calcium hydrogen phosphate	2.50	Coarse ash (%)	7.00
Sunflower meal	1.50	Calcium (%)	0.75
Stone powder	0.35	Total phosphorus (%)	0.45
Premix	1.35	Sulfur-containing amino acids (%)	0.50
Sodium chloride	0.3	Water (%)	0.50
Total	100.00	Total	

1. The premix involves the confidential formula of Xinjiang Tiankang Feed Technology Co., Ltd. (Urumqi, China), which will not be disclosed. 2. The nutritional levels are analytically guaranteed values tested by Xinjiang Tiankang Feed Technology Co., Ltd.

2.3. Sample Collection

The broiler chickens were euthanized under anesthesia after being raised for 28 days to assess the rate of ascites. Three chickens were randomly selected from the control group and 0.2% sodium chloride group, and the cecal contents were cryopreserved for the extraction of intestinal microbial genomes.

2.4. Microbial Total DNA Extraction

Total microbial DNA was extracted from broiler chicken cecal contents according to the protocol described previously [17]. Microbial genomic DNA was isolated with a commercial kit (catalog no. 51504; Qiagen, Shanghai, China; Shanghai Limin Industrial Co., Ltd., Shanghai, China) according to the manufacturer's description. Prior to high-throughput sequencing, all DNA samples underwent testing via 1.5% agarose gel electrophoresis to verify their integrity and quality. Additionally, the purity of the DNA was assessed using a 2000 NanoDrop (Thermo Fisher Scientific, Waltham, CA, USA) to ensure the reliability of subsequent experiments.

2.5. 16S rRNA Sequencing

The amplification of microbial DNA was carried out using specific barcode primers targeting the v3-v4 region of the 16S rRNA gene; the PCR primer sequence was F-5′-ACTCCTACGGGAGGCAGCA-3′), (R-5′-GGACTACHVGGGTWTCTAAT-3′). A commercial PCR kit (New England Biolabs, Ipswich, MA, USA) was used for PCR amplification. Amplifiers were quantified on a 1.5% agarose gel, and light bars between 400 and 450 bp were chosen for sequencing. The libraries were constructed using the NEB Next Ultra DNA Library Pre®Kit for Illumina, index codes were added, and the library was quantified using Qubit and Q-PCR. After library identification, computational sequencing was performed using NovaSeq 6000.

2.6. Bioinformatic Analysis

As per the quality control procedures outlined in QIIME (Version 1.7.0) (http://qiime.org/scripts/split_libraries_fastq.html) accessed on 21 July 2024 [18], the original tags were quality-filtered under certain filtering conditions to obtain high-quality clean tags [19]. OTU clustering and taxonomic annotation sequence analysis were performed on all valid tags using UPARSE software (Version 20.0.0.36) accessed on 24 July 2024 [20], and sequences with a similarity of ≥97% were assigned to the same OUT. For every representative sequence, the SSU rRNA database from the SILVA138 repository (Version SILVA 138.1) (http://www.arb-silva.de/, accessed on 1 February 2025) was utilized accessed on 10 August 2024 [21] and annotated using the parental method, and species annotation was performed at each taxonomic level (threshold: 0.8–1) [22]. To efficiently analyze various sequences, we conducted phylogenetic diversity and linkage assessments on all representative OTU sequences utilizing MUSCLE software (Version 3.8.31) [23]. The abundance information of OTUs was standardized by utilizing the default sequence count related to the sample with the fewest sequences. Subsequent alpha and beta diversity assessments were conducted on the standardized data. All computations were executed in QIIME (version 1.7.0) and represented visually using R software (version 2.15.3).

2.7. Statistical Analysis

The experimental data were analyzed using GraphPad Prism 8.4.3. Student's *t*-test was used to compare the mean values between the control group and the experimental group. If $p < 0.05$, the difference was considered significant. Experimental results are expressed as mean ± SEM.

3. Results

3.1. Construction of AS Model Induced by High-Salt Diet

To induce AS, sodium chloride concentrations of 0.1%, 0.2%, and 0.3% were added to the drinking water of broiler chickens beginning at 8 days of age. At 28 days of age, the broiler chickens were euthanized under anesthesia, and their growth performance and cecal microbial diversity were assessed (Figure 1A). There was no significant difference in the body weight of the 28-day-old chickens between the 0.1% sodium chloride group and the control group. However, the addition of 0.2% and 0.3% sodium chloride to the broilers' drinking water significantly impacted the weight gain of broiler chickens. The results indicated that the 0.2% sodium chloride group experienced a weight reduction of 22.57% compared to the control group. Furthermore, the 0.3% sodium chloride group demonstrated a more pronounced weight reduction of 37.10% (Figure 1B). This study further analyzed the incidence of ascites, revealing that the addition of 0.2% and 0.3% sodium chloride to drinking water resulted in ascites rates of 60% and 70%, respectively. In contrast, the inclusion of 0.1% sodium chloride did not induce ascites in broiler chickens (Figure 1C). Finally, the ascites heart index (AHI) was evaluated, revealing that 0.2% and 0.3% sodium chloride significantly increased the AHI compared to the control group. In contrast, 0.1% sodium chloride did not induce any changes in the broilers (Figure 1D). Therefore, when we subsequently analyzed the impact of AS on intestinal microorganisms, we used the 0.2% sodium chloride group as the experimental group.

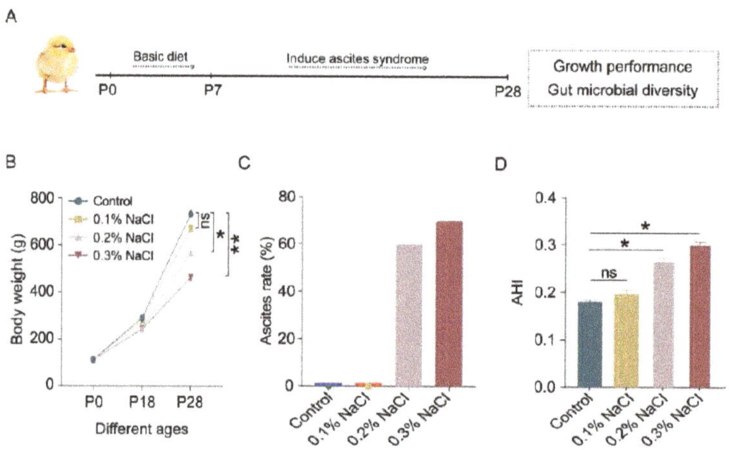

Figure 1. Construction of AS model induced by a high-salt diet. (**A**) Schematic of experimental design pattern. (**B**) Effects of different concentrations of sodium chloride on changes in body weight of broiler chickens. (**C**) Effects of different concentrations of sodium chloride on ascites rate in broiler chickens. (**D**) Effects of different concentrations of sodium chloride on AHI in broiler chickens. The experiment was performed in three biological replicates. All data are presented as mean ± SEM. The probability of significant differences was analyzed using Student's *t*-test. * $p \leq 0.05$, ** $p \leq 0.01$ compared with the control, ns means $p > 0.05$.

3.2. AS Reduces Cecal Microbial Diversity in Broiler Chickens

The rarefaction curves demonstrated a trend toward saturation, suggesting that at the 97% similarity threshold, the coverage of microbiome sequences in the two sample groups was adequate to assess the richness and diversity of the microbial community. The Chao index is a diversity estimation metric that relies on the abundance distribution of individual species within a sample. The Shannon index, grounded in information theory, serves as a measurement of species richness and evenness in an ecosystem. Faith's phylogenetic diversity (Faith-pd) is a diversity index derived from phylogenetic trees. Consequently, this study employed these three indices to analyze the rarefaction curve. The findings indicate that the rarefaction curve levels off and approaches stability, suggesting that the quantity of OTUs acquired for every sample is adequate (Figure 2A). To analyze the effects of AS on the microbial community, principal component analysis (PCA) and principal coordinate analysis (PCoA) were employed to assess the distribution of microorganisms between the two groups. The results showed that the composition of cecal microorganisms was significantly different between the AS group and the control group (Figure 2B). The differences in cecal microbiota abundance between the two groups were further analyzed, and the Venn results indicated that the AS group contained 2198 OTUs, while the control group had 3110 OTUs. Notably, there was a total of 542 OTUs shared between the two groups (Figure 2C). The above results indicate that AS significantly reduces the diversity of broiler cecal microbiota.

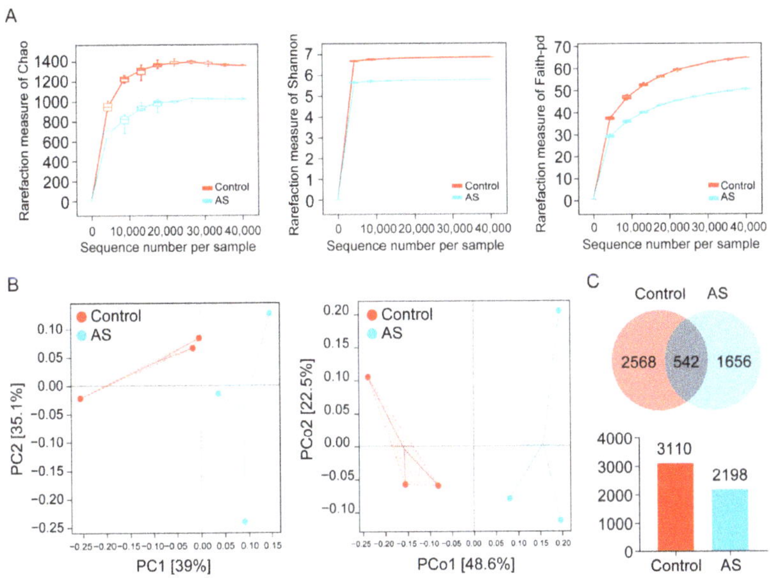

Figure 2. Analysis of the impact of AS on the number of cecal microorganisms. (**A**) Rarefaction curve plot of sequence number per sample group, such as Chao, Shannon, and Faith-pd. (**B**) Genus-level PCA of operational taxonomic units from sample groups and genus-level PCoA plots of microbial community dissimilarity (Bray–Curtis) in each sample-weighted uniFrac. Different colors represent different sample groups. (**C**) Venn diagrams on OTU level.

3.3. AS Reduces Alpha Diversity of Broiler Cecal Microbiota

Alpha diversity represents species richness, evenness, and diversity. This study selected the Shannon, Simpson, and Pielou-e indices for diversity analysis. The results showed that the alpha diversity of the cecal microbiota of the broiler chickens in the AS group was significantly reduced compared with that of the control group (Figure 3A–C).

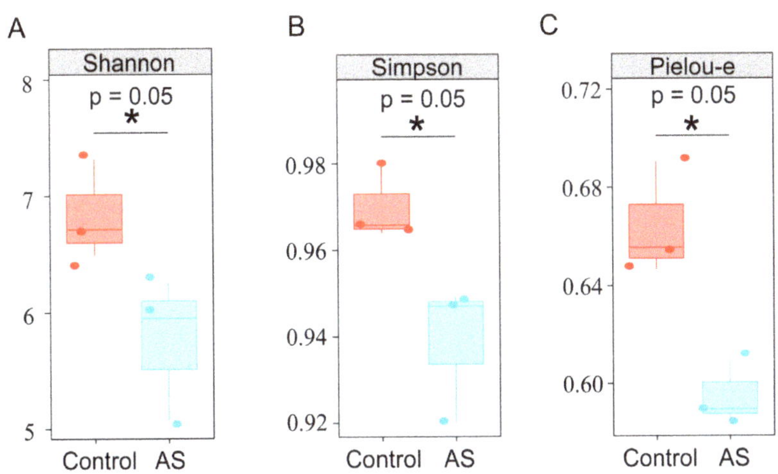

Figure 3. The alpha diversity in cecal microbiota. (**A**) Shannon index on OUT level. (**B**) Simpson index on OUT level. (**C**) Pielou-e index on OUT level. Statistical differences in the average values of each indicator were analyzed through independent *t*-tests. * indicates $p \leq 0.05$.

3.4. AS Affects the Composition of Cecal Microorganisms in Broiler Chickens

Changes in the composition of intestinal microbiota are among the primary causes of intestinal microbiota imbalance. Consequently, this study further investigated the effect of AS on the composition of cecal microorganisms in broiler chickens. Histogram analysis showed significant differences in the taxonomic composition of the cecal microorganisms when comparing the AS group to the control group at both the phylum (Figure 4A) and genus (Figure 4B) levels. The results indicated that Bacteroidetes and Firmicutes were the predominant bacterial phyla in the microbial communities of both the AS and control groups, with *Barnesiella* and *Lactobacillus* identified as the dominant genera. Further, LEfSe analysis was performed, and the results showed that Bacteroidetes was significantly increased in the AS group, while Actinobacteria was significantly decreased (Figure 4C). In addition, the abundance of *Dysgonomonas* and *Burkholderia* in the cecum of AS broiler chickens at the genus level was significantly increased, while the abundance of *Melissococcus, Oscillospira, Ruminococcus, Lysobacter, Christensenella, Anaerofustis, Slackia*, and *Olsenella* was significantly decreased (Figure 4D). This shows that AS significantly affects the composition of cecal microorganisms in broiler chickens.

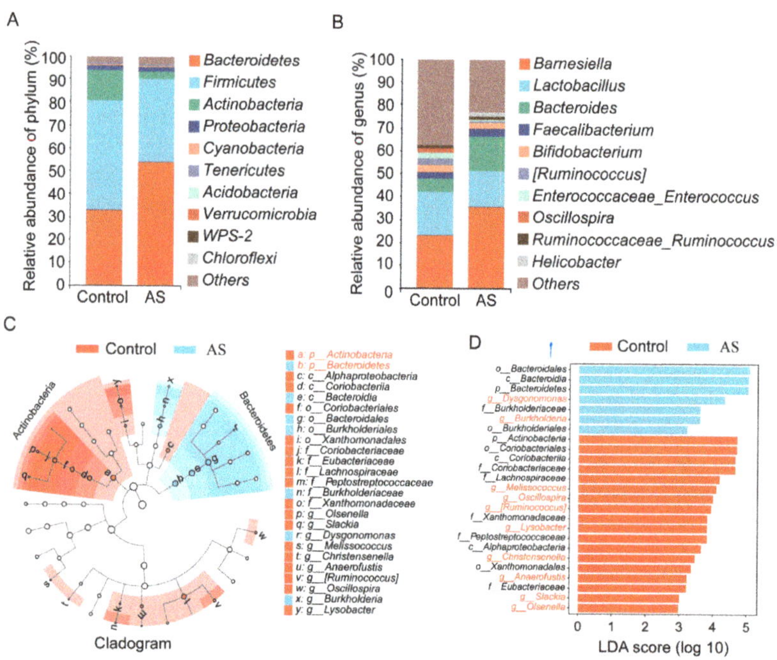

Figure 4. Analysis of the impact of AS on the composition of cecal microbiota. (**A**,**B**) Relative abundance of cecal microbiota at phylum and genus levels. (**C**) Phylogenetic profiles of specific bacterial taxa and dominant bacteria in two distinct groups were determined using LEfSe analysis. Biomarker groups are represented by colored circles and shaded areas. The diameter of each circle depends on the abundance of the taxon in the community. (**D**) Differential microbial enrichment between the control and AS groups is illustrated, with the red and blue horizontal color bars representing the microbial enrichment in the cecum of broiler chickens for the intact and AS groups, respectively.

3.5. Differential Metabolic Pathway Analysis

Based on the PICRUSt2 (Phylogenetic Investigation of Communities by Reconstruction of Unobserved States) differential microbiota function prediction method, the differential metabolic pathways of the two groups of cecal flora were analyzed. The results showed that the metabolites of AS broiler cecal microorganisms in novobiocin biosynthesis, polycyclic aromatic hydrocarbon degradation, the phosphotransferase system (PTS), bisphenol degradation, and linoleic acid metabolism pathways were significantly reduced, while the metabolites in beta-alanine metabolism, ubiquinone and another terpenoid–quinone biosynthesis, polyketide sugar unit biosynthesis, sphingolipid metabolism, and non-homologous end-joining pathways were significantly increased, especially the metabolites in novobiocin biosynthesis and non-homologous end-joining pathways (Figure 5). The above results indicate that AS affects intestinal health by changing the composition of the broiler cecal microbiota and affecting its metabolites.

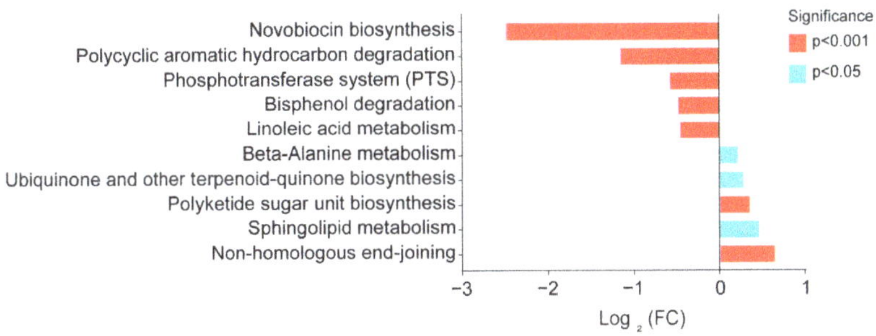

Figure 5. Analysis of differential metabolic pathways in the AS group compared with the control group. FC represents the change in metabolite abundance in the AS group compared with the control group.

4. Discussion

AS is a metabolic disease of broiler chickens caused by many factors, which seriously affect the growth and development of broiler chickens and cause significant economic losses in broiler chicken production [24]. Studies have shown that AS changes the normal histological structure of the heart, lungs, and liver of broiler chickens, thereby affecting their growth and development [25,26]. The intestinal microbiota, as the main component of the intestine, regulates normal digestion and absorption function [27]. However, the impact of AS on the intestinal health of broiler chickens, especially on the intestinal microbiota, requires further study.

To investigate the impact of AS on the composition of intestinal microbiota in broiler chickens, this study initially introduced varying concentrations of sodium chloride into the drinking water of the chickens to induce AS. Following this, 16S rRNA sequencing technology was employed to analyze the diversity of the intestinal microbiota. The results showed that sodium chloride concentrations of 0.2% and 0.3% significantly induced AS in broiler chickens. Guo et al. added 0.12% sodium chloride to drinking water to induce ascites syndrome in broiler chickens [28]. Consequently, this study further investigated the impact of AS induced by 0.2% sodium chloride on the cecal microbial diversity of broiler chickens.

The intestinal microbiota helps maintain the physiological structure and function of the intestine and is closely related to intestinal inflammation, barrier function, and host growth performance [29]. The composition of the intestinal microbiota is affected by different factors such as pathological conditions, genetics, environment, age, and diet, which are expected to play a crucial role in growth, immunity, and nutrient absorption [30]. The host diet significantly affects the intestinal microbiota, which in turn affects the host metabolism [31]. Therefore, this study further analyzed the effect of high-salt-diet-induced AS on the cecal intestinal microbiota of broiler chickens.

Numerous studies have shown that abnormal changes in the intestinal microbial community exist in AS patients and various animal models [32]. This study shows that AS significantly reduces the cecal microbial diversity of broiler chickens. Research shows that higher intestinal microbial diversity is beneficial to intestinal health [33]. This shows that AS affects intestinal health by reducing the microbial diversity of the cecum of broiler chickens. Alpha diversity is indicative of the distribution and abundance of microbial populations. The diversity, uniformity, and richness of microorganisms serve as important indicators influencing intestinal health [34]. Recent studies have demonstrated that AS increases the abundance of *Firmicutes, Proteobacteria, Actinobacteria, Firmicutes-Clostridium,* and *Gammaproteo* bacteria in the intestinal tract of rats while concurrently reducing the abundance of *Bacteroidetes, Spirochaetes, Bacteroidia, Spirochaetia,* and *Bacilli* [35]. This study demonstrates that AS reduces the alpha diversity of broiler cecal microorganisms, with the Shannon, Simpson, and Pielou-e indices showing marked reductions compared to the control group. The analysis of the cecal microbial composition found that AS changed the cecal microbial composition of broiler chickens at the phylum and genus levels. The LEfSe analysis results showed that AS reduced the abundance of *Actinobacteria* and increased the abundance of *Bacteroidetes* at the phylum level. There exists a gap between this result and the aforementioned findings, which may be attributed to the varying effects of AS on the intestinal flora across different species. *Firmicutes* and *Bacteroidetes* are the predominant bacterial phyla in the intestines of animals. *Bacteroidetes'* primarily function is to degrade carbohydrates, and their increased abundance is negatively correlated with the daily weight gain of animals [36]. AS may increase the abundance of cecal *Bacteroidetes*, thereby affecting the weight gain of broiler chickens. The intestinal microbiota regulates various physiological functions of the host by synthesizing and secreting specific metabolites [37]. Therefore, this study used bioinformatics prediction to further analyze differential metabolic pathways, and the results showed that AS significantly reduced metabolites related to polycyclic aromatic hydrocarbon degradation pathways. Polycyclic aromatic hydrocarbons are toxic organic pollutants that seriously affect the health of humans and animals, and bacteria have a strong ability to degrade polycyclic aromatic hydrocarbons [38]. It is suggested that AS may reduce bacterial abundance related to the degradation of PAH compounds in the cecum, thereby further affecting the intestinal health of broiler chickens.

This study shows that AS affects the diversity and abundance of broiler cecal microbiota. Thus, the gut microbiome can be used as a target to predict and improve AS in broiler chickens. These findings may provide valuable information for future studies on the impact of AS on intestinal microbial imbalance in broiler chickens. However, further research is needed on the interaction between cecal microorganisms and ascites syndrome.

5. Conclusions

In conclusion, this study demonstrated that AS significantly influences the diversity and composition of cecal microorganisms in broilers. This study provides foundational data for subsequent research on the interaction between intestinal microbiota and AS.

Author Contributions: Conceptualization, A.R. and X.F.; methodology, A.R.; software, Z.W.; validation, Z.W., R.G. and J.H. (Jiang He); formal analysis, A.R.; investigation, X.F.; resources, X.F.; data curation, J.H. (Juncheng Huang); writing—original draft preparation, A.R.; writing—review and editing, Z.W.; visualization, X.F.; supervision, C.S.; project administration, M.Q.; funding acquisition, X.F. All authors have read and agreed to the published version of the manuscript.

Funding: This work was supported by the National Natural Science Foundation of China [grant number 32060785] and a college student innovation and entrepreneurship training program project [grant number 202301].

Institutional Review Board Statement: The study was conducted in accordance with the Regulations for the Administration of Affairs Concerning Experimental Animals of the People's Republic of China and following the Research Ethics Committee of Tarim University (2023025).

Informed Consent Statement: Not applicable.

Data Availability Statement: The raw data supporting the conclusions of this article will be made available by the authors on request.

Conflicts of Interest: The authors declare no conflicts of interest.

References

1. Khajali, F. Managing broiler production challenges at high altitude. *Veter. Med. Sci.* **2022**, *8*, 1519–1527. [CrossRef] [PubMed]
2. Yu, J.; Liu, X.; Wang, K.; Wang, H.; Han, Y.; Kang, J.; Deng, R.; Zhou, H.; Duan, Z. Underlying mechanism of Qiling Jiaogulan Powder in the treatment of broiler ascites syndrome. *Poult. Sci.* **2023**, *102*, 102144. [CrossRef] [PubMed]
3. Wideman, R.F.; Rhoads, D.D.; Erf, G.F.; Anthony, N.B. Pulmonary arterial hypertension (ascites syndrome) in broilers: A review. *Poult. Sci.* **2013**, *92*, 64–83. [CrossRef] [PubMed]
4. Du, X.; Li, F.; Kong, F.; Cui, Z.; Li, D.; Wang, Y.; Zhu, Q.; Shu, G.; Tian, Y.; Zhang, Y.; et al. Altitude-adaption of gut microbiota in Tibetan chicken. *Poult. Sci.* **2022**, *101*, 101998. [CrossRef]
5. Wang, L.; Fu, G.; Liu, S.; Li, L.; Zhao, X. Effects of oxygen levels and a Lactobacillus plantarum strain on mortality and immune response of chickens at high altitude. *Sci. Rep.* **2019**, *9*, 16037. [CrossRef]
6. Bhagat, N.R.; Chauhan, P.; Verma, P.; Mishra, A.; Bharti, V.K. High-altitude and low-altitude adapted chicken gut-microbes have different functional diversity. *Sci. Rep.* **2023**, *13*, 20856. [CrossRef]
7. Zhang, M.; Li, D.; Yang, X.; Wei, F.; Wen, Q.; Feng, Y.; Jin, X.; Liu, D.; Guo, Y.; Hu, Y. Integrated multi-omics reveals the roles of cecal microbiota and its derived bacterial consortium in promoting chicken growth. *mSystems* **2023**, *8*, e0084423. [CrossRef]
8. Zhang, C.; Xi, Y.; Zhang, Y.; He, P.; Su, X.; Li, Y.; Zhang, M.; Liu, H.; Yu, X.; Shi, Y. Causal effects between gut microbiota and pulmonary arterial hypertension: A bidirectional Mendelian randomization study. *Heart Lung* **2024**, *64*, 189–197. [CrossRef]
9. Sharma, R.K.; Oliveira, A.C.; Yang, T.; Kim, S.; Zubcevic, J.; Aquino, V.; Lobaton, G.O.; Goel, R.; Richards, E.M.; Raizada, M.K. Pulmonary arterial hypertension-associated changes in gut pathology and microbiota. *ERJ Open Res.* **2020**, *6*, 00253-2019. [CrossRef]
10. Mootane, M.E.; Mafuna, T.; Ramantswana, T.M.; Malatji, D.P. Microbial community profiling in intestinal tract of indigenous chickens from different villages. *Sci. Rep.* **2024**, *14*, 21218. [CrossRef]
11. Śliżewska, K.; Markowiak-Kopeć, P.; Żbikowski, A.; Szeleszczuk, P. The effect of synbiotic preparations on the intestinal microbiota and her metabolism in broiler chickens. *Sci. Rep.* **2020**, *10*, 4281. [CrossRef] [PubMed]
12. Rychlik, I. Composition and Function of Chicken Gut Microbiota. *Animals* **2020**, *10*, 103. [CrossRef] [PubMed]
13. Chen, P.; Xu, T.; Zhang, C.; Tong, X.; Shaukat, A.; He, Y.; Liu, K.; Huang, S. Effects of Probiotics and Gut Microbiota on Bone Metabolism in Chickens: A Review. *Metabolites* **2022**, *12*, 1000. [CrossRef] [PubMed]
14. Kikusato, M. Phytobiotics to improve health and production of broiler chickens: Functions beyond the antioxidant activity. *Anim. Biosci.* **2021**, *34*, 345–353. [CrossRef] [PubMed]
15. Dittoe, D.K.; Olson, E.G.; Ricke, S.C. Impact of the gastrointestinal microbiome and fermentation metabolites on broiler performance. *Poult. Sci.* **2022**, *101*, 101786. [CrossRef]
16. Rinttilä, T.; Apajalahti, J. Intestinal microbiota and metabolites—Implications for broiler chicken health and performance. *J. Appl. Poult. Res.* **2013**, *22*, 647–658. [CrossRef]
17. Jian, C.; Luukkonen, P.; Yki-Järvinen, H.; Salonen, A.; Korpela, K. Quantitative PCR provides a simple and accessible method for quantitative microbiota profiling. *PLoS ONE* **2020**, *15*, e0227285. [CrossRef]
18. Caporaso, J.G.; Kuczynski, J.; Stombaugh, J.; Bittinger, K.; Bushman, F.D.; Costello, E.K.; Fierer, N.; Gonzalez Peña, A.; Goodrich, J.K.; Gordon, J.I.; et al. QIIME allows analysis of high-throughput community sequencing data. *Nat. Methods* **2010**, *7*, 335–336. [CrossRef]
19. Bokulich, N.A.; Subramanian, S.; Faith, J.J.; Gevers, D.; Gordon, J.I.; Knight, R.; Mills, D.A.; Caporaso, J.G. Quality-filtering vastly improves diversity estimates from Illumina amplicon sequencing. *Nat. Methods* **2013**, *10*, 57–59. [CrossRef]
20. Edgar, R.C. UPARSE: Highly accurate OTU sequences from microbial amplicon reads. *Nat. Methods* **2013**, *10*, 996–998. [CrossRef]

21. Wang, Q.; Garrity, G.M.; Tiedje, J.M.; Cole, J.R. Naïve Bayesian Classifier for Rapid Assignment of rRNA Sequences into the New Bacterial Taxonomy. *Appl. Environ. Microbiol.* **2007**, *73*, 5261–5267. [CrossRef] [PubMed]
22. Quast, C.; Pruesse, E.; Yilmaz, P.; Gerken, J.; Schweer, T.; Yarza, P.; Peplies, J.; Glöckner, F.O. The SILVA ribosomal RNA gene database project: Improved data processing and web-based tools. *Nucleic Acids Res.* **2012**, *41*, D590–D596. [CrossRef] [PubMed]
23. Edgar, R.C. MUSCLE: Multiple sequence alignment with high accuracy and high throughput. *Nucleic Acids Res.* **2004**, *32*, 1792–1797. [CrossRef] [PubMed]
24. Parveen, A.; Jackson, C.D.; Dey, S.; Tarrant, K.; Anthony, N.; Rhoads, D.D. Identification and validation of quantitative trait loci for ascites syndrome in broiler chickens using whole genome resequencing. *BMC Genet.* **2020**, *21*, 54. [CrossRef] [PubMed]
25. Sakumi, A.; Yamaguchi, R.; Tottori, J.; Uchida, K.; Tateyama, S. Liver capsule thickening characterized by mesothelial cell proliferation with vascularization in broilers ascites syndrome. *Avian Pathol.* **1996**, *25*, 147–153. [CrossRef]
26. Olkowski, A.A.; Rathgeber, B.M.; Sawicki, G.; Classen, H.L. Ultrastructural and molecular changes in the left and right ventricular myocardium associated with ascites syndrome in broiler chickens raised at low altitude. *J. Veter. Med.* **2001**, *48*, 1–14. [CrossRef]
27. Iqbal, Y.; Cottrell, J.J.; Suleria, H.A.R.; Dunshea, F.R. Gut Microbiota-Polyphenol Interactions in Chicken: A Review. *Animals* **2020**, *10*, 1391. [CrossRef]
28. Guo, D.; Zhang, J.; Han, Y.; Cui, L.; Wang, H.; Wang, K.; Li, P.; Deng, R.; Kang, J.; Duan, Z. Transcriptomic Study on the Lungs of Broilers with Ascites Syndrome. *Animals* **2023**, *13*, 175. [CrossRef]
29. Wang, W.; Li, Z.; Han, Q.; Guo, Y.; Zhang, B.; D'inca, R. Dietary live yeast and mannan-oligosaccharide supplementation attenuate intestinal inflammation and barrier dysfunction induced by *Escherichia coli* in broilers. *Br. J. Nutr.* **2016**, *116*, 1878–1888. [CrossRef]
30. Corrigan, A.; de Leeuw, M.; Penaud-Frézet, S.; Dimova, D.; Murphy, R.A. Phylogenetic and Functional Alterations in Bacterial Community Compositions in Broiler Ceca as a Result of Mannan Oligosaccharide Supplementation. *Appl. Environ. Microbiol.* **2015**, *81*, 3460–3470. [CrossRef]
31. Manor, O.; Dai, C.L.; Kornilov, S.A.; Smith, B.; Price, N.D.; Lovejoy, J.C.; Gibbons, S.M.; Magis, A.T. Health and disease markers correlate with gut microbiome composition across thousands of people. *Nat. Commun.* **2020**, *11*, 5206. [CrossRef] [PubMed]
32. Wu, P.; Zhu, T.; Tan, Z.; Chen, S.; Fang, Z. Role of Gut Microbiota in Pulmonary Arterial Hypertension. *Front. Cell. Infect. Microbiol.* **2022**, *12*, 812303. [CrossRef] [PubMed]
33. Clemente, J.C.; Ursell, L.K.; Parfrey, L.W.; Knight, R. The Impact of the Gut Microbiota on Human Health: An Integrative View. *Cell* **2012**, *148*, 1258–1270. [CrossRef] [PubMed]
34. Choi, K.Y.; Lee, T.K.; Sul, W.J. Metagenomic Analysis of Chicken Gut Microbiota for Improving Metabolism and Health of Chickens—A Review. *Asian-Australasian J. Anim. Sci.* **2015**, *28*, 1217–1225. [CrossRef]
35. Hong, W.; Mo, Q.; Wang, L.; Peng, F.; Zhou, Y.; Zou, W.; Sun, R.; Liang, C.; Zheng, M.; Li, H.; et al. Changes in the gut microbiome and metabolome in a rat model of pulmonary arterial hypertension. *Bioengineered* **2021**, *12*, 5173–5183. [CrossRef]
36. Ban-Tokuda, T.; Maekawa, S.; Miwa, T.; Ohkawara, S.; Matsui, H. Changes in faecal bacteria during fattening in finishing swine. *Anaerobe* **2017**, *47*, 188–193. [CrossRef]
37. Tilg, H.; Zmora, N.; Adolph, T.E.; Elinav, E. The intestinal microbiota fuelling metabolic inflammation. *Nat. Rev. Immunol.* **2019**, *20*, 40–54. [CrossRef]
38. Ghosal, D.; Ghosh, S.; Dutta, T.K.; Ahn, Y. Current State of Knowledge in Microbial Degradation of Polycyclic Aromatic Hydrocarbons (PAHs). *Front. Microbiol.* **2016**, *7*, 1369. [CrossRef]

veterinary sciences

MDPI

Article

Glycerol Monolaurate Complex Improved Antioxidant, Anti-Inflammation, and Gut Microbiota Composition of Offspring in a Sow–Piglet Model

Dan Li [1,†], Min Yang [2,†], Zhao Ma [1], Lianqiang Che [1], Bin Feng [1], Zhengfeng Fang [1], Shengyu Xu [1], Yong Zhuo [1], Jian Li [1], JiHhua Wang [3], Zhengfan Zhang [4], Zehui Wu [5], Tao Lin [1], De Wu [1] and Yan Lin [1,*]

1 Key Laboratory of Animal Disease-Resistance Nutrition and Feed Science, Institute of Animal Nutrition, Sichuan Agricultural University, Chengdu 611130, China; xiaomila99@outlook.com (D.L.); danlidadan@163.com (Z.M.); che.lianqiang@sicau.edu.cn (L.C.); fengbin@sicau.edu.cn (B.F.); zfang@sicau.edu.cn (Z.F.); shengyu_x@hotmail.com (S.X.); zhuoyong@sicau.edu.cn (Y.Z.); 14109@sicau.edu.cn (J.L.); lt04161211@163.com (T.L.); wude@sicau.edu.cn (D.W.)

2 Pet Nutrition and Health Research Center, Chengdu Agricultural College, Chengdu 611130, China; yang040101@yeah.net

3 Calid Biotech (Wuhan) Co., Ltd., Wuhan 430073, China; m18483617169@163.com

4 Hubei Key Laboratory of Animal Nutrition and Feed Science, Wuhan Polytechnic University, Wuhan 430023, China; zhangzhengfan@163.com

5 Sichuan Qiaozhu'er Breeding Co., Ltd., Neijiang 641100, China; wuzehui707@163.com

* Correspondence: linyan@sicau.edu.cn

† These authors contributed equally to this work.

Simple Summary: Antibiotic additives have significant effects in relieving the stress of sow delivery and weaning piglets and improving disease resistance in animals. However, the abuse of antibiotics led to bacterial resistance. The purpose of this study was to investigate the effects of GML on the reproductive performance of sows and the growth performance of piglets in late gestation and lactation diets and to preliminarily explore its possible mechanism to replace antibiotics. In conclusion, 0.2% GML supplementation improved the reproductive performance of sows and the growth performance of piglets during late gestation and lactation by improving the antioxidant capacity, anti-inflammatory action, and gut microflora balance.

Abstract: This study aimed to investigate the effects of maternal glycerol monolaurate complex (GML) and antibiotic (acetylisovaleryltylosin tartrate, ATLL) supplementation during late gestation and lactation on the reproductive performance of sows and the growth performance of piglets. In total, 64 pregnant sows were randomly divided into control, antibiotic, 0.1% GML, and 0.2% GML groups. The GML shortened their delivery interval and farrowing duration. The ATLL increased the level of malondialdehyde (MDA) in sows and piglets and enhanced glutathione peroxidase (GSH-Px) in piglets, while reducing the tumor necrosis factor-α (TNF-α) level in sows. The GML tended to increase milk protein in the colostrum and decreased the TNF-α of sows at lactation. Meanwhile, 0.2% GML increased the serum total superoxide dismutase (T-SOD) activity and interleukin-6 level in weaned piglets and decreased the TNF-α level in sows and weaned piglets. Furthermore, ATLL decreased the microbial diversity of sows, and GML tended to increase the microbial diversity of sows and piglets. The ATLL group had an increased relative abundance of Bacteroidota in weaned piglets. The GML decreased the relative abundance of Peptostreptococcales-Tissierellales, Proteobacteria, and the harmful bacteria *Romboutsia* in sows. Compared with the ATLL group, the 0.2% GML reduced the relative abundance of Bacteroidota in weaned piglets. Interestingly, both ATLL and GML supplementation decreased the relative abundance of harmful bacteria Peptostreptococcaceae in sows. Correlation analysis also found positive effects of ATLL and GML in anti-inflammatory and antioxidant aspects. In conclusion, GML enhanced reproductive and growth performance by improving antioxidant and anti-inflammatory status and maintaining intestinal flora balance, making it a promising alternative to ATLL in future applications.

Keywords: sows; piglets; glycerol monolaurate complex; fecal microbial; reproductive

1. Introduction

The reproductive performance of sows is directly related to production efficiency and economic benefits of the swine industry [1]. With the dramatic changes in hormones and metabolism in the body, especially in the late gestation and early lactation stages, sows often suffer from fatigue, anorexia, constipation, long labor, weight loss, and other symptoms during lactation [2]. Antibiotic additives have significant effects in relieving the stress of sow delivery and weaning piglets and improving disease resistance in animals. However, the abuse of antibiotics has led to bacterial resistance [3], high antibiotic levels in livestock and poultry, as well as environmental pollution [4], and is a threat to human health [5,6]. European food unions and countries such as China have enforced a total ban on using antibiotics in livestock feed. This has undoubtedly increased the pressure on livestock farmers [7] and has prompted researchers to focus attention on biologically active natural alternatives to improve the reproductive performance of sows, along with the growth and health of piglets, by improving their intestinal and immune functions [5,8].

Glycerol monolaurate (C15H30O4, GML), a lipophilic medium-chain fatty acid ester naturally found in breast milk, palm oil, and coconut oil [9]. There are a variety of biological activities [10] and nutritional effects of GML [11]. More importantly, whereas most antibiotics have only a single bacterial target, GML has multiple cellular targets; therefore, use of GML does not easily result in development of resistance to drugs [10]. Studies have shown that GML can improve the daily gain of fattening cows [12], reduce the number of methanobacteria and protozoa in the rumen fluid of sheep [13], stabilize the intestinal flora, improve the growth performance of weaned piglets [14], and improve the feed conversion ratio (FCR), meat quality, intestinal morphology, and barrier function of broilers by enhancing animal immunity [15] and antioxidant balance, as well as promoting their intestinal flora health [16,17]. Additionally, GML is widely used in broilers to maintain the intestinal microecological balance and improve growth performance [5]. and in weaned piglets to increase beneficial intestinal bacteria and reduce diarrhea [18].

As bioactive plant compounds, essential oils contain mainly thymol and cinnamaldehyde (CIN). Research has demonstrated that thymol has beneficial effects on the growth of beneficial bacteria and inhibits the growth of potentially harmful bacteria, including Escherichia coli and Clostridium perfringens [19,20]. Also, CIN possesses bacteriostatic, anti-bacterial, and other functions [21]. It has been used to treat diabetes and diseases caused by Helicobacter pylori [22] by enhancing gastrointestinal peristalsis, maintaining intestinal microecological balance [23], and enhancing antioxidation [24], because its mechanism of action is similar to that of antibiotics [25]. Dietary supplementation with 0.75% GML and oregano essential oil can improve the intestinal structure and morphology, as well as growth performance of broiler chickens [5]. Wang et al. [26] found that dietary supplementation of 0.1% and 0.2% GML complex (mixture including GLM, CIN, and thymol) improved the growth performance of weaned piglets, and the addition of 0.2% GML complex also improved the antioxidant capacity. Zhao et al. [27] showed that maternal supplementation with 0.1% GML improved the composition of sow milk. These results indicate that GML supplementation not only promotes piglet growth and health but also benefits maternal milk composition and the intestinal health of offspring. Although studies on supplementing GML have increased in recent years, few have reported the effect of GML on sow reproductive performance. We hypothesized that GML has a positive effect on improving sow reproductive and piglet growth performance, and can be used as an alternative to natural antibiotics. The aim of this study was to investigate the effects of GML on sow reproductive performance, suckling piglet growth performance, blood indexes, antioxidant capacity, and fecal microorganisms, and to initially explore the possible mechanisms.

2. Materials and Methods

2.1. Animals and Diets

In total, 64 healthy pregnant Landrace × Yorkshire sows with similar parity and body condition (90 d of gestation, back fat 15.08 ± 3.41 mm) were divided into four treatment groups randomly: control (CON), antibiotic (0.3% acetylisovaleryltylosin tartrate, ATLL), 0.1% GML, and 0.2% GML, with 16 replications per group. The experiment lasted 45 d, from 90 d of gestation (G 90) to 21 d of lactation (L 21). The GML complex contained GML (800 g/kg) and essential oils CIN (54 g/kg) and thymol (6 g/kg), provided by CALID Biotechnology Co., Ltd. (Wuhan, China). Diets were according to the experimental design with reference to the nutritional requirements of sows in the NRC 2012 [28]. The nutritional parameters of the gestation and lactation diets is listed in Supplementary Table S1.

2.2. Management

During gestation, sows were individually housed in gestation stalls, with the temperature maintained at 20 °C. The sows received daily feed at 8:30 and 14:30, and were fed 2.8 kg/d during 90–110 d of gestation. At 110 d of gestation, sows were moved from the gestation stalls to farrowing rooms and were kept in individual farrowing crates thereafter with temperature maintained at 23 °C. The litter sizes were standardized to 10–12 piglets each at 24 h post-farrowing by cross-fostering within the same treatment. On the day of delivery, no feed was provided to the sows. Starting the day after delivery, they were given a lactation diet at 2.0 kg/d, which increased by 1 kg/d over the first 5 days, after which they had unrestricted access to the food. During lactation, sow's milk was the sole source of nutrition for the suckling piglets. Both sows and piglets had continuous access to water throughout the study. All piglets were weaned at 21 days of lactation.

2.3. Growth Performance

The back fat thickness of the sows was measured at the P2 position using a B-mode ultrasonic device (Renco Lean Meater type 7, Minneapolis, MN, USA) at 90 d of gestation and 21 d of lactation.

Stool consistency and fecal morphology were recorded at 08:00 and 20:00 during gestation. Fecal scores were calculated according to Jing et al. [29], in which the scoring range was 0–4. A score of 3 or above was judged as constipated, and the constipation rate was calculated repeatedly.

2.4. Samples Collection

On the day of weaning, six piglets from each treatment group were randomly selected for collection of blood samples from the front cavity vein. At 1 and 21 d of lactation, six sows from each treatment group were randomly selected for collection of blood samples from the ear margin vein. Blood samples were centrifuged for 10 min at 3000 rpm at 4 °C after natural coagulation to obtain serum. Serum samples were immediately stored at −20 °C for further analysis.

At 20 d of lactation, six piglets, selected from different litters and weighing close to the average, were chosen per group, and feces were collected using rectal massage. The collection of feces of sows was carried out in the same way. Fecal samples were stored at −80 °C and tested for microbial diversity for sows and piglets.

Colostrum (10 mL) was collected from six sows per group from the 4th or 5th pair of nipples after the sow had delivered five piglets and was stored at −20 °C. Normal milk samples were collected from six sows from each group at 20 d of lactation and stored at −20 °C.

2.5. Determination

2.5.1. Determination of Antioxidant Capacity and Cytokines

Total superoxide dismutase (T-SOD) and glutathione peroxidase (GSH-Px) activities, total antioxidant capacity (T-AOC), and malondialdehyde (MDA) content were determined using commercial kits (Jiancheng Bioengineering Institute, Nanjing, China). Cytokine levels—interleukin-1β (IL-1β), interleukin-6 (IL-6), interleukin-10 (IL-10), and tumor necrosis factor-α (TNF-α)—were measured using ELISA kits (Jiancheng Bioengineering Institute, Nanjing, China). The IgA content in colostrum was determined using a commercial kit (Jiancheng Bioengineering Institute, Nanjing, China).

2.5.2. Determination of Milk Composition

The milk fat, lactose, milk protein, and total solid contents were determined using mid-infrared spectroscopy (Foss MilkoScan FT+, Fossomatic FC, Hillerød, Denmark).

2.5.3. Analysis of Fecal Microbial Diversity

The number of microorganisms in feces of piglets and sows was determined using 16s rRNA. Bacterial genomic DNA was isolated from fecal samples of sows and piglets using a DNA extraction kit (Omega Biotek, Norcross, GA, USA) and quantified using Nanodrop (Thermo Fisher Scientific Inc., Waltham, MA, USA). The PCR amplification, purification, and product analysis were performed as described by Dai et al. (2022) [30]. The samples were analyzed by Majorbio Bio-Pharm Technology Corporation Ltd. (Shanghai, China). High-throughput sequencing was performed on a MiSeq sequencer using the Reagent Kit v3 according to the manufacturer's instructions (Illumina Inc., San Diego, CA, USA) on the Genomic and Transcriptomic Platform (INRAE, Toulouse, France). Species composition was analyzed according to Le et al. [31].

3. Statistical Analysis

This trial was fully randomized with sow or litter size as the test unit, the sow model with diet treatment as the fixed effect, and estrus time as a random effect. The piglet model used sow diet treatment as a fixed effect and maternal difference as a random effect. All data were expressed as means, and data analysis was performed using one-way analysis of variance and statistical product and service solutions (SAS) statistical software (V9.4, SAS Institute Inc., Cary, NC, USA), followed by a generalized linear model. In the results, $p < 0.01$ was considered highly significant, $p < 0.05$ was significant, and $0.05 < p < 0.10$ was a trend.

4. Results

4.1. Effects of GML on Back Fat and Fecal Scores of Sows

Maternal supplementation with GML had no significant effect on back fat loss or fecal scores of sows ($p > 0.05$; Table 1).

Table 1. Effects of GML complex on back fat loss and fecal score of sows.

Items	CON	ATLL	0.1% GML	0.2% GML	SEM	p-Value
			Sow Back Fat Thickness, mm			
Day 90 of Gestation	14.81	15.06	15.07	15.38	0.43	0.98
Farrowing, within 24 h	16.00	15.30	15.14	13.80	0.36	0.56
Day 21 of Lactation	14.64	15.33	14.43	13.03	0.35	0.69
Total Back Fat Loss	−1.47	−1.08	−1.14	−1.68	0.22	0.89
Fecal Scores	1.66	1.42	1.41	1.58	0.08	0.64

Data are presented as mean (sows: n = 16). CON, control group (basal diets), ATLL, antibiotic group (basal diets + 0.3% acetylisovaleryltylosin tartrate), 0.1% GML, glycerol monolaurate group (basal diets + 0.1% GML complex), 0.2% GML, 0.2% glycerol monolaurate group (basal diets + 0.2% GML complex).

4.2. Effects of GML on Sow Reproductive Performances

Supplementation with GML shortened the delivery interval of sows ($p < 0.05$), and 0.1% GML shortened the farrowing duration of sows ($p < 0.05$), but had no significant effect on litter size, litter weight, or average body weight ($p > 0.05$; Table 2). Both ATLL and GML groups showed a trend of decreased incidence of lactating diarrhea in piglets ($p < 0.1$).

Table 2. Effects of GML complex on delivery performance of sows.

Items	CON	ATLL	0.1% GML	0.2% GML	SEM	p-Value
Farrowing Duration, min	195.4 ^{b}	154 ^{ab}	120.53 ^{a}	185.81 ^{b}	9.26	0.01
Delivery Interval, min	23.6 ^{b}	19.59 ^{ab}	13.1 ^{a}	12.75 ^{a}	1.48	0.02
		Litter Size, n				
Total Born	11.69	13.13	12.80	13.73	0.46	0.29
Born Alive	11.00	11.75	11.73	12.57	0.43	0.45
Weaning	9.28	10.81	10.09	11.25	0.32	0.60
		Litter Weight, kg				
Day 0 of Birth	15.18	14.82	14.86	17.11	0.56	0.24
Day of Weaning	62.82	59.42	60.11	66.73	1.65	0.33
		Average Pig Body Weight, kg				
Day 0 of Birth	1.39	1.30	1.36	1.31	0.04	0.37
Day of Weaning	6.29	6.09	6.39	6.50	0.13	0.46
Fecal Score of Piglets	0.57	0.56	0.52	0.36	0.01	0.22
Diarrhea Rate of Piglets, %	1.46	1.14	0.92	0.84	0.32	0.08

Values are presented as mean (sows: n = 16). CON, control group (basal diets), ATLL, antibiotic group (basal diets + 0.3% acetylisovaleryltylosin tartrate), 0.1% GML, glycerol monolaurate group (basal diets + 0.1% GML complex), 0.2% GML, 0.2% glycerol monolaurate group (basal diets + 0.2% GML complex).

4.3. Effects of GML on Sow Feed Intake During Lactation

Compared to the ATLL group, 0.2% GML increased the feed intake for 1–7 d of lactation. Meanwhile, there was a trend for ATLL to reduce the feed intake of sows at 15–21 d and 1–21 d of lactation ($p < 0.1$) (Table 3).

Table 3. Effects of GML complex on feed intake of sows during lactating.

Items	CON	ATLL	0.1% GML	0.2% GML	SEM	p-Value
G90-D0, kg	2.88	2.87	2.84	2.86	0.01	0.21
L1-L7, kg	4.1 ^{ab}	3.9 ^{a}	4.07 ^{ab}	4.19 ^{b}	0.03	0.03
L8-L14, kg	6.27	5.79	5.9	6.19	0.09	0.2
L15-L21, kg	6.34	5.8	5.88	6.19	0.08	0.06
L1-L21, kg	5.57	5.16	5.27	5.51	0.06	0.07

Abbreviations: G90 = day 90 of gestation; D0 = day 0 of delivery; L1 = day 1 of lactation. Data are presented as mean (sows: n = 16). CON, control group (basal diets), ATLL, antibiotic group (basal diets + 0.3% acetylisovaleryltylosin tartrate), 0.1% GML, glycerol monolaurate group (basal diets + 0.1% GML complex), 0.2% GML, 0.2% glycerol monolaurate group (basal diets + 0.2% GML complex.

4.4. Effects of GML on Composition of Colostrum and Normal Milk of Sows

Maternal supplementation with GML had no significant effects on fat, lactose, or total solids in the colostrum and milk of sows but tended to increase milk protein in colostrum ($p < 0.1$; Table 4). Moreover, 0.2% GML increased milk IgA content by 21.46% in colostrum and fat content by 9% in milk at 20 d of lactation.

Table 4. Effects of GML complex on the composition of colostrum and normal milk of sows.

Items	CON	ATLL	0.1% GML	0.2% GML	SEM	*p*-Value
			Colostrum			
Fat, %	6.65	7.75	7.33	6.92	0.31	0.17
Protein, %	14.55	16.29	16.99	17.59	0.57	0.08
Lactose, %	2.75	2.65	2.40	2.48	0.08	0.40
Total Solids, %	28.56	31.75	30.99	30.94	0.76	0.33
IgA, mg/ml	4.80	3.26	4.49	5.83	0.56	0.46
			Milk on Day 20 of Lactation			
Fat, %	6.65	6.86	6.99	7.25	0.29	0.22
Protein, %	5.02	5.08	5.05	5.09	0.12	0.35
Lactose, %	5.52	5.25	5.25	5.22	0.07	0.65
Total Solids, %	19.32	20.60	19.93	20.10	0.33	0.54

Data are presented as mean (sows: n = 6). CON, control group (basal diets), ATLL, antibiotic group (basal diets + 0.3% acetylisovaleryltylosin tartrate), 0.1% GML, glycerol monolaurate group (basal diets + 0.1% GML complex), 0.2% GML, 0.2% glycerol monolaurate group (basal diets + 0.2% GML complex).

4.5. Effects of GML on Antioxidant Capacity of Sows and Piglets

The ATLL increased the MDA levels of sows at 21 d of lactation and the serum GSH-Px levels in weaning piglets ($p < 0.05$; Table 5). The GML had no significant effect on the serum antioxidant indices of sows; however, 0.2% GML increased the serum T-SOD levels in weaning piglets ($p < 0.05$).

Table 5. Effects of GML complex on serum antioxidant indices of sows and piglets.

Items	CON	ATLL	0.1% GML	0.2% GML	SEM	*p*-Value
			Day 1 of lactation			
T-AOC, U/mL	2.89	2.73	2.73	2.67	0.10	0.95
T-SOD, U/mL	78.02	75.18	74.37	76.23	0.76	0.38
MDA, nmol/mL	4.94	4.15	5.09	5.20	0.38	0.48
GSH-Px, U/mL	785.53	739.42	779.27	676.22	19.70	0.18
			Day 21 of lactation			
T-AOC, U/mL	3.01	3.38	3.29	3.02	0.14	0.73
T-SOD, U/mL	54.27	50.38	54.28	53.78	1.41	0.75
MDA, nmol/mL	3.94 ^{a}	5.13 ^{b}	4.59 ^{ab}	4.55 ^{ab}	0.19	0.04
GSH-Px, U/mL	667.60	558.86	660.51	567.36	30.10	0.58
			Weaning piglets			
T-AOC, U/mL	3.38	3.40	4.07	5.154	0.14	0.43
T-SOD, U/mL	23.73 ^{a}	38.25 ^{ab}	39.80 ^{ab}	49.53 ^{b}	3.08	0.02
MDA, nmol/mL	7.31	8.315	8.07	7.09	0.41	0.27
GSH-Px, U/mL	140.73 ^{a}	210.70 ^{b}	145.86 ^{a}	187.91 ^{ab}	10.29	0.03

^{a,b} Values in the same row with different small letter superscripts mean significant difference ($p < 0.05$). T-AOC, total antioxidant capacity; T-SOD, total peroxide dismutase; MDA, malondialdehyde; GSH-Px, glutathione peroxidase. Data are presented as mean (sows: n = 6; piglets: n = 6). CON, control group (basal diets), ATLL, antibiotic group (basal diets + 0.3% acetylisovaleryltylosin tartrate), 0.1% GML, glycerol monolaurate group (basal diets + 0.1% GML complex), 0.2% GML, 0.2% glycerol monolaurate group (basal diets + 0.2% GML complex).

4.6. Effects of GML on Cytokine Secretion of Sows and Piglets

As shown in Table 6, ATLL and GML supplementation decreased the TNF-α of sows at 1 and 21 d of lactation ($p < 0.05$). The 0.1% GML increased the IL-10 of sows at 21 d of lactation ($p < 0.05$) and tended to increase serum IL-10 concentration of weaning piglets ($p < 0.1$). The 0.2% GML supplementation decreased the TNF-α level and increased the IL-6 concentration of weaning piglets (both $p < 0.05$).

Table 6. Effects of GML complex on cytokines and immunoglobulins in sows and piglets.

Items	CON	ATLL	0.1% GML	0.2% GML	SEM	*p*-Value
			Day 1 of lactation			
IL-1β, ng/L	202.18	188.38	186.77	199.66	6.88	0.76
TNF-α, ng/L	264.53 [b]	203.27 [a]	198.09 [a]	220.67 [a]	16.73	0.05
IL-6, ng/L	61.94	63.10	56.69	48.49	3.45	0.45
IL-10, ng/L	124.73	122.59	123.50	123.69	5.24	1.00
			Day 21 of lactation			
IL-1β, ng/L	181.64	185.91	172.90	179.62	10.07	0.98
TNF-α, ng/L	367.28 [c]	322.02 [b]	338.63 [b]	303.53 [a]	34.14	0.05
IL-6, ng/L	60.69	58.77	68.40	59.23	4.03	0.84
IL-10, ng/L	124.19 [a]	129.44 [a]	154.39 [b]	140.41 [ab]	6.54	0.05
			Weaning piglets			
IL-1β, ng/L	68.58	85.12	77.80	64.51	4.16	0.31
TNF-α, ng/L	115.54 [b]	99.191 [b]	110.35 [b]	76.76 [a]	12.45	0.01
IL-6, ng/L	40.85 [a]	57.70 [ab]	29.35 [a]	85.64 [b]	6.65	0.01
IL-10, ng/L	66.76	58.98	77.42	51.39	4.23	0.08

[a,b,c] Values in the same row with different small letter superscripts mean significant difference ($p < 0.05$). Data are presented as mean (sows: n = 6; piglets: n = 6). CON, control group (basal diets), ATLL, antibiotic group (basal diets + 0.3% acetylisovaleryltylosin tartrate), 0.1% GML, glycerol monolaurate group (basal diets + 0.1% GML complex), 0.2% GML, 0.2% glycerol monolaurate group (basal diets + 0.2% GML complex).

4.7. Effects of GML on Gut Microbial Composition of Sows and Piglets

The Sobs index difference (Figure 1) showed that dietary supplementation with ATLL decreased the microbial diversity of sows compared with the CON group ($p < 0.05$; Figure 1A), whereas this increased in the 0.1% GML group ($p < 0.01$). However, this difference was not significant in piglets ($p > 0.05$; Figure 1B).

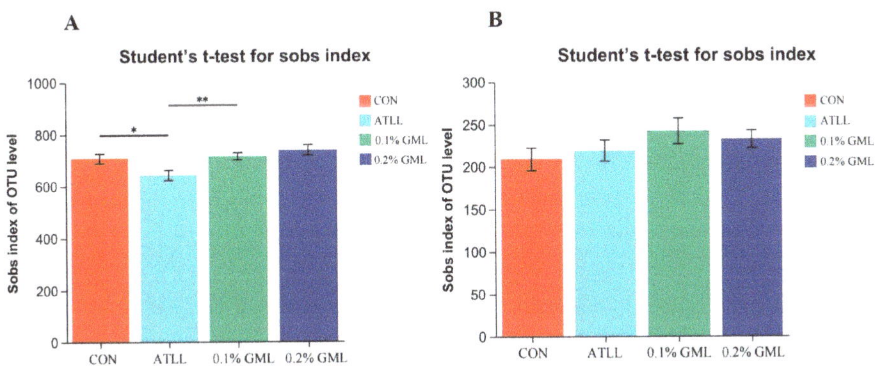

Figure 1. Analysis of microbial diversity of sows (**A**) and piglets (**B**). (**A**) Sows on day 20 of lactation. (**B**) Piglets on day of weaning. * $p < 0.05$, ** $p < 0.01$.

Principal coordinate analysis based on Bray–Curtis distance showed a different trend in sows and piglets ($p < 0.1$; Figure 2). The 0.1% and 0.2% GML treatments both tended to increase the microbial diversity of sows and piglets. This indicated that GML addition contributed to increased gut microbial diversity.

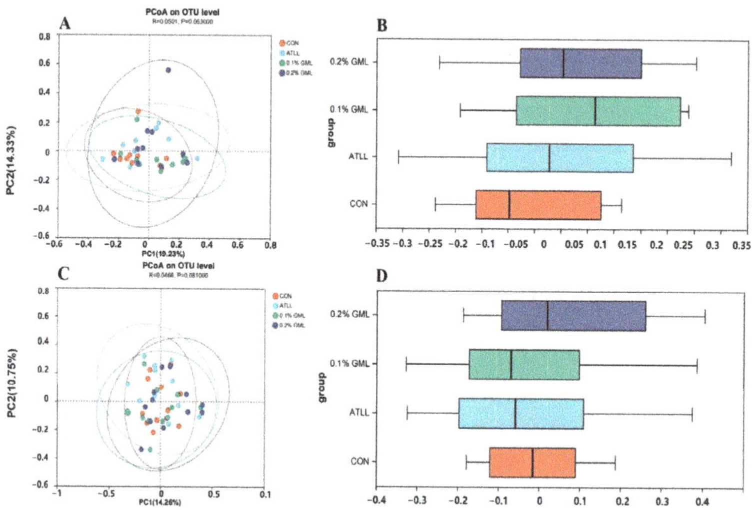

Figure 2. Effect of GML on intestinal microbial diversity of sows (**A**,**B**) and piglets (**C**,**D**). The different color legends represent different feeding feeds. CON, control group (basal diets), ATLL, antibiotic group (basal diets + 0.3% acetylisovaleryltylosin tartrate), 0.1% GML, glycerol monolaurate group (basal diets + 0.1% GML complex), 0.2% GML, 0.2% glycerol monolaurate group (basal diets + 0.2% GML complex).

Bacterial community heat map analysis and sample cluster tree analysis at the genus level for the species composition of sows and piglets in the treatment groups are shown in Figure 3. At the genus level, composition of microbes in sows in the CON and ATLL groups was similar but differed from that in the GML group. The dominant genera in the piglet fecal microbiota in the ATLL group differed from those of other treatment groups.

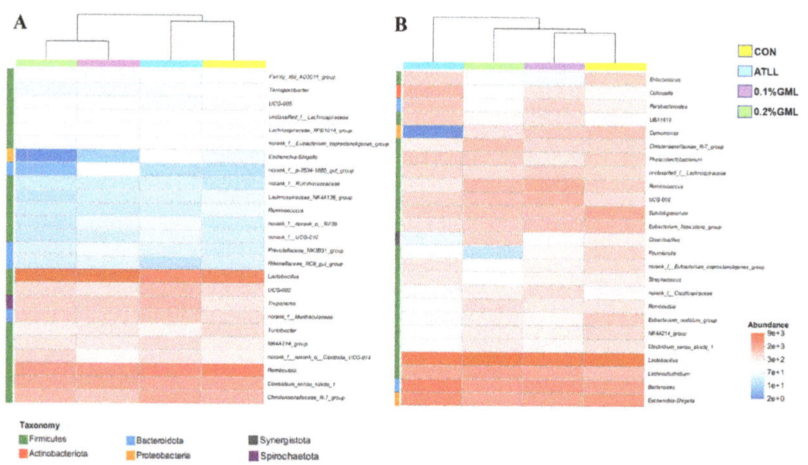

Figure 3. Bacterial community heat map analysis and sample cluster tree of genus level in sows (**A**) and piglets (**B**). CON, control group (basal diets), ATLL, antibiotic group (basal diets + 0.3% acetylisovaleryltylosin tartrate), 0.1% GML, glycerol monolaurate group (basal diets + 0.1% GML complex), 0.2% GML, 0.2% glycerol monolaurate group (basal diets + 0.2% GML complex). The abundance change of different species in the sample is displayed by the color gradient.

Significant differences among the top 10 dominant phylum (A), orders (B), families (C), and genera (D) in fecal samples of sows from different groups were determined (Figure 4). At the phylum level, addition of 0.1% and 0.2% GML decreased the relative abundance of Peptostreptococcales-Tissierellales ($p < 0.05$) and Proteobacteria ($p < 0.01$), respectively (Figure 4A). At the order level (Figure 4B), ATLL supplementation decreased the relative abundance of Lactobacillales ($p > 0.1$). However, the addition of 0.1% GML increased the relative abundance of Lactobacillales ($p > 0.1$)) and significantly decreased the relative abundance of Peptostreptococcales-Tissierellales ($p < 0.01$). At the family level, ATLL and GML supplementation both decreased the relative abundance of Peptostreptococcaceae ($p < 0.05$; Figure 4C). At the genus level, GML supplementation decreased the relative abundance of *Romboutsia* ($p < 0.05$; Figure 4D).

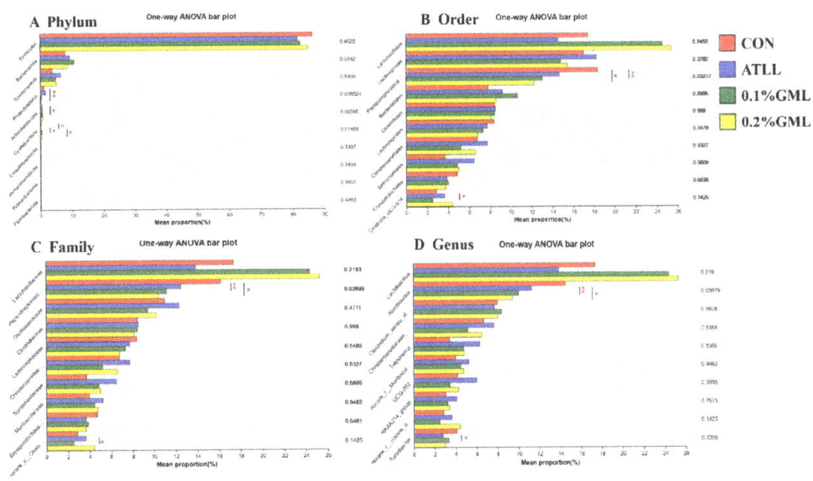

Figure 4. Analysis of group significant differences test among the top ten dominant phyla (**A**), orders (**B**), families (**C**), and genera (**D**) in fecal samples of sows. CON, control group (basal diets), ATLL, an-tibiotic group (basal diets + 0.3% acetylisovaleryltylosin tartrate), 0.1% GML, glycerol monolaurate group (basal diets + 0.1% GML complex), 0.2% GML, 0.2% glycerol monolaurate group (basal diets + 0.2% GML complex). The species names at the taxonomic level are shown on the *y* axis, while the *x* axis represents the average relative abundance. * $p < 0.05$, ** $p < 0.01$.

Changes in the gut microbiota of mothers may affect the microbial composition of their offspring. As shown in Figure 5A, at the phylum level, the changes in microorganisms in piglets were similar to those in sows. The ATLL group significantly increased relative abundance of Bacteroidota ($p < 0.05$). Treatment with 0.2% GML significantly decreased the relative abundance of Actinobacteria ($p < 0.05$). It is worth mentioning that compared with the ATLL group, the 0.2% GML treatment significantly reduced the relative abundance of Bacteroidota ($p < 0.05$). At the order level (Figure 5B), the ATLL group showed a significant increase in the relative abundance of Bacteroidales ($p < 0.05$), however, the 0.2% GML treatment significantly reduced the relative abundance of Bacteroidales ($p = 0.018$). At the family level (Figure 5C), the ATLL treatment significantly increased the relative abundance of Bacteroidaceae ($p < 0.05$). At the genus level (Figure 5D), the ATLL treatment significantly increased the relative abundance of *Bacteroides* ($p < 0.05$). whereas GML supplementation significantly increased the relative abundance of *Ruminococcus* ($p < 0.05$), compared with the ATLL group.

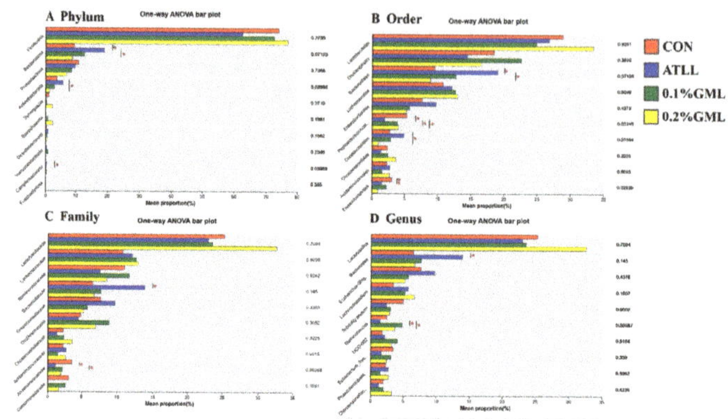

Figure 5. Analysis of group significant differences test among the top ten dominant phyla (**A**), orders (**B**), families (**C**), and genera (**D**) in fecal samples of piglets. CON, control group (basal diets), ATLL, antibiotic group (basal diets + 0.3% acetylisovaleryltylosin tartrate), 0.1% GML, glycerol monolaurate group (basal diets + 0.1% GML complex), 0.2% GML, 0.2% glycerol monolaurate group (basal diets + 0.2% GML complex). The y axis shows species names at the taxonomic level, while the x axis displays the average relative abundance of the species. * $p < 0.05$ and ** $p < 0.01$.

The Spearman correlation analysis between microbial composition and the antioxidant/inflammatory cytokine levels in sows is shown in Figure 6. This analysis revealed distinct correlations between bacterial genera and the body's antioxidant and inflammatory cytokine levels, highlighting the intricate role these bacteria play in regulating both antioxidant activity and inflammation. Specifically, the relative abundance of *Lactobacillus* was positively correlated with levels of T-SOD and IL-10 ($R > 0.4$, $p < 0.05$). In contrast, the relative abundance of *UCG-002* showed a negative correlation with T-SOD and TNF-α ($R < -0.4$, $p < 0.05$). Similarly, *norank_f_UCG-010* was negatively correlated with T-SOD, IL-6, and IL-1β ($R < -0.4$, $p < 0.05$). Finally, the *NK4A214_group* demonstrated a positive correlation with T-AOC ($R > 0.4$, $p < 0.05$) while being negatively correlated with IL-6 and IL-1β ($R < -0.4$, $p < 0.05$).

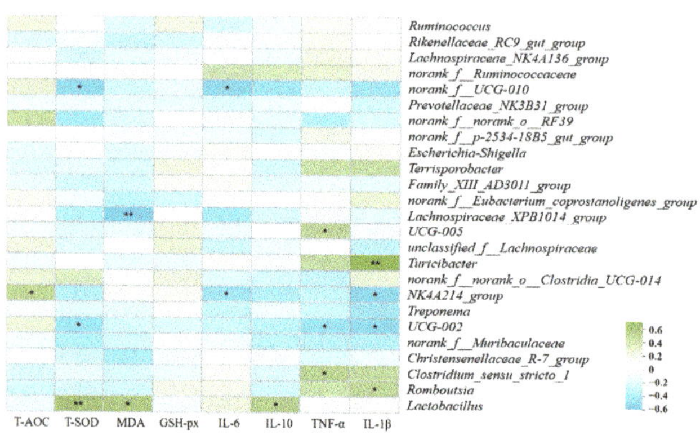

Figure 6. Heat map of correlation analysis between microbial composition and the antioxidant/inflammatory cytokine levels in sows at the genus level. The Spearman's correlation coefficient r, ranging from −1 to 1, represents the value shown on the middle heat map. A value of $R < 0$ indicates a negative correlation, while $R > 0$ indicates a positive correlation. * $p < 0.05$ and ** $p < 0.01$.

5. Discussion

Long labor can increase the risk of hypoxia and death in piglets, especially for the lastborn or lightest piglets [32]. In the present study, GML supplementation shortened the delivery interval and farrowing duration of sows. The addition of GML increased the weaning litter weight of piglets. Further analysis showed that 0.2% GML increased the IgA in colostrum and thus reduced diarrhea incidence in piglets during lactation. Changes in milk composition are conductive to piglet growth [33]. Li et al. [34] found that the addition of α -GML to the diet improved the milk immunoglobulin of sows, thus improving the health of piglets, and the IgA and IgG levels increased in sows along with the diet α -GML. GML maintained sow body condition and increased weight gain in piglets [35], while maternal 0.1% GML supplementation significantly increased fat and protein in colostrum and normal milk of sows [27]. However, GML can enhance the performance of weaned piglets by stabilizing the intestinal microbiota [14]. The concentrations of polyunsaturated and monounsaturated fatty acids in chicken egg yolk were increased with the addition of 0.3 g/kg GML. [16]. It was reported that, as a single medium-chain fatty acid glycerate, GML could be absorbed quickly and directly by piglets without emulsification, thereby reducing energy consumption [36]. However, medium-chain fatty acids in sows found higher colostrum protein concentrations [37]. Supplementation of GML in sows resulted in piglets with improved apparent total tract digestibility of dry matter, crude protein, and ether extract [34], implying that GML supplementation may benefit nutrient absorption by changing milk composition and piglet growth.

Before and after delivery, sows usually experience severe inflammatory reactions and oxidative stress that damage tissues and cells [38]; at the same time, this also damages the health of piglets [39]. Serum SOD and GSH-Px, which are antioxidant enzymes, can remove excess free radicals and play an important role in maintaining the oxidation balance in the body [40], and MDA as an oxidative product is considered a marker of oxidative stress [41]. In this study, addition of antibiotics increased the MDA level in sows in late lactation. Many studies have shown that antibiotic treatment induces the oxidative damage commonly caused by bacteria, leading to the production of reactive oxygen species, and causing oxidative stress that leads to cell death [42]. In contrast, ATLL and 0.2% GML contribute to activating the antioxidant enzyme system in weaned piglets, thus protecting them from excessive oxidative damage. Similar findings were reported in poultry [43].

Previous studies have indicated that modulation of the intestinal microbiota by antibiotics may suppress the immune response and alleviate inflammation [44,45]. In mouse experiments, the use of antibiotics can reduce the levels of the pro-inflammatory cytokine TNF-α [46,47]. Furthermore, in this study, application of ATLL and GML both significantly decreased the secretion of the pro-inflammatory cytokine TNF-α in sows, suggesting that GML may have a similar effect to ATLL in mitigating inflammatory responses. It has been shown that GML can effectively regulate pro-inflammatory factors, and inflammatory and immune responses by inhibiting the expression of inflammatory factors, chemokines, and cytokines [17,48]. Additionally, it has been found that GML can inhibit pro-inflammatory factors such as IL-2, IFN-γ, TNF-α, and IL-10 in a dose-dependent manner [15]. The levels of TNF-α and IL-10 in lactating sows and weaned piglets supplemented with different doses of GML corroborated these findings. However, we found in piglets given 0.1% and 0.2% GML that the low doses reduced the IL-6 level, but this was not dose-dependent. We speculate that this discrepancy may be related to the pleiotropic effects of IL-6, which can induce B cell differentiation and immunoglobulin production to mitigate inflammation [49], although the specific mechanism remains unclear. The above conclusions suggest that both ATLL and GML can effectively regulate inflammation and immune responses, and that maternal GML supplementation has more beneficial effects on offspring after birth than ATLL.

The intestinal microbiota is widely regarded as a key regulator of host physiology [50], oxidative stress [51], immune homeostasis [52], and reproductive performance [53]). Additionally, mother-to-infant microbial transmission has been observed in humans [54] and pigs [47]. Previous studies indicated that the use of antibiotics has a negative impact on the gut microbiota, leading to a reduction in species diversity, thus affecting stability of the gut microbiota and adversely affecting the host's gut health [55]. In contrast, research has demonstrated that GML helps maintain the integrity of the intestinal barrier function, promoting gut health [16,17]. Additionally, GML can improve the growth performance of weaned piglets and stabilize the gut microbiota [14]. The results of our study are consistent with previous findings, indicating that the use of antibiotics alters the composition of the gut microbiota, leading to a decrease in microbial diversity, while GML helps increase microbial diversity.

Gut microbes can affect the antioxidant response of the gut, and once the gut microbiota is unbalanced, it will destroy the intestinal barrier, increase the oxidative damage and reduce the antioxidant capacity [51]. It was found that 18 microbial genera were associated with antioxidant capacity of growing pigs [56]. Probiotics can play a positive role in regulating the oxidative stress response by scavenging free radicals and enhancing the activity of antioxidant enzymes [57,58]. In addition to participating in antioxidant responses, microorganisms also reduce intestinal inflammation by promoting the expression of anti-inflammatory factors and inhibiting the production of proinflammatory cytokines [59,60]. Tang et al. [61] observed that the relative abundance of *Romboutsia* and *Peptostreptococcaceae* was negatively correlated with immune characteristics. Wang et al. [47] found that *Bacteroides* play an important role in immunity and anti-inflammation. Based on the correlation analysis between microbiota and antioxidants, our study suggests that *Lachnospiraceae_XPB1014_group*, *NK4A214_group*, *Lactobacillus*, and *UCG-002* may play a positive role in producing antioxidant and anti-inflammation effects. Furthermore, research suggests that antibiotics can influence the host's inflammatory immune response by altering the microbial community [62,63]. GML can upregulate the favorable microbial abundance and reduce the occurrence of inflammation [17]. Maternal GML supplementation enhanced intestinal oxidative stability and barrier function in offspring and attenuated intestinal inflammatory responses in offspring [64]. Our study showed that adding antibiotics and GML improved the body immunity by adjusting the abundance of microorganisms associated with immunity. Meanwhile, the correlation analysis also found the positive effects of ATLL and GML in anti-inflammatory and antioxidant aspects. These findings strongly support the hypothesis that the antibiotic and GML groups may reduce inflammation by improving composition of the intestinal microbiota.

Energy intake was negatively correlated with an increase in Bacteroidetes' relative abundance [65]. Our study showed that the relative abundance of Bacteroidetes increased in the antibiotic group, but decreased in the 0.2% GML group. In this 0.2% GML group, host growth increased by promoting energy intake, while antibiotics had the opposite effect. This may be because GML is conducive to reducing pH in the intestinal tract and this has a strong bactericidal effect [66], which can reduce the relative abundance of *Escherichia coli* [67] in the intestinal tract and increase the number of beneficial bacteria such as *Lactobacillus* [68]. Additionally, the effect of GML, which encourages the competitive growth of good bacteria but is detrimental to proliferation of pathogenic bacteria, may be associated with improved intestinal microbiota structure in sows and piglets. Studies have shown that 450 mg/kg GML can increase beneficial microorganisms, such as intestinal *Bifidobacteria* and *Lactobacillus* [69]. Consequently, the improved reproductive performance of sows and the growth performance of piglets were also attributed to the improved intestinal microbiota due to GML. This finding is important for understanding the potential role of antibiotic alternatives in maintaining gut health and their application in the livestock industry.

Overall, GML presents numerous advantages over traditional antibiotics, such as reduced oxidative stress, improved immune function, enhanced gut health, better growth performance, and beneficial effects on reproductive performance. However, despite these promising benefits, its full potential has yet to be fully explored. Further research is essential to better understand the underlying mechanisms of GML's effects and to optimize its application in livestock production.

6. Conclusions

Maternal ATLL supplementation increased maternal oxidative stress responses by interfering with gut microbial composition while enhancing autoimmunity through synergistic effects. Dietary supplementation with GML in late-gestation and lactating sows can enhance reproductive performance by reducing the sow–piglet inflammatory response and increasing the piglets' antioxidant capacity. However, maternal GML supplementation improved sow–piglet gut microbial diversity and balance through maternal microbial transmission, thereby improving growth of piglets and reducing the rate of piglet diarrhea by reducing the sow–piglet inflammatory response.

Supplementary Materials: The following supporting information can be downloaded at https://www.mdpi.com/article/10.3390/vetsci12010024/s1, Table S1: The nutritional parameters of the gestation and lactation diet [70].

Author Contributions: Conceptualization, Y.L.; data curation, Y.L., D.L., Z.M., B.F., Z.F., J.L., Z.Z., Z.W., T.L. and D.W.; formal analysis, D.L., Z.M., L.C., B.F., S.X., Y.Z., Z.W. and T.L.; investigation, B.F., Y.Z. and D.W.; methodology, Y.L., M.Y., L.C., Z.F., S.X., Y.Z., J.L., J.W. and Z.Z.; resources, Z.W.; validation, Y.L.; writing—original draft, Y.L., M.Y., D.L. and T.L.; writing—review and editing, Y.L. and D.W. All authors have read and agreed to the published version of the manuscript.

Funding: This research was funded by the Sichuan Province "14th Five-Year" Sichuan Pig Major Science and Technology Project (2021ZDZX0009), the Natural Science Foundation of Sichuan Province (2022NSFSC0058), and National Modern Agricultural Industry Technology System Sichuan Pig innovation team SCSZTD-2024-08.

Institutional Review Board Statement: The animal study protocol was approved by The Institutional Animal Care and Use Committee of Sichuan Agricultural University (SCAUAC202108-3).

Informed Consent Statement: Not applicable.

Data Availability Statement: The datasets presented in this study can be found in online repositories. The names of the repository/repositories and accession number(s) can be found at https://www.ncbi.nlm.nih.gov/ accessed on 20 December 2024, PRJNA913793.

Acknowledgments: This work was supported by Calid Biotech (Wuhan) Co., Ltd.

Conflicts of Interest: Author JiHhua Wang is a member of Calid Biotech (Wuhan) Co., Ltd. and author Zehui Wu is a member of Sichuan Qiaozhu'er Breeding Co., Ltd. The remaining authors declare that the research was conducted in the absence of any commercial or financial relationships that could be construed as a potential conflict of interest.

References

1. Zhang, S.; Zhang, J.; Olasege, B.S.; Ma, P.; Qiu, X.; Gao, H.; Wang, C.; Wang, Y.; Zhang, Q.; Yang, H.; et al. Estimation of Genetic Parameters for Reproductive Traits in Connectedness Groups of Duroc, Landrace and Yorkshire Pigs in China. *J. Anim. Breed. Genet.* **2020**, *137*, 211–222. [CrossRef]
2. Gu, X.L.; Song, Z.H.; Li, H.; Wu, S.; Wu, S.S.; Ding, Y.N.; He, X.; Yin, Y.L.; Fan, Z.Y. Effects of Dietary Isomaltooligosaccharide and *Bacillus* Spp. Supplementation During Perinatal Period on Lactational Performance, Blood Metabolites, and Milk Composition of Sows. *J. Sci. Food Agric.* **2019**, *99*, 5646–5653. [CrossRef] [PubMed]
3. Sánchez-Cordón, P.J.; Montoya, M.; Reis, A.L.; Dixon, L.K. African Swine Fever: A Re-Emerging Viral Disease Threatening the Global Pig Industry. *Vet. J.* **2018**, *233*, 41–48. [CrossRef] [PubMed]
4. Danner, M.C.; Robertson, A.; Behrends, V.; Reiss, J. Antibiotic Pollution in Surface Fresh Waters: Occurrence and Effects. *Sci. Total Environ.* **2019**, *664*, 793–804. [CrossRef] [PubMed]
5. Amer, S.A.; Tolba, S.A.; AlSadek, D.M.M.; Abdel Fattah, D.M.; Hassan, A.M.; Metwally, A.E. Effect of Supplemental Glycerol Monolaurate and Oregano Essential Oil Blend on the Growth Performance, Intestinal Morphology, and Amino Acid Digestibility of Broiler Chickens. *BMC Vet. Res.* **2021**, *17*, 312. [CrossRef] [PubMed]
6. McAllister, T.A.; Wang, Y.; Diarra, M.S.; Alexander, T.; Stanford, K. Challenges of a One-Health Approach to the Development of Alternatives to Antibiotics. *Anim. Front.* **2018**, *8*, 10–20. [CrossRef] [PubMed]
7. Anadón, A. WS14 The EU Ban of Antibiotics as Feed Additives (2006): Alternatives and Consumer Safety. *J. Vet. Pharmacol. Ther.* **2006**, *29*, 41–44. [CrossRef]
8. Kogut, M.H. Issues and Consequences of Using Nutrition to Modulate the Avian Immune Response. *J. Appl. Poult. Res.* **2017**, *26*, 605–612. [CrossRef]
9. Schlievert, P.M.; Kilgore, S.H.; Seo, K.S.; Leung, D.Y.M. Glycerol Monolaurate Contributes to the Antimicrobial and Anti-inflammatory Activity of Human Milk. *Sci. Rep.* **2019**, *9*, 14550. [CrossRef]
10. Thormar, H.; Hilmarsson, H.; Bergsson, G. Stable Concentrated Emulsions of the 1-Monoglyceride of Capric Acid (Monocaprin) with Microbicidal Activities Against the Food-Borne Bacteria *Campylobacter jejuni*, *Salmonella* spp., and *Escherichia coli*. *Appl. Environ. Microbiol.* **2006**, *72*, 522–526. [CrossRef]
11. Fortuoso, B.F.; Dos Reis, J.H.; Gebert, R.R.; Barreta, M.; Griss, L.G.; Casagrande, R.A.; de Cristo, T.G.; Santiani, F.; Campigotto, G.; Rampazzo, L.; et al. Glycerol Monolaurate in the Diet of Broiler Chickens Replacing Conventional Antimicrobials: Impact on health, Performance and Meat Quality. *Microb. Pathog.* **2019**, *129*, 161–167. [CrossRef] [PubMed]
12. Jordan, E.; Lovett, D.K.; Monahan, F.J.; Callan, J.; Flynn, B.; O'Mara, F.P. Effect of Refined Coconut Oil or Copra Meal on Methane Output and on Intake and Performance of Beef Heifers. *J. Anim. Sci.* **2006**, *84*, 162–170. [CrossRef] [PubMed]
13. Liu, H.; Vaddella, V.; Zhou, D. Effects of Chestnut Tannins and Coconut Oil on Growth Performance, Methane Emission, Ruminal Fermentation, and Microbial Populations in Sheep. *J. Dairy Sci.* **2011**, *94*, 6069–6077. [CrossRef] [PubMed]
14. Decuypere, J.A.; Dierick, N.A. The Combined Use of Triacylglycerols Containing Medium-Chain Fatty Acids and Exogenous Lipolytic Enzymes as an Alternative to in-Feed Antibiotics in Piglets: Concept, Possibilities and Limitations. An Overview. *Nutr. Res. Rev.* **2003**, *16*, 193–210. [CrossRef] [PubMed]
15. Zhang, M.S.; Sandouk, A.; Houtman, J.C. Glycerol Monolaurate (GML) Inhibits Human T Cell Signaling and Function by Disrupting Lipid Dynamics. *Sci. Rep.* **2016**, *6*, 30225.
16. Kong, L.; Wang, Z.; Xiao, C.; Zhu, Q.; Song, Z. Glycerol Monolaurate Ameliorated Intestinal Barrier and Immunity in Broilers by Regulating Intestinal Inflammation, Antioxidant Balance, and Intestinal Microbiota. *Front. Immunol.* **2021**, *12*, 713485. [CrossRef] [PubMed]
17. Mo, Q.; Fu, A.; Deng, L.; Zhao, M.; Li, Y.; Zhang, H.; Feng, F. High-dose Glycerol Monolaurate Up-Regulated Beneficial Indigenous Microbiota without Inducing Metabolic Dysfunction and Systemic Inflammation: New Insights into Its Antimicrobial Potential. *Nutrients* **2019**, *11*, 1981. [CrossRef] [PubMed]
18. Castillo, M.; Martín-Orúe, S.M.; Roca, M.; Manzanilla, E.G.; Badiola, I.; Perez, J.F.; Gasa, J. The Response of Gastrointestinal Microbiota to Avilamycin, Butyrate, and Plant Extracts in Early-Weaned Pigs. *J. Anim. Sci.* **2006**, *84*, 2725–2734. [CrossRef]
19. Hashemipour, H.; Kermanshahi, H.; Golian, A.; Khaksar, V. Effects of Carboxy Methyl Cellulose and Thymol + Carvacrol on Performance, Digesta Viscosity and Some Blood Metabolites of Broilers. *J. Anim. Physiol. Anim. Nutr.* **2014**, *98*, 672–679. [CrossRef] [PubMed]
20. Hashemipour, H.; Khaksar, V.; Rubio, L.A.; Veldkamp, T.; van Krimpen, M.M. Effect of Feed Supplementation with a Thymol Plus Carvacrol Mixture, in Combination or Not with an NSP-Degrading Enzyme, on Productive and Physiological Parameters of Broilers Fed on Wheat-Based Diets. *Anim. Feed Sci. Technol.* **2016**, *211*, 117–131. [CrossRef]

21. Yin, L.; Chen, J.; Wang, K.; Geng, Y.; Lai, W.; Huang, X.; Chen, D.; Guo, H.; Fang, J.; Chen, Z.; et al. Study the Antibacterial Mechanism of Cinnamaldehyde Against Drug-Resistant Aeromonas Hydrophila In Vitro. *Microb. Pathog.* **2020**, *145*, 104208. [CrossRef]
22. Farrokhfall, K.; Khoshbaten, A.; Zahediasl, S.; Mehrani, H.; Karbalaei, N. Improved Islet Function is Associated with Anti-inflammatory, Antioxidant and Hypoglycemic Potential of Cinnamaldehyde on Metabolic Syndrome Induced by High Tail Fat in Rats. *J. Funct. Foods* **2014**, *10*, 397–406. [CrossRef]
23. Tiihonen, K.; Kettunen, H.; Bento, M.H.; Saarinen, M.; Lahtinen, S.; Ouwehand, A.C.; Schulze, H.; Rautonen, N. The Effect of Feeding Essential Oils on Broiler Performance and Gut Microbiota. *Br. Poult. Sci.* **2010**, *51*, 381–392. [CrossRef]
24. Valentini, J.; Da Silva, A.S.; Fortuoso, B.F.; Reis, J.H.; Gebert, R.R.; Griss, L.G.; Boiago, M.M.; Lopes, L.Q.S.; Santos, R.C.V.; Wagner, R.; et al. Chemical Composition, Lipid Peroxidation, and Fatty Acid Profile in Meat of Broilers Fed with Glycerol Monolaurate Additive. *Food Chem.* **2020**, *330*, 127187. [CrossRef] [PubMed]
25. Kroismayr, A.; Sehm, J.; Pfaffl, M.W.; Schedle, K.; Plitzner, C.; Windisch, W. Effects of Avilamycin and Essential Oils on mRNA Expression of Apoptotic and Inflammatory Markers and Gut Morphology of Piglets. *Czech J. Anim. Sci.* **2008**, *53*, 377–387. [CrossRef]
26. Wang, Y.; Li, J.; Wang, H.; Mi, Y.; Xue, Y.; Li, J.; Ma, Y. Effects of Essential Oil Coated with Glycerol Monolaurate on Growth Performance, Intestinal Morphology, and Serum Profiles in Weaned Piglets. *Anim. Biosci.* **2023**, *36*, 753–760. [CrossRef]
27. Zhao, H.; Tian, M.; Xiong, L.; Lin, T.; Zhang, S.; Yue, X.; Liu, X.; Chen, F.; Zhang, S.; Guan, W. Maternal Supplementation with Glycerol Monolaurate Improves the Intestinal Health of Suckling Piglets by Inhibiting the NF-κB/MAPK Pathways and Improving Oxidative Stability. *Food Funct.* **2023**, *14*, 3290–3303. [CrossRef]
28. NRC. *Nutrient Requirements of Swine*; National Academies Press: Washington, DC, USA, 2012.
29. Hu, J.; Park, J.H.; Kim, I.H. Effect of Dietary Supplementation with Lactobacillus Plantarum on Growth Performance, Fecal Score, Fecal Microbial Counts, Gas Emission and Nutrient Digestibility in Growing Pigs. *Anim. Feed. Sci. Technol.* **2022**, *290*, 115295. [CrossRef]
30. Dai, F.; Lin, T.; Cheng, L.; Wang, J.; Zuo, J.; Feng, D. Effects of Micronized Bamboo Powder on Growth Performance, Serum Biochemical Indexes, Cecal Chyme Microflora and Metabolism of Broilers Aged 1–22 days. *Trop. Anim. Health Prod.* **2022**, *54*, 166. [CrossRef] [PubMed]
31. Le Floc'h, N.; Achard, C.S.; Eugenio, F.A.; Apper, E.; Combes, S.; Quesnel, H. Effect of Live Yeast Supplementation in Sow Diet During Gestation and Lactation on Sow and Piglet Fecal Microbiota, Health, and Performance. *J. Anim. Sci.* **2022**, *100*, skac209. [CrossRef] [PubMed]
32. Langendijk, P.; Plush, K. Parturition and Its Relationship with Stillbirths and Asphyxiated Piglets. *Animals* **2019**, *9*, 885. [CrossRef]
33. Theil, P.K.; Lauridsen, C.; Quesnel, H. Neonatal Piglet Survival: Impact of Sow Nutrition Around Parturition on Fetal Glycogen Deposition and Production and Composition of Colostrum and Transient Milk. *Animal* **2014**, *8*, 1021–1030. [CrossRef]
34. Li, L.; Wang, H.; Zhang, N.; Zhang, T.; Ma, Y. Effects of α-Glycerol Monolaurate on Intestinal Morphology, Nutrient Digestibility, Serum Profiles, and Gut Microbiota in Weaned Piglets. *J. Anim. Sci.* **2022**, *100*, skac046. [CrossRef]
35. Risley, C.; Lopez, J.; Cordts, C.; von Brandestein, F. PSV-17 Glycerol Monolaurate Improves Litter Performance and Modifies Milk Composition when Fed to Lactating Sows. *J. Anim. Sci.* **2024**, *102* (Suppl. S2), 314–315. [CrossRef]
36. Cui, Z.; Wang, X.; Liao, S.; Qi, M.; Zha, A.; Zuo, G.; Liao, P.; Chen, Y.; Guo, C.; Tan, B. Effects of Medium-Chain Fatty Acid Glycerides on Nutrient Metabolism and Energy Utilization in Weaned Piglets. *Front. Vet. Sci.* **2022**, *9*, 938888. [CrossRef]
37. Swanson, A.J.; Hesse, A.; Levesque, C.L.; Pierre, B.S. PSVII-7 The Effects on Supplementation of Medium Chain Fatty Acids on Piglet Health and Immune Markers. *J. Anim. Sci.* **2022**, *100* (Suppl. S2), 174–175. [CrossRef]
38. Berchieri-Ronchi, C.B.; Kim, S.W.; Zhao, Y.; Correa, C.R.; Yeum, K.J.; Ferreira, A.L. Oxidative Stress Status of Highly Prolific Sows During Gestation and Lactation. *Animal* **2011**, *5*, 1774–1779. [CrossRef]
39. Prater, M.R.; Laudermilch, C.L.; Liang, C.; Holladay, S.D. Placental Oxidative Stress Alters Expression of Murine Osteogenic Genes and Impairs Fetal Skeletal Formation. *Placenta* **2008**, *29*, 802–808. [CrossRef]
40. Ren, W.; Yin, Y.; Liu, G.; Yu, X.; Li, Y.; Yang, G.; Li, T.; Wu, G. Effect of Dietary Arginine Supplementation on Reproductive Performance of Mice with Porcine Circovirus Type 2 Infection. *Amino Acids* **2012**, *42*, 2089–2094. [CrossRef]
41. Lykkesfeldt, J.; Svendsen, O. Oxidants and Antioxidants in Disease: Oxidative Stress in Farm Animals. *Vet. J.* **2007**, *173*, 502–511. [CrossRef]
42. Kohanski, M.A.; Dwyer, D.J.; Hayete, B.; Lawrence, C.A.; Collins, J.J. A Common Mechanism of Cellular Death Induced by Bactericidal Antibiotics. *Cell* **2007**, *130*, 797–810. [CrossRef]
43. Fortuoso, B.F.; Galli, G.M.; Griss, L.G.; Armanini, E.H.; Silva, A.D.; Fracasso, M.; Mostardeiro, V.; Morsch, V.M.; Lopes, L.Q.S.; Santos, R.C.V.; et al. Effects of Glycerol Monolaurate on Growth and Physiology of Chicks Consuming Diet Containing Fumonisin. *Microb. Pathog.* **2020**, *147*, 104261. [CrossRef] [PubMed]

44. Seifert, H.A.; Benedek, G.; Nguyen, H.; Gerstner, G.; Zhang, Y.; Kent, G.; Vandenbark, A.A.; Bernhagen, J.; Offner, H. Antibiotics Protect Against EAE by Increasing Regulatory and Anti-Inflammatory Cells. *Metab. Brain Dis.* **2018**, *33*, 1599–1607. [CrossRef]
45. de Oliveira, G.L.V.; Leite, A.Z.; Higuchi, B.S.; Gonzaga, M.I.; Mariano, V.S. Intestinal Dysbiosis and Probiotic Applications in Autoimmune Diseases. *Immunology* **2017**, *152*, 1–12. [CrossRef] [PubMed]
46. Gómez-Hurtado, I.; Gimenez, P.; García, I.; Zapater, P.; Francés, R.; González-Navajas, J.M.; Manichanh, C.; Ramos, J.M.; Bellot, P.; Guarner, F.; et al. Norfloxacin is More Effective than Rifaximin in Avoiding Bacterial Translocation in an Animal Model of Cirrhosis. *Liver Int.* **2018**, *38*, 295–302. [CrossRef] [PubMed]
47. Wang, J.; Xiang, Q.; Gu, S.; Gu, Y.; Yao, M.; Huang, W.; Gao, W.; Tang, L.L. Short- and Long-Term Effects of Different Antibiotics on the Gut Microbiota and Cytokines Level in Mice. *Infect. Drug Resist.* **2022**, *15*, 6785–6797. [CrossRef] [PubMed]
48. Haase, A.T.; Rakasz, E.; Schultz-Darken, N.; Nephew, K.; Weisgrau, K.L.; Reilly, C.S.; Li, Q.; Southern, P.J.; Rothenberger, M.; Peterson, M.L.; et al. Glycerol Monolaurate Microbicide Protection against Repeat High-Dose SIV Vaginal Challenge. *PLoS ONE* **2015**, *10*, e0129465. [CrossRef] [PubMed]
49. Hirano, T.; Yasukawa, K.; Harada, H.; Taga, T.; Watanabe, Y.; Matsuda, T.; Kashiwamura, S.; Nakajima, K.; Koyama, K.; Iwamatsu, A.; et al. Complementary DNA for a Novel Human Interleukin (BSF-2) that Induces B Lymphocytes to Produce Immunoglobulin. *Nature* **1986**, *324*, 73–76. [CrossRef] [PubMed]
50. Brown, E.M.; Clardy, J.; Xavier, R.J. Gut Microbiome Lipid Metabolism and Its Impact on Host Physiology. *Cell Host Microbe* **2023**, *31*, 173–186. [CrossRef]
51. Tang, J.; Xu, L.; Zeng, Y.; Gong, F. Effect of Gut Microbiota on LPS-Induced Acute Lung Injury by Regulating the TLR4/NF-kB Signaling Pathway. *Int. Immunopharmacol.* **2021**, *91*, 107272. [CrossRef] [PubMed]
52. Graham, D.B.; Xavier, R.J. Conditioning of the Immune System by the Microbiome. *Trends Immunol.* **2023**, *44*, 499–511. [CrossRef] [PubMed]
53. Liu, Y.; Cheng, X.; Zhen, W.; Zeng, D.; Qu, L.; Wang, Z.; Ning, Z. Yeast Culture Improves Egg Quality and Reproductive Performance of Aged Breeder Layers by Regulating Gut Microbes. *Front. Microbiol.* **2021**, *12*, 633276. [CrossRef] [PubMed]
54. Ferretti, P.; Pasolli, E.; Tett, A.; Asnicar, F.; Gorfer, V.; Fedi, S.; Armanini, F.; Truong, D.T.; Manara, S.; Zolfo, M.; et al. Mother-to-Infant Microbial Transmission from Different Body Sites Shapes the Developing Infant Gut Microbiome. *Cell Host Microbe* **2018**, *24*, 133–145.e135. [CrossRef] [PubMed]
55. Costa, M.C.; Bessegatto, J.A.; Alfieri, A.A.; Weese, J.S.; Filho, J.A.; Oba, A. Different Antibiotic Growth Promoters Induce Specific Changes in the Cecal Microbiota Membership of Broiler Chicken. *PLoS ONE* **2017**, *12*, e0171642. [CrossRef]
56. Li, Z.; Zhao, Y.; Wang, H.; Zhang, W.; Zhang, C.; Xie, J.; Ma, X. High-Fibre Diets Regulate Antioxidative Capacity and Promote Intestinal Health by Regulating Bacterial Microbiota in Growing Pigs. *J. Anim. Physiol. Anim. Nutr.* **2024**, *108*, 357–365. [CrossRef]
57. Cao, X.; Tang, L.; Zeng, Z.; Wang, B.; Zhou, Y.; Wang, Q.; Zou, P.; Li, W. Effects of Probiotics BaSC06 on Intestinal Digestion and Absorption, Antioxidant Capacity, Microbiota Composition, and Macrophage Polarization in Pigs for Fattening. *Front. Vet. Sci.* **2020**, *7*, 570593. [CrossRef] [PubMed]
58. Wang, Y.; Wu, Y.; Wang, Y.; Xu, H.; Mei, X.; Yu, D.; Wang, Y.; Li, W. Antioxidant Properties of Probiotic Bacteria. *Nutrients* **2017**, *9*, 521. [CrossRef] [PubMed]
59. Xiong, Y.; Brinkman, C.C.; Hittle, L.E.; Fricke, W.F.; Mongodin, E.F.; Bromberg, J.S. Gut Microbiota Determines Graft Survival Outcome Through Regulation of Anti-Inflammatory and Pro-Inflammatory Responses. *Transplantation* **2018**, *102*, S333. [CrossRef]
60. Yu, X.; Cui, Z.; Qin, S.; Zhang, R.; Wu, Y.; Liu, J.; Yang, C. Effects of *Bacillus licheniformis* on Growth Performance, Diarrhea Incidence, Antioxidant Capacity, Immune Function, and Fecal Microflora in Weaned Piglets. *Animals* **2022**, *12*, 1609. [CrossRef] [PubMed]
61. Tang, Y.; Pu, Q.; Zhao, Q.; Zhou, Y.; Jiang, X.; Han, T. Effects of Fucoidan Isolated from Laminaria japonica on Immune Response and Gut Microbiota in Cyclophosphamide-Treated Mice. *Front. Immunol.* **2022**, *13*, 916618. [CrossRef]
62. He, B.; Liu, Y.; Hoang, T.K.; Tian, X.; Taylor, C.M.; Luo, M.; Tran, D.Q.; Tatevian, N.; Rhoads, J.M. Antibiotic-Modulated Microbiome Suppresses Lethal Inflammation and Prolongs Lifespan in Treg-Deficient Mice. *Microbiome* **2019**, *7*, 145. [CrossRef]
63. Yoon, S.; Lee, G.; Yu, J.; Lee, K.; Lee, K.; Si, J.; You, H.J.; Ko, G. Distinct Changes in Microbiota-Mediated Intestinal Metabolites and Immune Responses Induced by Different Antibiotics. *Antibiotics* **2022**, *11*, 1762. [CrossRef]
64. Zhao, M.; Zhang, J.; Liu, T.; Wang, J.; Cai, H.; Zhang, X.; Quay Huai Xia, D.; Feng, F.; Tang, J. Differential Modulations of Lauric Acid and Its Glycerides on High Fat Diet-Induced Metabolic Disorders and Gut Microbiota Dysbiosis. *Food Res. Int.* **2022**, *157*, 111437. [CrossRef]
65. Jumpertz, R.; Le, D.S.; Turnbaugh, P.J.; Trinidad, C.; Bogardus, C.; Gordon, J.I.; Krakoff, J. Energy-Balance Studies Reveal Associations Between Gut Microbes, Caloric Load, and Nutrient Absorption in Humans. *Am. J. Clin. Nutr.* **2011**, *94*, 58–65. [CrossRef] [PubMed]

66. Wang, Y.; Abdullah; Zhang, C.; Li, Y.; Zhang, H.; Wang, J.; Feng, F. Effects of Dietary Glycerol Monolaurate on the Growth Performance, Digestive Enzymes, Body Composition and Non-Specific Immune Response of White Shrimp (*Litopenaeus vannamei*). *Aquac. Rep.* **2020**, *18*, 100535. [CrossRef]
67. Jang, I.S.; Ko, Y.H.; Kang, S.Y.; Lee, C.Y. Effect of a Commercial Essential Oil on Growth Performance, Digestive Enzyme Activity and Intestinal Microflora Population in Broiler Chickens. *Anim. Feed Sci. Technol.* **2007**, *134*, 304–315. [CrossRef]
68. Kovanda, L.; Zhang, W.; Wei, X.; Luo, J.; Wu, X.; Atwill, E.R.; Vaessen, S.; Li, X.; Liu, Y. In Vitro Antimicrobial Activities of Organic Acids and Their Derivatives on Several Species of Gram-Negative and Gram-Positive Bacteria. *Molecules* **2019**, *24*, 3770. [CrossRef]
69. Schlievert, P.M.; Peterson, M.L. Glycerol Monolaurate Antibacterial Activity in Broth and Biofilm Cultures. *PLoS ONE* **2012**, *7*, e40350. [CrossRef]
70. Lin, Y.; Li, D.; Ma, Z.; Che, L.; Feng, B.; Fang, Z.; Xu, S.; Zhuo, Y.; Li, J.; Hua, L.; et al. Maternal Tributyrin Supplementation in Late Pregnancy and Lactation Improves Offspring Immunity, Gut Microbiota, and Diarrhea Rate in a Sow Model. *Front Microbiol.* **2023**, *14*, 1142174. [CrossRef]

veterinary sciences

MDPI

Article

Effects of Long-Term Serum Starvation on Autophagy, Metabolism, and Differentiation of Porcine Skeletal Muscle Satellite Cells

Yi Wang [1,†], Juan Gao [1,†], Bojun Fan [1,†], Yuemin Hu [1], Yuefei Yang [1], Yajie Wu [1], Jiaqiao Zhu [1], Junwei Li [1], Feng Li [2] and Huiming Ju [1,*]

[1] College of Veterinary Medicine, Yangzhou University/Jiangsu Co-Innovation Center for Prevention and Control of Important Animal Infectious Diseases and Zoonosis, Yangzhou 225009, China; wyi@yzu.edu.cn (Y.W.); mz120201425@stu.yzu.edu.cn (J.G.); mx120211006@stu.yzu.edu.cn (B.F.); mx120190778@stu.yzu.edu.cn (Y.H.); yfyang@yzu.edu.cn (Y.Y.); mx120211004@stu.yzu.edu.cn (Y.W.); jqzhu1998@163.com (J.Z.); lijunwei@yzu.edu.cn (J.L.)

[2] Department of Reproductive Medicine Center, Northern Jiangsu People's Hospital Affiliated to Yangzhou University/Clinical Medical College, Yangzhou University, Yangzhou 225009, China; minprince@163.com

* Correspondence: hmju@yzu.edu.cn; Tel.: +86-13912125712; Fax: +86-5147972218

† These authors contributed equally to this work.

Simple Summary: This study investigated the effects of long-term serum starvation on autophagy, metabolism, and differentiation of porcine skeletal muscle satellite cells (SMSCs) and elucidated the role of autophagy in skeletal muscle development to provide a theoretical basis for improving meat production in domestic pigs. The results showed that long-term serum starvation induced autophagy through the AMPK/mTOR signaling pathway, accelerated cell metabolism and apoptosis, exacerbated reactive oxygen species accumulation, and inhibited myogenic and adipogenic differentiation of SMSCs. In addition, serum starvation-induced autophagy moderately promoted the myogenic and adipogenic differentiation of SMSCs. However, these effects were insufficient to counteract the inhibition of cell differentiation by long-term serum starvation. This study provides insight into leveraging serum starvation as a stressor to regulate muscle growth and metabolism in domestic pigs.

Abstract: This study investigated the effects of long-term serum starvation on autophagy, metabolism, and differentiation of porcine skeletal muscle satellite cells (SMSCs) and elucidated the role of autophagy in skeletal muscle development. Our findings provide a theoretical basis for improving meat production in domestic pigs. The SMSCs isolated and preserved in our laboratory were revived and divided into six groups based on the culture medium serum concentration to simulate varying levels of serum starvation: 20% serum (control group), 15% serum (mild serum starvation group), 5% serum (severe serum starvation group), and their autophagy inhibition groups supplemented with 3-methyladenine. After 96 h of culture, the apoptosis rate, mitochondrial membrane potential, reactive oxygen species, and ATP were measured to evaluate the effects of serum starvation on the SMSCs' metabolism. Additionally, the levels of autophagy-related proteins, autophagosomes, and autolysosomes were measured to investigate the impact of long-term serum starvation on autophagy. The expression of proteins associated with myogenic and adipogenic differentiation (MHC, MyoD1, peroxisome proliferator-activated receptor γ, and lipoprotein lipase) as well as lipid content were also determined to investigate the effects of long-term serum starvation on SMSC differentiation. The results showed that long-term serum starvation induced autophagy through the AMPK/mTOR signaling pathway, accelerated cell metabolism and apoptosis, exacerbated reactive oxygen species accumulation, and inhibited myogenic and adipogenic differentiation of SMSCs. Moreover, these effects were positively correlated with the level of serum starvation. In addition, serum starvation-induced autophagy moderately promoted the myogenic and adipogenic differentiation of SMSCs; however, these effects were insufficient to counteract the inhibition of cell differentiation by long-term serum starvation. This study provides insight into leveraging serum starvation as a stressor to regulate muscle growth and metabolism in domestic pigs.

Keywords: serum starvation; skeletal muscle satellite cell; autophagy; cell metabolism; differentiation

1. Introduction

Skeletal muscle satellite cells (SMSCs), which were discovered by Mauro in 1961, are undifferentiated and quiescent myogenic precursor cells located between the basal lamina and sarcolemma of adult muscle tissue [1]. When muscle growth is impaired or stressed by degenerative muscular diseases, SMSCs can play a crucial role in muscle development, maintenance, and regeneration by forming myofibers through myogenic differentiation. These cells can also directly influence the growth of porcine skeletal muscle and meat production [2–4]. Autophagy involves the formation of autophagosomes, which facilitate the recycling of biological macromolecules such as proteins and organelles like mitochondria into the lysosome [5]. Within the lysosome, these components are degraded into small molecules such as amino acids and monosaccharides, enabling the reuse of energy and materials. Autophagy has recently received great attention because of its role in skeletal muscle metabolic homeostasis and disease progression. Skeletal muscle differentiation is a stress-related process, and the growth and differentiation of skeletal muscle can be promoted by activation of the AMPK signaling pathway, caspase activation, excessive reactive oxygen species (ROS) accumulation, mitochondrial fission, and changes in autophagic flux [6]. Studies have shown that autophagic flux increases in skeletal muscle during hypoxia, disuse atrophy, starvation, and lack of exercise [7–9]. We previously confirmed that short-term (24 h) and mild (15% serum concentration) serum starvation promoted SMSC metabolism, induced autophagy via the AMPK/mTOR signaling pathway, and promoted myogenic and adipogenic differentiation in SMSCs [10]. We hypothesized that serum starvation plays an important role in the growth and metabolism of animal skeletal muscle. To further investigate the effects of serum starvation on porcine SMSCs, we induced long-term mild and severe serum starvation in SMSCs by reducing serum concentration for an extended period (96 h) and then evaluated the effects on the autophagy, metabolism, and differentiation of SMSCs. This study contributes to a comprehensive understanding of the effects of serum starvation on skeletal muscle growth and development, and the results lay a theoretical foundation for optimizing the use of serum starvation-induced autophagy to improve feed efficiency in animal production.

2. Materials and Methods

2.1. Cell Grouping and Treatment

Porcine SMSC cell lines were preserved in our laboratory. We resuscitated the cells and divided them into six groups when they reached 70–80% confluence. The 20% S group (the control group), the 15% S group, and the 5% S group refer to the cells grown in culture media containing 20%, 15%, and 5% serum (FBS, Ref#10270-106, Gibco, Emeryville,

CA, USA). We subjected the above groups to autophagy inhibition via treatment with 5 mM of the autophagy inhibitor 3-methyladenine (3-MA, HY-19312, MCE, Plymouth, MN, USA) and named them 20% S + 3-MA group, 15% S + 3-MA group, and 5% S + 3-MA group, respectively. Autophagy-related proteins, cell metabolic indices, and differentiation abilities were detected after 96 h of culture under the corresponding conditions.

2.2. Detection of Cell Metabolism

The apoptosis detection kit (C1062S, Beyotime, Shanghai, China), ROS detection kit (S0033S, Beyotime, Shanghai, China), and MMP detection kit (C2006, Beyotime, Shanghai, China) were used to detect cell apoptosis rate, mitochondrial active oxygen (ROS) level, and mitochondrial membrane potential (MMP) level, respectively, by flow cytometry (Beckman Coulter, Fullerton, CA, USA). The concentration of ATP was determined by the ATP detection kit (S0026, Beyotime, Shanghai, China). The specific methods were similar to those in reference [10].

2.3. Detection of Autophagy

2.3.1. Detection of Lysosomes and Autophagosomes

Mitochondria and lysosomes were stained with Mito-Tracker (C1048, Beyotime, Shanghai, China) and Lyso-Tracker (C1046, Beyotime, Shanghai, China); then, the cells were photographed by fluorescence microscope, and the fluorescence intensity was analyzed by ImageJ software 1.8.0_172 (NIH, Bethesda, MD, USA).

Lipofectamine™ 3000 (L3000150, Thermo, Waltham, MA, USA) diluted in Opti-MEM™ (31985070, Thermo, Waltham, MA, USA) was combined with premixed LC3B plasmid (842-pmCherry-EGFP-LC3B, Beyotime, Shanghai, China) solution at a 1:1 ratio for autophagosome detection. After incubation with DNA–lipid complex for 24 h, samples were observed under a fluorescence microscope.

2.3.2. Detection of Autophagy-Related Protein Expression by Western Blot

Total protein was extracted from the cells, and tubulin (11224-1-AP, Proteintech, Rosemont, IL, USA) was used as the internal reference protein. The ratios of phospho-mTOR/mTOR (67778-1-lg and 66888-1-lg, Proteintech, Rosemont, IL, USA), phospho-AMPK/AMPK (GTX03702, GeneTex, Irvine, CA, USA and 66536-1-lg, Proteintech, Rosemont, IL, USA), LC3B-II/LC3B-I (ab229327, Abcam, Waltham, MA, USA) and p62/Tubulin (ab233207, Abcam, Waltham, MA, USA) were detected by Western blot. The gray values of hybrid bands were measured by ImageJ software 1.8.0_172 (NIH, Bethesda, MD, USA) The expression of the proteins was quantified and calculated as reported by [10,11].

2.4. Detection of Cell Differentiation

2.4.1. Myogenesis and Lipogenesis Induction

Myogenesis differentiation medium and lipogenesis induction solution were used to induce myogenic and lipogenesis differentiation of SMSCs, respectively. The specific induction methods were similar to those in reference [10].

2.4.2. Detection of Myogenic Differentiation Markers by Western Blot

MHC and MyoD1 antibodies (sc-53088 and sc-32758, Santa Cruz, TX, USA) were used to detect myogenesis. The specific methods used to detect myogenic differentiation markers are described in Section 2.3.2.

2.4.3. Detection of Myogenic Differentiation Markers by Cellular Immunofluorescence

MHC, MyoD1 antibodies, and the FITC fluorescent secondary antibody (SA00003-1, Proteintech, Rosemont, IL, USA) were incubated after the SMSCs were fixed and permeabilized. DAPI was used to stain the cell nucleus, and laser confocal microscope (TCS SP8 STED, Leica, Wetzlar, Germany) was used to detect MHC and MyoD1 protein expression. ImageJ software 1.8.0_172 (NIH, Bethesda, MD, USA) was used to analyze the fluorescence intensity. Myogenic differentiation markers were detected as previously described [12].

2.4.4. Detection of Adipogenic Differentiation Markers by qRT-PCR

After lipogenesis induction, we isolated total mRNA from the cells and reverse transcribed them into cDNA. We detected the expression of PPARγ and LPL mRNA by qRT-PCR with GAPDH gene as the internal reference and operated the reaction systems according to TB Green® Premix Ex Taq™ II (RR82LR, Takara, Takatsuki, Japan). The related primers and the expression of the proteins was quantified as reported in [13].

2.4.5. Detection of Fat Content with Oil Red O Staining

After lipogenesis induction, the cells were incubated with Oil Red O (G1262, Solarbio, Beijing, China) fixative, then stained with the Oil Red O staining solution. We observed the results under a microscope. Fat droplets were extracted with isopropanol and OD values at 510 nm were quantitatively detected.

2.4.6. Detection of TG Levels

The supernatant of the cells after lipogenesis induction, digesting, and lysing was used as the test solution, and the absorbance value at 420 nm was detected by TG detection kit (BC0620, Solarbio, Beijing, China). The TG content of each group was quantitatively calculated according to the statistical cell number.

2.5. Statistical Analysis

All experiments were carried out with a minimum of three independent repeats. Data were presented as the x ± SD. Statistical differences between groups were analyzed using one-way ANOVA with GraphPad Prism7 software (GraphPad Software, San Diego, CA, USA), and differences were considered significant when $p < 0.05$.

3. Results and Analysis

The following data are presented in the order of the results from the 20% S, 20% S + 3-MA, 15% S, 15% S + 3-MA, 5% S, and 5% S + 3-MA groups.

3.1. Cell Metabolism

3.1.1. Apoptosis Rate

The apoptosis rates in the six groups were 6.52 ± 1.33, 5.17 ± 1.10, 7.06 ± 0.67, 5.08 ± 0.40, 8.77 ± 0.45, and 6.43 ± 0.56, respectively, shown in Figure 1A,D. The apoptosis rates increased gradually in the 20% S, 15% S, and 5% S groups, and the apoptosis rate of 5% S was significantly higher than that of 20% S ($p < 0.01$). The protein expression levels were lower in groups with 3-MA than in the corresponding groups without 3-MA. These results indicate that serum starvation elevated the rate of apoptosis, and this effect was positively correlated with the level of serum starvation. Serum concentrations lower than those in the control group adversely affected cell metabolism. Additionally, 3-MA effectively mitigated the increase in apoptosis rates across all groups.

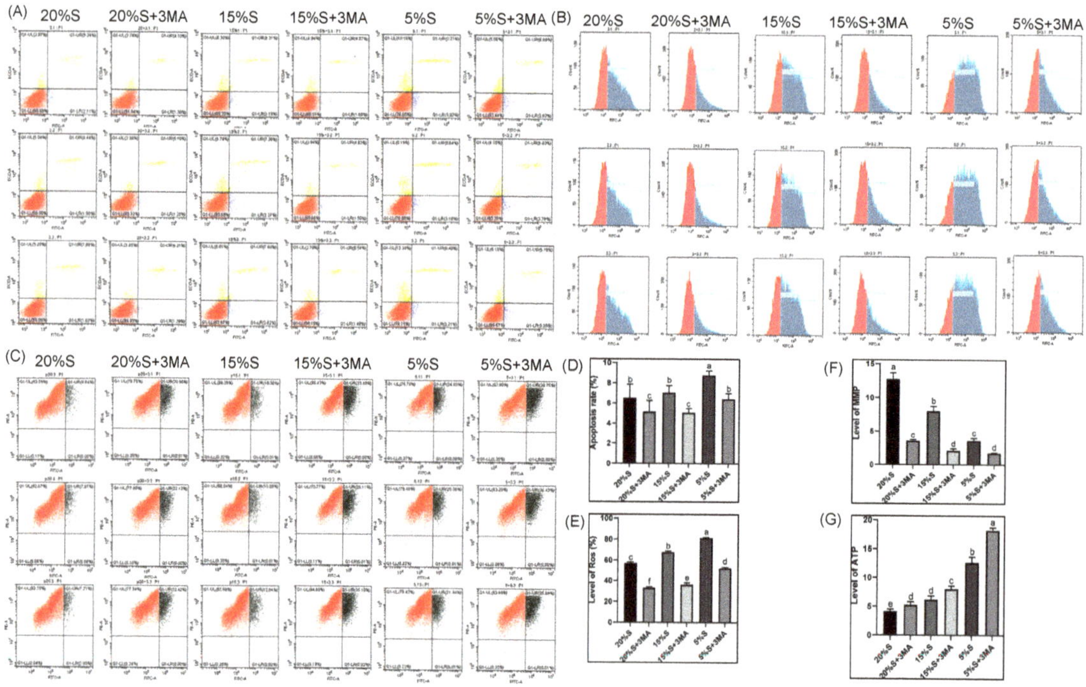

Figure 1. Detection and analysis of cellular metabolic levels. (**A**) Detection of apoptosis rate. Red spots represents the normal cells, yellow spots represents the late apoptotic cells and the mechanically damaged cells, and blue spots represents the early apoptotic cells. (**B**) Detection of ROS levels. (**C**) Detection of MMP. The red spots represents cells with high membrane potential, and the green ones represents cells with low membrane potential. (**D**) Analysis of apoptosis rate. (**E**) Analysis of ROS levels. (**F**) Analysis of MMP by JC-1 red-green fluorescence ratio. (**G**) Detection and analysis of intracellular ATP levels. The data are expressed as the mean ± SD (n = 3) of three experiments. The statistical significance of differences between groups was analyzed using one-way ANOVA, the same letters indicate the difference is not significant, while different letters indicate the difference is significant ($p < 0.05$).

3.1.2. ROS Level

The ROS levels in the six groups were 56.99 ± 1.11, 33.07 ± 1.1, 67.77 ± 0.9, 36.31 ± 1.7, 81.4 ± 0.53, and 52.55 ± 0.7, respectively, shown in Figure 1B,E. The ROS levels increased gradually in the 20% S, 15% S, and 5% S groups ($p < 0.01$). The protein expression levels were significantly lower in groups with 3-MA than in the corresponding groups without 3-MA ($p < 0.01$). The results suggest that increasing serum starvation was correlated with a gradual escalation in cellular oxidative stress and a significant increase in ROS levels. Moreover, the severity of cell damage was positively correlated with the level of serum starvation. Treatment with 3-MA effectively attenuated this increase in ROS levels.

3.1.3. MMP Level

The MMP in each group was 12.72 ± 0.96, 3.57 ± 0.16, 8.02 ± 0.67, 2.08 ± 0.31, 3.57 ± 0.38, and 1.75 ± 0.04, respectively, shown in Figure 1C,F. The MMP decreased gradually in the 20% S, 15% S, and 5% S groups ($p < 0.01$). The protein expression levels were significantly lower in groups with 3-MA than in the corresponding groups without 3-MA ($p < 0.01$). Long-term nutritional deficiency hindered the SMSCs' mitochondrial metabolism. Additionally, a significant decrease in MMP was observed upon inhibition of autophagy, suggesting that autophagy mitigated mitochondrial dysfunction.

3.1.4. ATP Level

The ATP level in each group was 4.13 ± 0.41, 5.26 ± 0.54, 6.18 ± 0.62, 8.01 ± 0.52, 12.65 ± 0.97, and 18.22 ± 0.51, respectively, shown in Figure 1G. Compared with the 20% S group, the ATP levels in the remaining five groups gradually increased, suggesting that long-term serum starvation accelerated cell metabolism and induced compensatory production of ATP to compensate for the lack of cellular energy. However, the ATP levels further increased following autophagy inhibition, indicating that autophagy may play a positive role in maintaining normal cell metabolism by inhibiting intracellular energy consumption through specific pathways.

The results of Section 3.1 indicate that long-term serum starvation (mild and severe) showed inhibitory effects on cell growth and metabolism.

3.2. *Autophagy*

3.2.1. Content of Autophagosomes and Lysosomes

The number of autophagosomes in the six groups were 2.33 ± 0.52, 0.83 ± 0.75, 4.60 ± 0.89, 2.75 ± 1.04, 6.25 ± 0.96, and 3.13 ± 0.83, shown in Figure 2A,B. The fluorescence intensity of lysosomes in each group was 23.46 ± 1.72, 22.95 ± 0.75, 25.16 ± 0.74, 21.16 ± 0.68, 32.8 ± 2.18, and 23.87 ± 0.06, respectively, shown in Figure 2C,D. The fluorescence intensity of the lysosomes was significantly higher in the 5% S group than in the 20% S group ($p < 0.05$). Additionally, the 5% S + 3-MA group showed significantly reduced fluorescence ($p < 0.05$) compared with the 5% S group; however, fluorescence did not differ significantly among the other groups. The detection results of autophagosomes were basically consistent with those of lysosomes. These results indicate that increased serum starvation was correlated with an elevation in cellular autophagy, which was mitigated by 3-MA.

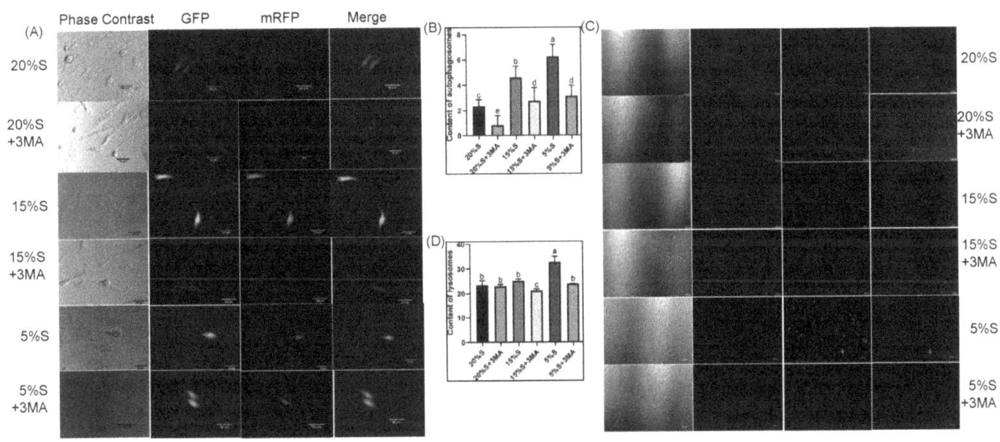

Figure 2. Detection and analysis of autophagosomes and lysosomes. (**A**) Detection of intracellular autophagic vesicles by fluorescence microscopy; (**B**) analysis of the average fluorescence value of Lyso-Tracker Red; (**C**) fluorescent probe localization of mitochondria (green) and lysosomes (red) in living cells; (**D**) analysis of the average fluorescence values of Lyso-Tracker Red. The data are expressed as the mean ± SD (n= 3) of three experiments. The statistical significance of differences between groups was analyzed using one-way ANOVA, the same letters indicate the difference is not significant, while different letters indicate the difference is significant ($p < 0.05$).

3.2.2. Ratios of p-mTOR/mTOR and p-AMPK/AMPK

The gray value ratios of p-mTOR and mTOR proteins (p-mTOR/mTOR) related to the AMPK/mTOR signaling pathway in the six groups were 0.69 ± 0.04, 1.72 ± 0.07, 0.61 ± 0.07, 1.04 ± 0.20, 0.39 ± 0.05, and 0.97 ± 0.04, respectively, shown in Figure 3A,B. The ratios decreased gradually in the 20% S, 15% S, and 5% S groups ($p < 0.01$). The protein expression levels were significantly higher in groups with 3-MA than in the corresponding groups without 3-MA ($p < 0.01$). The gray value ratios of p-AMPK and AMPK proteins (p-AMPK/AMPK) in each group were 1.50 ± 0.17, 1.58 ± 0.46, 4.96 ± 0.08, 3.77 ± 0.47, 13.31 ± 1.15, and 6.33 ± 0.52, respectively, shown in Figure 3A,C. The ratios increased gradually in the 20% S, 15% S, and 5% S groups ($p < 0.01$). The protein expression levels were significantly lower in the 15% S and 5% S group with 3-MA than in the corresponding group without 3-MA ($p < 0.01$). The results suggest that serum starvation may induce autophagy through the AMPK/mTOR signaling pathway.

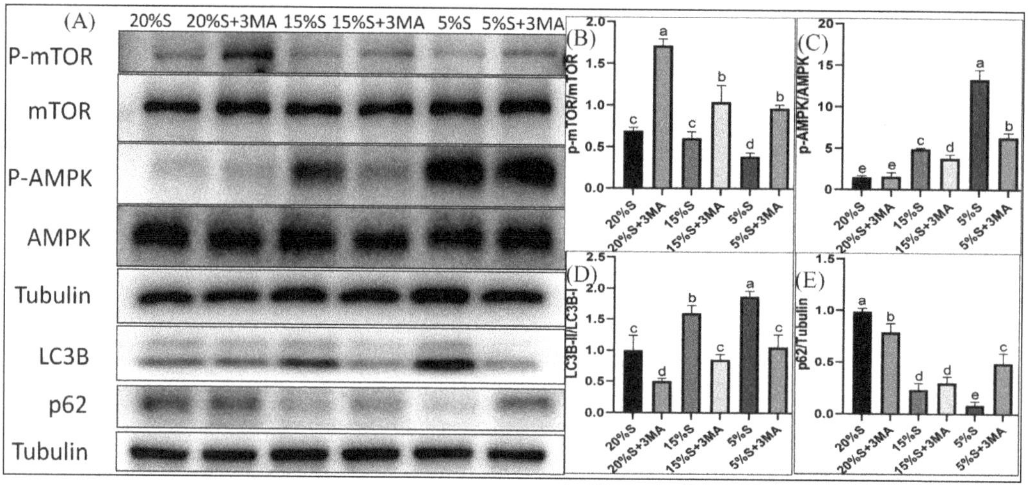

Figure 3. Detection and analysis of mTOR, p-mTOR, AMPK, p-AMPK, LC3B, and p62 expression by Western blot. (**A**) WB results of mTOR, p-mTOR, AMPK, p-AMPK, LC3B, and p62; (**B**) grayscale analysis of p-mTOR/mTOR; (**C**) grayscale analysis of p-AMPK/AMPK; (**D**) grayscale analysis of LC3B-II/LC3B-I; (**E**) grayscale analysis of p62. The data are expressed as the mean ± SD ($n = 3$) of three experiments. The statistical significance of differences between groups was analyzed using one-way ANOVA, the same letters indicate the difference is not significant, while different letters indicate the difference is significant ($p < 0.05$) (Supplementary File S1).

3.2.3. Relative Expression of LC3B and p62

The expression ratios of the autophagy marker proteins LC3B-II and LC3B-I (LC3B-II/LC3B-I) in the six groups were 1.00 ± 0.24, 0.50 ± 0.04, 1.60 ± 0.13, 0.85 ± 0.09, 1.87 ± 0.10, and 1.06 ± 0.20, respectively, shown in Figure 3A,D. The ratios increased gradually in the 20% S, 15% S, and 5% S groups. The ratios were significantly lower in groups with 3-MA than in the corresponding groups without 3-MA ($p < 0.01$). The relative expression of the p62 protein in each group was 0.99 ± 0.03, 0.79 ± 0.09, 0.23 ± 0.06, 0.30 ± 0.06, 0.09 ± 0.04, and 0.49 ± 0.10, respectively, shown in Figure 3A,E. The expression of the p62 protein decreased gradually in the 20% S, 15% S, and 5% S groups ($p < 0.01$). The expression levels in the 3-MA groups were higher than those in the corresponding groups without 3-MA. The results suggest that as the degree of serum starvation increases, the degree of cellular autophagy strengthens.

By measuring the indicators of Section 3.2, we observed that long-term serum starvation induced autophagy via the AMPK/mTOR pathway, with the level of autophagy showing a positive correlation with the severity of serum starvation.

3.3. Cell Differentiation

3.3.1. Relative Expression of Myogenic Differentiation Markers

The relative expression levels of MHC protein in each group were 1.03 ± 0.10, 0.60 ± 0.03, 0.63 ± 0.06, 0.16 ± 0.01, 0.09 ± 0.02, and 0.10 ± 0.03, respectively, shown in Figure 4A,B. The relative expression levels of MyoD1 protein in the six groups were 1.38 ± 0.11, 1.05 ± 0.11, 1.02 ± 0.09, 0.49 ± 0.11, 0.39 ± 0.12, and 0.26 ± 0.09, respectively, shown in Figure 4A,C. The expression of both proteins decreased gradually in the 20% S, 15% S, and 5% S groups ($p < 0.01$). The protein expressions were lower in groups with 3-MA than in the corresponding groups without 3-MA.

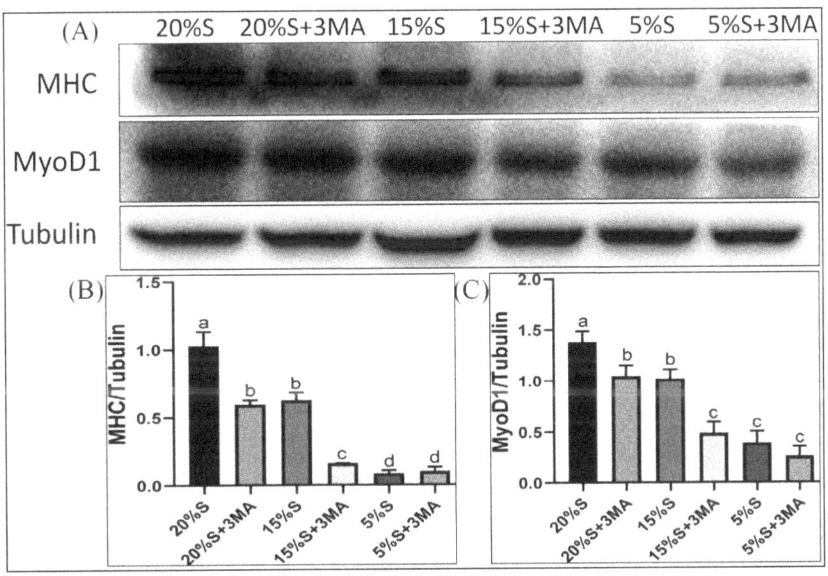

Figure 4. Detection and analysis of MHC and MyoD1 by Western blot. (**A**) WB results of MHC and MyoD1; (**B**) grayscale analysis of MHC; (**C**) grayscale analysis of MyoD1. The data are expressed as the mean ± SD ($n = 3$) of three experiments. The statistical significance of differences between groups was analyzed using one-way ANOVA, the same letters indicate the difference is not significant, while different letters indicate the difference is significant ($p < 0.05$) (Supplementary File S1).

The fluorescence intensities of MHC protein in the six groups were 27.12 ± 0.83, 18.7 ± 0.93, 22.3 ± 0.95, 16.23 ± 0.32, 18.19 ± 0.78, and 14.31 ± 0.26, respectively, shown in Figure 5A,B. Those of MyoD1 protein in each group were 18.89 ± 0.25, 11.65 ± 0.34, 16.19 ± 0.74, 9.4 ± 1.07, 12.74 ± 1.27, and 6.74 ± 1.06, shown in Figure 5C,D. The expression of both proteins decreased gradually in the 20% S, 15% S, and 5% S groups ($p < 0.01$). The protein expression levels were lower in groups with 3-MA than in the corresponding groups without 3-MA ($p < 0.01$). The results were basically consistent with those of the Western blot analysis, which indicate that long-term serum starvation inhibited the myogenic differentiation of SMSCs, while autophagy mitigated the decline in myogenic differentiation during serum starvation.

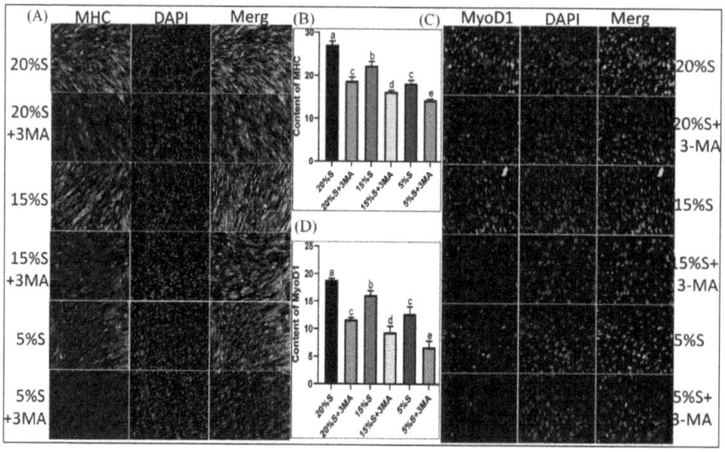

Figure 5. Detection and analysis of MHC and MyoD1 by cellular immunofluorescence. (**A**) Immunofluorescence staining results of MHC. The green part represents the positive cells that can express MHC, the blue part represents the nucleus; (**B**) quantitative analysis of MHC fluorescence; (**C**) immunofluorescence staining results of MyoD1. The green part represents the positive cells that can express MyoD1, the blue part represents the nucleus; (**D**) quantitative analysis of MyoD1 fluorescence. The data are expressed as the mean ± SD (n = 3) of three experiments. The statistical significance of differences between groups was analyzed using one-way ANOVA, the same letters indicate the difference is not significant, while different letters indicate the difference is significant ($p < 0.05$).

3.3.2. Relative Expression of Adipogenic Differentiation Markers

The relative mRNA expression of PPARγ in the six groups were 1.08 ± 0.27, 0.69 ± 0.05, 0.79 ± 0.06, 0.69 ± 0.04, 0.59 ± 0.02, and 0.35 ± 0.03, respectively. Those of the LPL protein in the six groups were 1.00 ± 0.01, 0.60 ± 0.04, 0.75 ± 0.05, 0.58 ± 0.12, 0.46 ± 0.21, and 0.32 ± 0.02, respectively, shown in Figure 6. The expression of the two mRNA levels decreased gradually in the 20% S, 15% S, and 5% S groups. The protein expressions were lower in groups with 3-MA than in the corresponding groups without 3-MA.

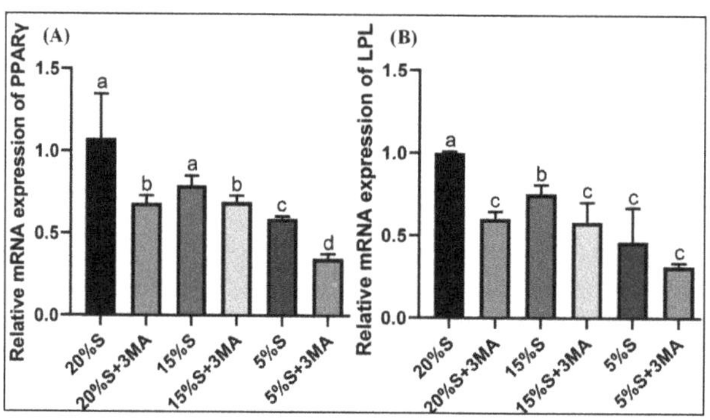

Figure 6. Detection and analysis of PPARγ and LPL by qRT-PCR. (**A**) Expression of PPARγ mRNA; (**B**) expression of LPL mRNA. The data are expressed as the mean ± SD (n = 3) of three experiments. The statistical significance of differences between groups was analyzed using one-way ANOVA, the same letters indicate the difference is not significant, while different letters indicate the difference is significant ($p < 0.05$).

3.3.3. Content of Fat

The cells in all groups contained orange-red lipid droplets, with the highest number and largest area of lipid droplets being observed in the 20% S group and the lowest number and smallest area being found in the 5% S group. The addition of 3-MA significantly reduced the number and area of lipid droplets across all groups, shown in Figure 7A. The optical density at 510 nm was measured using an Oil Red O staining kit, and the values for the six groups were 0.19 ± 0.01, 0.12 ± 0.01, 0.18 ± 0.01, 0.11 ± 0.01, 0.12 ± 0.01, and 0.09 ± 0.01, respectively, shown in Figure 7B. The values decreased gradually in the 20% S, 15% S, and 5% S groups ($p < 0.01$). The values were lower in groups with 3-MA than in the corresponding groups without 3-MA ($p < 0.01$). These findings were consistent with those obtained upon microscopic observation following Oil Red O staining.

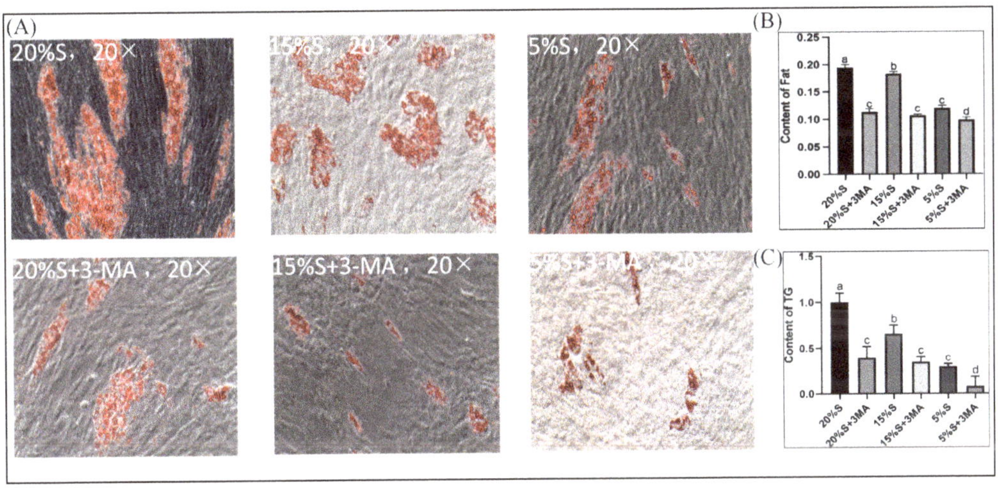

Figure 7. Detection and analysis of fat content and TG levels. (**A**) Detection of fat content (20×), the red part is stained fat, and the unstained surrounding is muscle; (**B**) quantitative determination of the OD value; (**C**) detection and analysis of TG levels. The data are expressed as the mean ± SD (n= 3) of three experiments. The statistical significance of differences between groups was analyzed using one-way ANOVA, the same letters indicate the difference is not significant, while different letters indicate the difference is significant ($p < 0.05$).

3.3.4. Content of TG

The TG contents in the six groups were 1.00 ± 0.10, 0.39 ± 0.10, 0.66 ± 0.09, 0.35 ± 0.06, 0.29 ± 0.03, and 0.14 ± 0.01, respectively, shown in Figure 7C. The contents decreased gradually in the 20% S, 15% S, and 5% S groups ($p < 0.01$). The contents were lower in groups with 3-MA than in the corresponding groups without 3-MA ($p < 0.05$ or $p < 0.01$).

The results of Sections 3.3.2–3.3.4 indicate that serum starvation reduced adipogenic differentiation and that this reduction was positively correlated with the level of serum starvation. Additionally, autophagy played a role in alleviating the decline in adipogenic differentiation during serum starvation.

Overall, the results of Section 3.3 indicate that serum starvation ultimately led to the inhibition of SMSC differentiation. However, the results obtained from the autophagy inhibition groups indicate that autophagy significantly promoted SMSC differentiation during long-term serum starvation.

4. Discussion

SMSCs are undifferentiated myogenic precursors in adult muscle tissue that undergo myogenic differentiation to form muscle fibers when stimulated by specific signals [1]. We previously found that short-term mild serum starvation (15% S for 24 h) promoted SMSCs' metabolism and enhanced their myogenic and adipogenic differentiation through the autophagic pathway [1,2]. These findings underscore the potential for the application of serum starvation to promote the growth of animal skeletal muscle. To better understand the impact of serum starvation on skeletal muscle growth, SMSCs were subjected to different levels of serum starvation simulated by reducing the serum concentration in culture media over an extended period (96 h). This approach allowed the investigation of the effects of serum starvation on SMSC metabolism, autophagy, and differentiation.

We measured the apoptosis rate, ROS levels, MMP, and ATP levels [14–17], which are commonly used to reflect cellular metabolic levels, to comprehensively evaluate the effects of long-term serum starvation on cell metabolism. We found that long-term serum starvation accelerated cell metabolism and apoptosis, negatively impacting cell growth, but that cells compensated for the energy deficiency by increasing ATP production and activating autophagy to mitigate these adverse effects. In contrast, our previous studies showed that short-term mild serum starvation increased ATP production in cells, while the apoptosis rate, ROS, and MMP were not significantly different from those in normal cells, thus promoting cell metabolism without affecting cell growth [10]. However, long-term serum starvation (mild and severe) showed inhibitory effects on cell growth and metabolism.

Starvation is a well-known stressor that triggers autophagy, and the quantity of autolysosomes and autophagosomes is commonly used to detect the extent of autophagy in cells [18,19]. Additionally, the autophagy marker proteins p62 and LC3B are frequently employed to measure the levels of autophagy [20–22]. The AMPK/mTOR signaling pathway is a well-established mechanism for autophagy, and the activation of AMPK and the inhibition of mTOR collaboratively initiate the autophagy process [23–25]. By measuring these indicators, we observed that long-term serum starvation induced autophagy via the AMPK/mTOR pathway, with the level of autophagy showing a positive correlation with the severity of serum starvation. This finding aligns with the previously observed effects of short-term serum starvation on autophagy in SMSCs [11]. Since multiple signaling pathways can cause autophagy, other molecular pathways that may modulate autophagy and muscle differentiation in response to serum starvation need further verification.

Low concentrations of horse serum in the culture medium have been shown to induce the differentiation of SMSCs into myocytes [26], while the medium used for induction of adipogenic differentiation promoted their differentiation into adipocytes [27–30]. The genes encoding MyoD1 and MHC are used as markers for myogenic differentiation [31,32], while PPARγ and lipoprotein lipase are the primary transcription factors regulating adipogenesis [33,34]. Measurement of these indicators enables evaluation of the impact of long-term serum starvation on SMSCs' differentiation at the molecular level. Our findings indicated that long-term serum starvation significantly reduced both myogenic and adipogenic differentiation of SMSCs, and that these changes were accompanied by notable decreases in lipid and triglyceride levels following adipogenic differentiation. Additionally, the extent of this reduction was positively correlated with the level of serum starvation. The results obtained from the autophagy inhibition groups indicated that autophagy significantly promoted SMSC differentiation during long-term serum starvation. However, the adverse effects of excessive ROS accumulation, reduced mitochondrial membrane potential, and increased apoptosis rates on SMSC differentiation during long-term serum starvation appeared to be more pronounced, ultimately leading to the inhibition of SMSC differentiation. Our previous research showed that short-term mild serum starvation (15% S for 24 h) promoted SMSCs' differentiation, while short-term severe serum starvation (5% S for 24 h) inhibited their differentiation [10].

Based on the results of this experiment and our previous findings, short-term mild serum starvation significantly enhances the metabolism and differentiation of SMSCs. While autophagy induced by long-term serum starvation moderately promoted myogenic and adipogenic differentiation of SMSCs, this effect was insufficient to counteract the overall inhibitory impact of long-term serum starvation on cell differentiation. Therefore, we are currently exploring strategies to promote muscle development through controlled starvation. Specifically, we are using an intermittent mild fasting mouse model to investigate the effects of starvation on skeletal muscle growth. The results of mouse experiments indicate that intermittent mild fasting induced autophagy via the AMPK/mTOR signaling pathway. Compared with mice on a standard feeding protocol, the fasting regimen led to an increase in skeletal muscle mass and a decrease in visceral fat content, despite a reduction in food intake (data to be published).

Supplementary Materials: The following supporting information can be downloaded at: https://www.mdpi.com/article/10.3390/vetsci12010011/s1, File S1: The Original WB Figures.

Author Contributions: Conceptualization, H.J.; methodology, H.J., Y.W. (Yi Wang), Y.H., and J.G.; data curation, J.G. and Y.W. (Yajie Wu); formal analysis, B.F. and Y.W. (Yajie Wu); investigation, Y.Y.; resources, J.L. and H.J.; writing—original draft, Y.W. (Yi Wang) and H.J.; writing—review and editing, J.G. and H.J.; supervision, F.L. and J.Z. All authors have read and agreed to the published version of the manuscript.

Funding: This work was supported by the Priority Academic Program Development of Jiangsu Higher Education Institutions (PAPD) and Jiangsu Co-innovation Center for Prevention, the National Natural Science Foundation of China (31872323).

Institutional Review Board Statement: The animal study protocol was approved by the Institutional Ethics Committee of Yangzhou University (No. 202104831).

Informed Consent Statement: Not applicable.

Data Availability Statement: The datasets used during the current study are available from the corresponding author on reasonable request.

Conflicts of Interest: The authors declare no conflicts of interest. The funders had no role in the design of the study; in the collection, analyses, or interpretation of data; in the writing of the manuscript; or in the decision to publish the results.

References

1. Mauro, A. Satellite cell of skeletal muscle fibers. *J. Biophys. Biochem. Cytol.* **1961**, *9*, 493–495. [CrossRef]
2. Mukund, K.; Subramaniam, S. Skeletal muscle: A review of molecular structure and function, in health and disease. *Wiley Interdiscip. Rev. Syst. Biol. Med.* **2020**, *12*, e1462. [CrossRef] [PubMed]
3. Giuliani, G.; Rosina, M.; Reggio, A. Signaling pathways regulating the fate of fibro/adipogenic progenitors (FAPs) in skeletal muscle regeneration and disease. *FEBS J.* **2022**, *289*, 6484–6517. [CrossRef]
4. Molina, T.; Fabre, P.; Dumont, N.A. Fibro-adipogenic progenitors in skeletal muscle homeostasis, regeneration and diseases. *Open Biol.* **2021**, *11*, 210110. [CrossRef]
5. Ohsumi, Y. Historical landmarks of autophagy research. *Cell Res.* **2014**, *24*, 9–23. [CrossRef]
6. Baechler, B.L.; Bloemberg, D.; Quadrilatero, J. Mitophagy regulates mitochondrial network signaling, oxidative stress, and apoptosis during myoblast differentiation. *Autophagy* **2019**, *15*, 1606–1619. [CrossRef] [PubMed]
7. Call, J.A.; Nichenko, A.S. Autophagy: An essential but limited cellular process for timely skeletal muscle recovery from injury. *Autophagy* **2020**, *16*, 1344–1347. [CrossRef]

8. Fiacco, E.; Castagnetti, F.; Bianconi, V.; Madaro, L.; De Bardi, M.; Nazio, F.; D'Amico, A.; Bertini, E.; Cecconi, F.; Puri, P.L.; et al. Autophagy regulates satellite cell ability to regenerate normal and dystrophic muscles. *Cell Death Differ.* **2016**, *23*, 1839–1849. [CrossRef] [PubMed]
9. Paolini, A.; Omairi, S.; Mitchell, R.; Vaughan, D.; Matsakas, A.; Vaiyapuri, S.; Ricketts, T.; Rubinsztein, D.C.; Patel, K. Attenuation of autophagy impacts on muscle fibre development, starvation induced stress and fibre regeneration following acute injury. *Sci. Rep.* **2018**, *8*, 9062. [CrossRef] [PubMed]
10. Wang, Y.; Gao, J.; Fan, B.; Hu, Y.; Yang, Y.; Wu, Y.; Li, F.; Ju, H. Different levels of autophagy induced by transient serum starvation regulate metabolism and differentiation of porcine skeletal muscle satellite cells. *Sci. Rep.* **2023**, *13*, 13153. [CrossRef]
11. Wang, Y.; Gao, J.; Hu, Y.; Yang, Y.; Fan, B.; Wu, Y.; Ju, H. Effect of Transient Serum Starvation on Metabolism and Autophagy of Porcine Skeletal Muscle Satellite Cells. *Chin. J. Anim. Vet. Sci.* **2024**, *55*, 3408–3417.
12. Wang, H.; Yang, Y.; Wang, Y.; Peng, Y.; Xu, L.; Hu, Y.; Ju, H. Effect of the Porcine STC-1 Gene on Autophagy and Mitochondrial Function as Induced by Serum Starvation. *Biochem. Genet.* **2022**, *60*, 2533–2551. [CrossRef]
13. Ju, H.; Yang, Y.; Sheng, A.; Qi, Y. MicroRNA-378 promotes myogenic differentiation by targeting BMP4. *Mol. Med. Rep.* **2016**, *13*, 2194–2200. [CrossRef]
14. Alsayyah, C.; Ozturk, O.; Cavellini, L.; Belgareh-Touze, N.; Cohen, M.M. The regulation of mitochondrial homeostasis by the ubiquitin proteasome system. *Biochim. Biophys. Acta Bioenerg.* **2020**, *1861*, 148302. [CrossRef] [PubMed]
15. Sztalryd, C.; Brasaemle, D.L. The perilipin family of lipid droplet proteins: Gatekeepers of intracellular lipolysis. *Biochim. Biophys. Acta Mol. Cell Biol. Lipids.* **2017**, *1862 Pt B*, 1221–1232. [CrossRef]
16. Kishikawa, J.-I.; Inoue, Y.; Fujikawa, M.; Nishimura, K.; Nakanishi, A.; Tanabe, T.; Imamura, H.; Yokoyama, K. General anesthetics cause mitochondrial dysfunction and reduction of intracellular ATP levels. *PLoS ONE.* **2018**, *13*, e0190213. [CrossRef]
17. Wood, J.D.; Enser, M.; Fisher, A.V.; Nute, G.R.; Sheard, P.R.; Richardson, R.I.; Whittington, F.M.; Hughes, S.I. Fat deposition, fatty acid composition and meat quality: A review. *Meat Sci.* **2008**, *78*, 343–358. [CrossRef] [PubMed]
18. Zhao, Y.G.; Codogno, P.; Zhang, H. Machinery, regulation and pathophysiological implications of autophagosome maturation. *Nat. Rev. Mol. Cell Biol.* **2021**, *22*, 733–750. [CrossRef] [PubMed]
19. Nakatogawa, H. Mechanisms governing autophagosome biogenesis. *Nat. Rev. Mol. Cell Biol.* **2020**, *21*, 439–458. [CrossRef]
20. Shvets, E.; Abada, A.; Weidberg, H.; Elazar, Z. Dissecting the involvement of LC3B and GATE-16 in p62 recruitment into autophagosomes. *Autophagy* **2011**, *7*, 683–688. [CrossRef] [PubMed]
21. Bjorkoy, G.; Lamark, T.; Johansen, T. p62/SQSTM1: A missing link between protein aggregates and the autophagy machinery. *Autophagy* **2006**, *2*, 138–139. [CrossRef]
22. Shvets, E.; Fass, E.; Scherz-Shouval, R.; Elazar, Z. The N-terminus and Phe52 residue of LC3 recruit p62/SQSTM1 into autophagosomes. *J. Cell Sci.* **2008**, *121 Pt 16*, 2685–2695. [CrossRef] [PubMed]
23. Sancak, Y.; Bar-Peled, L.; Zoncu, R.; Markhard, A.L.; Nada, S.; Sabatini, D.M. Ragulator-Rag complex targets mTORC1 to the lysosomal surface and is necessary for its activation by amino acids. *Cell* **2010**, *141*, 290–303. [CrossRef] [PubMed]
24. Shaw, R.J. LKB1 and AMP-activated protein kinase control of mTOR signalling and growth. *Acta Physiol.* **2009**, *196*, 65–80. [CrossRef] [PubMed]
25. Jia, J.; Abudu, Y.P.; Claude-Taupin, A.; Gu, Y.; Kumar, S.; Choi, S.W.; Peters, R.; Mudd, M.H.; Allers, L.; Salemi, M.; et al. Galectins control MTOR and AMPK in response to lysosomal damage to induce autophagy. *Autophagy* **2019**, *15*, 169–171. [CrossRef] [PubMed]
26. Starkey, J.D.; Yamamoto, M.; Yamamoto, S.; Goldhamer, D.J. Skeletal muscle satellite cells are committed to myogenesis and do not spontaneously adopt nonmyogenic fates. *J. Histochem. Cytochem.* **2011**, *59*, 33–46. [CrossRef] [PubMed]
27. Beloor, J.; Kang, H.K.; Moon, Y.S. Serum lipids can convert bovine myogenic satellite cells to adipo cytes. *Asian Australas. J. Anim. Sci.* **2010**, *23*, 1519–1526. [CrossRef]
28. Phua, W.W.T.; Wong, M.X.Y.; Liao, Z.; Tan, N.S. An apparent functional consequence in skeletal muscle physiology via peroxisome proliferator-activated receptors. *Int. J. Mol. Sci.* **2018**, *19*, 1425. [CrossRef]
29. Asakura, A.; Komakim, M.; Rudnicki, M. Muscle satellite cells are multipotential stem cells that exhibit myogenic, osteogenic, and adipogenic differentiation. *Differentiation* **2010**, *68*, 245–253. [CrossRef]
30. Wada, M.R.; Inagawa-Ogashiwa, M.; Shimizu, S.; Yasumoto, S.; Hashimoto, N. Generation of different fates from multipotent muscle stem cells. *Development* **2022**, *129*, 2987–2995. [CrossRef] [PubMed]
31. Lahmann, I.; Zhang, Y.; Baum, K.; Wolf, J.; Birchmeier, C. An oscillatory network controlling self-renewal of skeletal muscle stem cells. *Exp. Cell Res.* **2021**, *409*, 112933. [CrossRef] [PubMed]
32. Yamanouchi, K.; Nakamura, K.; Takeuchi, S.; Hosoyama, T.; Matsuwaki, T.; Nishihara, M. Suppression of MyoD induces spontaneous adipogenesis in skeletal muscle progenitor cell culture. *Anim. Sci. J.* **2021**, *92*, e13573. [CrossRef]

33. Gu, H.; Zhou, Y.; Yang, J.; Li, J.; Peng, Y.; Zhang, X.; Miao, Y.; Jiang, W.; Bu, G.; Hou, L.; et al. Targeted overexpression of PPARgamma in skeletal muscle by random insertion and CRISPR/Cas9 transgenic pig cloning enhances oxidative fiber formation and intramuscular fat deposition. *FASEB J.* **2021**, *35*, e21308. [CrossRef]
34. Hu, X.; Tang, J.; Hu, X.; Bao, P.; Pan, J.; Chen, Z.; Xian, J. MiR-27b Impairs Adipocyte Differentiation of Human Adipose Tissue-Derived Mesenchymal Stem Cells by Targeting LPL. *Cell Physiol. Biochem.* **2018**, *47*, 545–555. [CrossRef] [PubMed]

MDPI AG
Grosspeteranlage 5
4052 Basel
Switzerland
Tel.: +41 61 683 77 34

Veterinary Sciences Editorial Office
E-mail: vetsci@mdpi.com
www.mdpi.com/journal/vetsci

www.ingramcontent.com/pod-product-compliance
Lightning Source LLC
Chambersburg PA
CBHW061848210926
51850CB00006B/1003
* 9 7 8 3 7 2 5 8 5 2 4 1 3 *